Lecture Notes in Networks and Systems

Volume 251

The series "Lecture Notes in Networks and Systems" publishes the latest developments in Networks and Systems—quickly, informally and with high quality. Original research reported in proceedings and post-proceedings represents the core of LNNS.

Volumes published in LNNS embrace all aspects and subfields of, as well as new challenges in, Networks and Systems.

The series contains proceedings and edited volumes in systems and networks, spanning the areas of Cyber-Physical Systems, Autonomous Systems, Sensor Networks, Control Systems, Energy Systems, Automotive Systems, Biological Systems, Vehicular Networking and Connected Vehicles, Aerospace Systems, Automation, Manufacturing, Smart Grids, Nonlinear Systems, Power Systems, Robotics, Social Systems, Economic Systems and other. Of particular value to both the contributors and the readership are the short publication timeframe and the world-wide distribution and exposure which enable both a wide and rapid dissemination of research output.

The series covers the theory, applications, and perspectives on the state of the art and future developments relevant to systems and networks, decision making, control, complex processes and related areas, as embedded in the fields of interdisciplinary and applied sciences, engineering, computer science, physics, economics, social, and life sciences, as well as the paradigms and methodologies behind them.

Indexed by SCOPUS, INSPEC, WTI Frankfurt eG, zbMATH, SCImago.

All books published in the series are submitted for consideration in Web of Science.

More information about this series at http://www.springer.com/series/15179

Phayung Meesad · Sunantha Sodsee ·
Watchareewan Jitsakul ·
Sakchai Tangwannawit
Editors

Recent Advances in Information and Communication Technology 2021

Proceedings of the 17th International Conference on Computing and Information Technology (IC2IT 2021)

Springer

Editors
Phayung Meesad
Faculty of Information Technology
and Digital Innovation
King Mongkut's University of Technology
North Bangkok
Bangkok, Thailand

Sunantha Sodsee
Faculty of Information Technology
and Digital Innovation
King Mongkut's University of Technology
North Bangkok
Bangkok, Thailand

Watchareewan Jitsakul
Faculty of IT and Digital Innovation
King Mongkut's University of Technology
North Bangkok
Bangkok, Thailand

Sakchai Tangwannawit
Faculty of Information Technology
and Digital Innovation
King Mongkut's University of Technology
North Bangkok
Bangkok, Thailand

ISSN 2367-3370 ISSN 2367-3389 (electronic)
Lecture Notes in Networks and Systems
ISBN 978-3-030-79756-0 ISBN 978-3-030-79757-7 (eBook)
https://doi.org/10.1007/978-3-030-79757-7

This Springer imprint is published by the registered company Springer Nature Switzerland AG
The registered company address is: Gewerbestrasse 11, 6330 Cham, Switzerland

Preface

To keep up with the emerging trends in Information Technology and Digital Innovation, the International Conference on Computing and Information Technology (IC^2IT) has been organized for 17 years to serve the IT and digital communities as a platform to share new research and innovation.

In 2021, the $IC^2IT2021$ is held on May13th–14th, 2021, in Bangkok, Thailand. The main contributions in the conference proceedings are in the areas of machine learning, natural language processing, image processing, intelligent systems and algorithms, as well as network and cloud computing. All submitted papers from 17 countries were peer reviewed by at least two, usually three members of the international program committee (IPC). Only 34 submissions have been accepted based on originality, important issue and clarification for presentation at the conference, showing research directions that reflecting digital disruption in the world. All accepted papers are included in the conference proceedings, which are published in the well-established and worldwide-distributed series on Advances in Intelligent Systems and Computing edited by Janusz Kacprzyk for eight years now.

Herein, we would like to thank the authors, as well as the members of the program committee, and all participating university collaborates in both Thailand and overseas for their outstanding support and academic cooperation. We also would like to say thank you to all staff members of the Faculty of Information Technology and Digital Innovation at King Mongkut's University of Technology North Bangkok, who have carried out many technical and organizational tasks. Finally, we wish all participants pleasure the conference and fulfill the academic ideas.

February 2021

Phayung Meesad
Sunantha Sodsee
Watchareewan Jitsakul
Sakchai Tangwannawit

Organization

Program Committee

M. Aiello	UNI-Stuttgart, Germany
S. Auwatanamongkol	NIDA, Thailand
T. Bernard	Li-Parad, France
P. Boonyopakorn	KMUTNB, Thailand
A. Bunteong	UBU, Thailand
J. H. Chan	KMUTT, Thailand
N.H.H. Cuong	Danang Uni, Vietnam
N. Ditcharoen	UBU, Thailand
T. Eggendorfer	HS Weingarten, Germany
K. Hemmi	Uni of Nagasaki, Japan
S. Hiranpongsin	UBU, Thailand
U. Inyaem	RMUTT, Thailand
W. Janratchakool	RMUTT, Thailand
W. Jitsakul	KMUTNB, Thailand
N. Jongsawat	RMUTT, Thailand
M. Kaenampornpan	MSU, Thailand
M. Ketcham	KMUTNB, Thailand
A. Kongthon	NECTEC, Thailand
P. Kropf	Uni NE, Switzerland
P. Kuacharoen	NIDA, Thailand
M. Kubek	FernUni, Germany
P. Kunakornvong	RMUTT, Thailand
U. Lechner	UniBW, Germany
M. Maliyaem	KMUTNB, Thailand
P. Meesad	KMUTNB, Thailand
C. Namman	UBU, Thailand
A. Nanthaamornphong	PSU, Thailand
K. Nimkerdphol	RMUTT, Thailand

K. Pasupa	KMITL, Thailand
W. Pongpech	NIDA, Thailand
N. Porrawatpreyakorn	KMUTNB, Thailand
P. Saengsiri	TISTR, Thailand
T. Siriborvornratanakul	NIDA, Thailand
S. Smanchat	KMUTNB, Thailand
S. Sodsee	KMUTNB, Thailand
W. Sriurai	UBU, Thailand
T. Sucontphunt	NIDA, Thailand
A. Taguchi	TCU, Japan
S. Tangwannawit	KMUTNB, Thailand
C. Thaenchaikun	PSU, Thailand
K. Thongglin	PSU, Thailand
J. Thongkam	MSU, Thailand
N. Tongtep	PSU, Thailand
D. H. Tran	HNUE, Vietnam
K. Treeprapin	UBU, Thailand
D. Tutsch	Uni-Wuppertal, Germany
N. Utakrit	KMUTNB, Thailand
S. Valuvanathorn	UBU, Thailand
D. Van Hieu	TGU, Vietnam
V. Wattanasoontorn	PSU, Thailand
N. Wisitpongphan	KMUTNB, Thailand
K. Woraratpanya	KMITL, Thailand
P. Wuttidittachotti	KMUTNB, Thailand

Organizing Partners

In Cooperation with
King Mongkut's University of Technology North Bangkok (KMUTNB)
FernUniversitaet in Hagen, Germany (FernUni)
Chemnitz University of Technology, Germany (CUT)
Oklahoma State University, USA (OSU)
Edith Cowan University, Western Australia (ECU)
Hanoi National University of Education, Vietnam (HNUE)
Mahasarakham University (MSU)
National Institute of Development Administration (NIDA)
Rajamangala University of Technology Thanyaburi (RMUTT)
Rajamangala University of Technology Krungthep (RMUTK)
Kanchanaburi Rajabhat University (KRU)
Phetchaburi Rajabhat University (PBRU)

Gesellschaft für Informatik (GI)
Council of IT Deans of Thailand (CITT)
The Thailand Convention and Exhibition Bureau (TCEB)

Contents

DB2SM: An Efficient Resampling Technique for Imbalanced Data Classification

Pharnuphon Jiraamphorn(✉) and Eakasit Pacharawongsakda

Big Data Engineering Program, College of Innovative Technology and Engineering, Dhurakij Pundit University, Bangkok, Thailand
{625162020002,eakasit.pac}@dpu.ac.th

Abstract. Data classification is the main task of Machine Learning. However, it is not easy to build a good classification model because it requires well-prepared data and efficient classification techniques. One of the data issues that we inevitably have to face is the imbalanced data problem. This problem typically refers to a situation that the classes are not represented equally, e.g., one class is more occur than another class. Therefore, the classifier will be biased to the majority class and give high performance for that class. To deal with this issue, various techniques have been proposed. In this work, we introduce an improvement technique, called DB2SM. Our method employs 2 times DBSCAN clustering to find a proper area and SMOTE for oversample minority instances. For more detail, the first clustering task removes the negative instances (e.g., majority class), which are located close to the positive classes. Then the second clustering task generates clusters that have only positive or negative instances. Finally, SMOTE was applied to create new positive instances. The experimental results with 20 imbalanced datasets showed that our proposed method gives a better performance in terms of accuracy, AUC and F-measure, compared to SMOTE, DBCS and DBSM.

Keywords: Imbalanced data · Resampling · DBSCAN · SMOTE

1 Introduction

Handling imbalanced data is one of the most challenging tasks in Machine Learning. In real-world use-cases, for example, fraud detection, intrusion detection, and disease diagnostics have an imbalanced data issue. There are a few fraudulent cases compared to normal cases and a few patients compared to normal people. A large number of instances in the dataset are called the majority class and the fewer are called the minority class. In this situation, the classification model will be biased to the majority class and give high accuracy. To handle these issues, there are two main approaches: data level and algorithm level [1]. The methods in the first approach are focused on sampling data. It tends to resample instances to a balanced dataset, for example, random undersampling (RUS) or random oversampling (ROS). The second approaches adjust the machine learning techniques to handle the imbalanced dataset. The techniques in this approach including one class SVM, cost sensitive learning.

P. Meesad et al. (Eds.): IC2IT 2021, LNNS 251, pp. 1–12, 2021.
https://doi.org/10.1007/978-3-030-79757-7_1

Each solution has its own limitation. For example, the quality of synthetic instances using SMOTE technique is dependent on the k-nearest neighbor data. If they contain noises or outliers, it will reduce the potential of new instances too [2]. This paper introduces an improvement technique by combining DBSCAN and SMOTE to solve some existing problems and improve oversampling quality on the imbalanced data. By the assumption to generate new instances in the high-density area or largest cluster of minority class will receive a good instance quality as the original data does.

2 Related Works

In this section, we reviewed previously research works which aimed to address the imbalanced data.

M. Ester et al. [3] presented DBSCAN (A Density-Based Algorithm for Discovering Clusters in Large Spatial Database with Noise) which is a technique that clustered data by exploring every instance. If they have the number of neighbors equal to or greater than the minimum number of points (*minpts*) and located within a constant radius called epsilon (ε) then the algorithm will form a cluster for it and extend the cluster area by applying the same play to their neighbors. The DBSCAN was mostly used because it can handle unlabeled data and can cluster to various shapes instead of an oval.

The Random Over-Sampling (ROS) technique is the simplest method that duplicates the minority class instances. However, this might lead to an over-fitting issue since the new instances are the same as previously. To overcome this problem, Chawla et al. [4] proposed another way by synthetic new instances near the previous minority class instances. This technique is called SMOTE (Synthetic Minority Over-sampling TEchnique) and it's widely used to do over-sampling in the imbalanced dataset.

Y. Sanguanmak and A. Hanskunatai [5] proposed a hybrid technique called DBSM that combining DBSCAN and SMOTE in the same process. After applying DBSCAN to the imbalanced dataset, DBSM removes 50% of negative (majority class) instances in each cluster which have the smallest distance from their centroid. This task is similar to undersampling but it does not random. Then DBSM employs SMOTE to oversampling positive (minority class) instances. This method gave improved performance compared to base-line techniques.

M.K. Verma et al. [6] presented DBCS method which applies DBSCAN to group data into clusters. In the next step, DBCS computes the ratio of positive and negative instances and suggests resampling techniques e.g., undersampling or oversampling to each cluster. For oversampling, it generates the synthetic positive instances within an epsilon (ε) value of each cluster. This method was examined with eleven datasets from KEEL data repository with C4.5 classifier. The results showed a better average AUC and accuracy if model via compared with the using of original data, SMOTE, ROS, and RUS techniques.

X. Xiaolong et al. [7] used the DBSCAN to separate positive instances into three groups including an explicit group of minority class called the core samples, a combination group of majority class, and minority class called the borderline samples, and the isolated minority that located among majorities called the noise samples. The algorithm removed all noise samples before applied different oversampling techniques for remaining groups. The first is applying SMOTE within the core samples, the second is calculating the cluster center in borderline samples and then generated synthetic instances between the cluster center and other instances in the cluster. This work is called DSMOTE. The experiment was done on four UCI datasets which contained two-class and multi-class datasets. The results showed better performances including precision, recall, and F-value while compared with original data, SMOTE, and borderline-SMOTE with J48 decision tree classifier.

M. Lv et al. [8] combined the SMOTE, Boderline SMOTE and boosting techniques such as AdaBoost and cost-sensitive AdaBoost to handle imbalanced credit card clients datasets from UCI. A comparison result show the performance of over-sampled SMOTE-AdaBoost and Borderline SMOTE AdaBoost is slightly worse than the traditional AdaBoost. They noticed on a possibility of using SMOTE with high-dimensional imbalanced data.

N. Netirungroj and E. Pacharawongsakda [9] proposed two steps framework which employs both undersampling and oversampling techniques. This work is called TOP (TwO-levels of Positive Resampling Framework). TOP creates two areas: inner area and outer area. The first area is located near the positive instances and the second area is far from those positive instances. They applied the undersampling technique in the inner area and the oversampling technique to the outer area. An experiment was testing on many UCI imbalanced datasets and evaluated the model performance by using F1, GM, AUROC, and AUPRC. The experimental results were compared with many techniques such as Baseline-SMOTE, ROSE and DBSM. The technique yielded F1 score better than an average improvement across all datasets.

3 DB2SM Methodology

The goal of this work is to improve an oversampling quality by trying to generate a new instance in the proper area of minority class. The process contains two main function: the first is finding the high density area or largest cluster of minority instances (assumed that this is the best area for the minority class) by applying two rounds of DBSCAN, then the second is creating new instances by applying SMOTE in that area. The process was explained in these following steps:

Step 1. The first round of DBSCAN is applied with parameter: epsilon = e, minimum point = minpnts, and add the result clusters in C. The C contains positive clusters (all instances have minority class), negative clusters (all instances have majority class), and mixed cluster (included positive instances and negative instances). All members in each cluster connected to their neighbors within distance $< \; =$ e by DBSCAN algorithm.

Step 2. For each mixed cluster in C, we calculate Euclidean distance between all positive members and all negative members in C.

Step 3. Remove negative instances that have distance less than epsilon (e) from C. This condition reveals a positive group in clusters and separates them from connected by negative instances. (We keep the removed negative instances in an empty dataset called delInstance for later use).

Step 4. Combine all instances from the 3rd step and re-apply the second DBSCAN with the same parameters again. By assumption, C might contain the positive cluster, the negative cluster, and some isolated instances.

Step 5. Choose the largest positive cluster from the 4th step and apply SMOTE with parameter: number of instances = LI from Eq. (2). Then remove duplicated instances by comparing new instances with members in delInstance. (We assume that positive members will less than negative members to prevent bias and over-fitting from new positive members).

Step 6. Appending original data with new instances as the new training data and use this dataset to build a classification model. The processes are shown in Fig. 1.

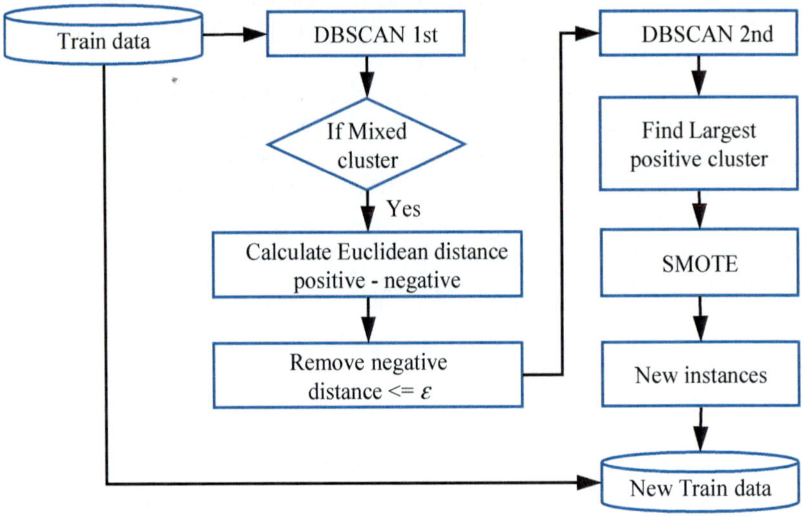

Fig. 1. The methodology

From these steps, the psudo-code was constructed as below.

Algorithm: DB2SM

Input : T : Train dataset contains positive & negative,
ε : Epsilon,
minpnts : Minimum Number of Point
LI : Lack of Information for the minority class (LI)
Output : *New_T* : New Balanced train data

```
C =DBSCAN)T, ε, minpnts // (C : set of cluster
for i =1 to number of cluster in C do
   if C_i is mixed cluster then
      for j =1 to number of positive in C_i do
         for k =1 to number of negative in C_i do
            d_jk =calEuclideanDistance)p_j, n_k(   //p_j :positive, n_k : negative member
            if d_jk <=ε then
               moveInstance)n_k, delInstance(   //delInstance : empty dataset
            end if
         end for
      end for
      appendInstance)D, C_i // (D : empty dataset
   end if
   delCluster)C, C_i( // delete C_i from C
end for
E =DBSCAN)D, ε, minpnts( // E:  set of cluster
C =C + E
Lp =findLargestPositiveCluster)C(
newInstance =SMOTE)Lp, LI // (newInstance : dataset
newInstance =delDuplicate)newInstance, delInstance(
New_T =T + new_instance
return(New_T )
```

4 Experiment and Result

4.1 Datasets

In the experiment, we tested with 20 imbalanced datasets from UCI Machine Learning Repository which characteristics were showed as in Table 1. Each dataset contains only two classes (positive, negative) and have no missing values. Imbalanced Ratio (*IR*) and Lack of Information (*LI*) for the minority class are indicators of the relationship between positive and negative in data which are calculated by these equations.

$$IR = \frac{number\ of\ negative}{number\ of\ positive} \tag{1}$$

$$LI = number\ of\ negative - number\ of\ positive \tag{2}$$

Table 1. Datasets characteristics

Dataset	Attribute	Example	IR	LI
ecoli1	8	336	3.36	182
ecoli2	8	336	5.46	232
ecoli3	8	336	8.60	266
glass0	10	214	2.06	74
glass1	10	214	1.82	62
glass6	10	214	6.38	156
haberman	4	306	2.78	254
iris0	5	150	2.00	50
new-thyroid1	6	215	5.14	145
new-thyroid2	6	215	5.14	145
page-blocks0	11	5,472	8.79	4,354
pima	9	768	1.87	232
segment0	20	2,308	6.02	1,650
vehicle0	19	846	3.25	448
vehicle1	19	846	2.90	412
vehicle2	19	846	2.88	410
vehicle3	19	846	2.99	422
wiscosin	10	683	1.86	205
yeast1	9	1,484	2.46	626
yeast3	9	1,484	8.10	1,158

4.2 Experimental Design and Evaluation

In the experiment, we used RapidMiner Studio version 9.8 as standard tool. All datasets were split into two parts as train data and test data by ratio 70:30. We used a simple Decision Tree as a classification model and evaluated performance with three widely used resampling techniques which are SMOTE, DBCS, DBSM. The experimental procedure is shown in the Fig. 2.

Classification performances are evaluated using statistical measurement tools based on confusion matrix (Table 2.) which are accuracy, AUC and F-measure. The calculation showed in Eqs. (3)–(9).

$$Accuracy = \frac{TP + TN}{TP + TN + FP + FN} \tag{3}$$

$$TP_{rate} = \frac{TP}{(TP + FN)} \tag{4}$$

$$FP_{rate} = \frac{FP}{(FP + TN)} \tag{5}$$

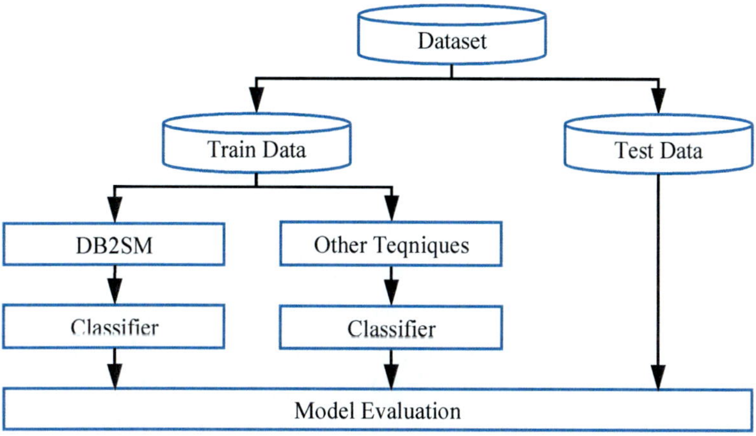

Fig. 2. Design of the experiment

Table 2. Confusion Matrix

	Pridicted as *positive*	Pridicted as *negative*
Actually *positive*	TP	FN
Actually *negative*	FP	TN

$$AUC = \frac{1 + TP_{rate} + FP_{rate}}{2} \quad (6)$$

$$Precision = \frac{TP}{(TP + FP)} \quad (7)$$

$$Recall = \frac{TP}{(TP + FN)} \quad (8)$$

$$F - measure = \frac{2*Recall*Precision}{Recall + Precision} \quad (9)$$

4.3 Results

The results showed in Tables 3, 4 and 5. The best results displayed in bold.

Table 3. Accuracy result comparison

Dataset	Original	SMOTE	DBCS	DBSM	DB2SM
ecoli1	**0.9109**	0.8911	0.8812	0.8911	0.8317
ecoli2	0.9208	0.9307	0.9406	0.9307	**0.9703**
ecoli3	0.8812	0.8614	0.8911	0.7822	**0.9307**
glass0	**0.8438**	0.7500	0.8125	0.7812	**0.8438**
glass1	0.7188	0.6250	0.6875	0.5938	**0.7344**
glass6	0.9844	**1.0000**	0.9688	0.9844	**1.0000**
haberman	**0.7065**	0.6196	0.6522	0.6087	0.6739
iris0	1.0000	1.0000	1.0000	1.0000	1.0000
new-thyroid1	**0.9688**	0.9531	**0.9688**	0.9531	0.9531
new-thyroid2	0.9219	0.9531	0.9531	0.9531	**0.9688**
page-blocks0	0.9604	0.9683	0.9287	0.9562	**0.9714**
pima	0.6957	0.6696	0.5957	0.6696	**0.7000**
segment0	**0.9928**	0.9884	0.9855	0.9884	0.9913
vehicle0	0.9055	0.8031	0.7205	0.8031	**0.9252**
vehicle1	0.7323	0.3858	0.3937	0.3858	**0.7677**
vehicle2	0.8976	0.8858	0.9134	0.8858	**0.9409**
vehicle3	**0.7598**	0.3819	0.4213	0.3819	0.7047
wiscosin	0.9463	0.9512	**0.9659**	0.9610	0.9512
yeast1	0.7326	0.7348	0.4629	0.7348	**0.7551**
yeast3	**0.9618**	0.9528	0.9169	0.9483	**0.9618**

Table 4. AUC result comparison

Dataset	Original	SMOTE	DBCS	DBSM	DB2SM
ecoli1	0.8640	**0.9340**	0.8960	0.8590	0.8270
ecoli2	0.8410	0.8590	0.8910	0.9160	**0.9330**
ecoli3	0.7380	0.8130	0.8840	0.8250	**0.8510**
glass0	0.8390	0.8110	0.7950	0.7820	**0.8890**

(*continued*)

Table 4. *(continued)*

Dataset	Original	SMOTE	DBCS	DBSM	DB2SM
glass1	0.6760	0.6320	0.6930	0.6540	**0.7180**
glass6	0.8890	**1.0000**	0.9640	0.9820	**1.0000**
haberman	0.5200	0.5520	**0.5980**	0.5470	0.5710
iris0	0.5000	0.5000	0.5000	0.5000	0.5000
new-thyroid1	0.9750	0.8450	**0.9810**	0.8450	0.8910
new-thyroid2	0.8440	0.9270	0.8850	0.9270	**0.9730**
page-blocks0	0.8230	**0.9460**	0.9330	0.9420	0.9140
pima	0.7040	**0.7460**	0.7360	**0.7460**	0.7340
segment0	**0.9720**	0.9690	0.9680	0.9690	0.9710
vehicle0	0.9170	0.8630	0.8050	0.8630	**0.9390**
vehicle1	0.5000	0.6050	0.5790	0.6050	**0.6370**
vehicle2	0.8870	0.9070	0.8850	0.9070	**0.9180**
vehicle3	**0.6530**	0.5960	0.6400	0.5960	0.6460
wiscosin	0.9280	0.9470	**0.9630**	0.9540	0.9380
yeast1	0.6950	**0.7610**	0.6180	**0.7610**	0.7030
yeast3	0.9050	0.9430	0.8830	**0.9450**	0.9250

Table 5. F-measure result comparison

Dataset	Original	SMOTE	DBCS	DBSM	DB2SM
ecoli1	**0.8000**	0.7843	0.7500	0.7660	0.7018
ecoli2	0.7778	0.8205	0.8500	0.8108	**0.9231**
ecoli3	0.2500	0.4615	**0.5600**	0.3889	0.3636
glass0	**0.7619**	0.6190	0.7273	0.6667	0.7368
glass1	0.5909	0.5556	0.6154	0.5517	**0.6667**
glass6	0.9412	**1.0000**	0.9000	0.9474	**1.0000**
haberman	0.4000	0.4068	0.4286	0.4000	**0.4444**
iris0	1.0000	1.0000	1.0000	1.0000	1.0000
new-thyroid1	0.9000	0.8421	**0.9091**	0.8421	0.8421
new-thyroid2	0.7619	0.8696	0.8696	0.8696	**0.9167**
page-blocks0	0.7735	0.8452	0.7221	0.8125	**0.8554**
pima	0.4068	0.6346	0.6235	0.6346	**0.6634**

(continued)

Table 5. (*continued*)

Dataset	Original	SMOTE	DBCS	DBSM	DB2SM
segment0	**0.9701**	0.9529	0.9419	0.9529	0.9643
vehicle0	0.8065	0.7024	0.6502	0.7024	0.8652
vehicle1	**0.8110**	0.4730	0.4797	0.4730	0.3059
vehicle2	0.7937	0.7914	0.8333	0.7914	0.8855
vehicle3	0.2469	0.4530	**0.4806**	0.4530	0.4681
wiscosin	0.9209	0.9265	**0.9510**	0.9429	0.9254
yeast1	0.4848	0.6289	0.5286	**0.6312**	0.5856
yeast3	0.8172	0.7879	0.6783	0.7723	**0.8350**

Table 6. Parameters and Results

Dataset	ε	*minpts*	Accuracy	AUC	F-measure
ecoli1	0.50	16	0.8317	0.8275	0.7018
ecoli2	0.15	14	0.9703	0.9327	0.9231
ecoli3	0.10	1	0.9307	0.8508	0.3636
glass0	0.50	5	0.8438	0.8886	0.7368
glass1	0.50	3	0.7344	0.7179	0.6667
glass6	0.35	2	1.0000	1.0000	1.0000
harberman	0.10	2	0.6739	0.5711	0.4444
iris0	0.20	2	1.0000	0.5000	1.0000
new-thyroid1	0.10	11	0.9531	0.8907	0.8421
new-thyroid2	0.10	16	0.9688	0.9734	0.9167
page-blocks0	0.10	8	0.9714	0.9142	0.8554
pima	0.10	14	0.7000	0.7425	0.6634
segment0	0.85	5	0.9913	0.9711	0.9643
vehicle0	0.10	7	0.9252	0.9394	0.8652
vehicle1	0.10	2	0.7677	0.6374	0.3059
vehicle2	0.10	13	0.9409	0.9185	0.8855
vehicle3	0.11	11	0.7047	0.6458	0.4681
wiscosin	0.10	17	0.9512	0.9376	0.9254
yeast1	0.05	1	0.7551	0.7026	0.5856
yeast3	0.10	7	0.9618	0.9248	0.8350

5 Conclusion and Discussion

In this paper, we proposed DB2SM as one of the good choices to handle imbalanced data. Our method employs two rounds of DBSCAN to find the proper area of minority class and then applying SMOTE technique to generate new minority instances in that area. The experimental results show that AUC and F-measure for all resampling techniques are better than the original train data in various datasets. Moreover, the DB2SM won 10 cases compared to the baseline, SMOTE, DBCS, and DBSM.

Along the process, we changed parameters such as an epsilon and minpnts to see how each dataset forming its appearances. The epsilon was set in the range 0.01–1.00 step by 0.01 from a normalized distance. The minpnts prefered from 1 to the number of positive instance in cluser. We found that changing the sensitive parameters affected performance. This might from the ambiguity between minority instances, noise, and outlier. However, some parameters can caused to generate a lower implicit instance. For example in Table 6, ecoli3 and yeast1 datasets used small epsilon and minpnts that allowed DBSCAN to create a tiny cluster and forced SMOTE to upsampling in this limited area. The result gained a good accuracy but F-measure was poor. It is possible that the model predicted the negative in high corrective rate but very low for positive.

6 Future Work

In the experiment, we used simple SMOTE as an oversampling method, but there are more variations of SMOTE including: SMOTE-NC, Borderline-SMOTE, SVM-SMOTE, ADASYN [10] that can increase the quality of result data. Furthermore, the experiment was settled on small datasets which consisted of tiny attributes then we can use the simply Euclidean DBSCAN [11] for gaining not much time by the average run-time complexity of DBSCAN is about O(nlog n). [12] In other scenarios such as a high dimensional data or more than two-class problem, these will effect to the computational time then we must discover an additional clustering technique to resolve this kind of problem.

References

1. Longadge, R., Dongre, S., Malik, L.: Class imbalance problem in data mining: review. Int. J. Comput. Sci. Netw. (IJCSN) **2**(1) (2013)
2. Ali, H., Salleh, M., Saedudin, R., Hussain, K., Mushtaq, M.: Imbalance class problems in data mining: a review. Indonesian J. Electr. Eng. Comput. Sci. **14**(3) (2019)
3. Ester, M., Kriegel, H., Sander, J., Xu, X.: A density-based algorithm for discovering clusters in large spatial database with noise. In: KDD, vol. 96, no. 34, pp. 226–231 (1996)
4. Chawla, N.V., Bowyer, K.W., Hall, L.O., Kegelmeyer, W.P.: SMOTE: synthetic minority over-sampling technique. J. Artif. Intell. Res. **16**, 321–357 (2002)
5. Sanguanmak, Y., Hanskunatai, A.: DBSM: the combination of DBSCAN and SMOTE for imbalanced data classification. In: 2016 13th International Joint Conference on Computer Science and Software Engineering, JCSSE 2016 (2016)
6. Verma, M.K., Xaxa, M.K., Verma, S.: DBCS: density based cluster sampling for solving imbalanced classification problem. In: 2017 International conference of Electronics, Communication and Aerospace Technology (ICECA) (2017)

7. Xiaolong, X., Wen, C., Yenfei, S.: Over-sampling algorithm for imbalanced data classification. J. Syst. Eng. Electron. **30**(6), 1182–1191 (2019)
8. Lv, M., Ren, Y., Chen, Y.: Research on imbalanced data: based on SMOTE-AdaBoost algorithm. In: The 3rd International Conference on Electronic Information Technology and Computer Engineering (EITCE), pp. 1165–1170 (2019)
9. Netirungroj, N., Pacharawongsakda, E.: TOP: an efficient two-levels of positive resampling framework for class imbalanced data. In: 2018 IEEE International Conference on Big Data, Cloud Computing, Data Science & Engineering (BCD) (2018)
10. Wijaya, C.: 5 SMOTE Techniques for Oversampling your Imbalance Data. https://towardsdatascience.com/5-smote-techniques-for-oversampling-your-imbalance-data-b8155bdbe2b5. Accessed 14 Spet 2020
11. Gan, J., Tao, Y.: On the hardness and approximation of Euclidean DBSCAN. ACM Trans. Datab. Syst. **42**(3), Article 14 (2017)
12. Schubert, E., Sander, J., Ester, M., Kriegel, H., Xu, X.: DBSCAN revisited, revisited: why and how you should (still) use DBSCAN. ACM Trans. Datab. Syst. **42**(3), Article 19 (2017)

Intelligent Fruit Recognition System Using Deep Learning

Ha Huy Cuong Nguyen[1(✉)], Anh Tuan Luong[2], Trung Hai Trinh[3], Phan Hieu Ho[1], Phayung Meesad[4], and Thanh Thuy Nguyen[5]

[1] The University of Danang, Danang, Vietnam
nhhcuong@sdc.udn.vn, hophanhieu@ac.udn.vn
[2] Dong A University, Da Nang City, Vietnam
tuanla@donga.edu.vn
[3] Vietnam-Korea University of Information and Communication Technology, Danang, Vietnam
tthai@vku.udn.vn
[4] Faculty of Information Technology and Digital Innovation, King Mongkut's University of Technology North Bangkok, Bangkok, Thailand
phayung.m@itd.kmutnb.ac.th
[5] Faculty of Computer Science, VNU University of Engineering and Technology, Hanoi, Vietnam
nguyenthanhthuy@vnu.edu.vn

Abstract. Industrial Revolution 4.0 has made us people more professional, automating all production stages from office work to project work on farms. In the precision agriculture, it is very urgent to bring new and effective solutions to using artificial intelligence for people to use and improve the manual steps gradually, and increase the automation feature. So, automatic fruit recognition technique is the latest trend and effective technique in precision agriculture. This paper proposes a technical solution for fruit classification using deep learning. Automatic fruit identification using computer vision is considered a challenging task. This is because there are similarities between fruits and changes in the external environment such as light affect the fruit recognition model. Most previously implemented techniques have some limitations since their testing and evaluation is done using a limited set of data sets. Some implementations, does not consider changes to the external environment for the image are considered in this implementation. In this paper, exploring part of the deep learning algorithms was achieved and discovered strengths and weaknesses for these algorithms. The knowledge was gained on deep learning and a model was built that could recognize fruits from images.

Keywords: Fruit recognition · Computer vision · Image processing · Image classification · Convolution Neural Network

1 Introduction

Artificial Intelligence (AI) is a system of machines or application software programs created by humans that can automatically perform intelligent behaviors

P. Meesad et al. (Eds.): IC2IT 2021, LNNS 251, pp. 13–22, 2021.
https://doi.org/10.1007/978-3-030-79757-7_2

such as the ability to communicate with humans, can learn and adapt intelligently to many situations. The narrow areas of AI include: Machine Learning; Deep Learning; Virtual Reality; Augmented Reality; Mixed Reality and Computer Vision. In the content of the topic, we mention the field of deep learning (Deep Learning). Deep Learning models, such as Convolutions Neural Networks (CNNs), have been successfully applied in image classification, text and speech recognition problems.

In recent years, we have witnessed many outstanding achievements in Computer Vision. Large image processing systems such as Facebook, Google or Amazon have included in their products intelligent functions such as facial recognition, self driving car development or automatic delivery.

Researchers in the country recently are focusing on the field of Machine Learning. Machine Learning is a method of analyzing data from which to automate the construction of analytical models. It can be said that this technology is very promising to bring optimal support for businesses with many real world applications, such as voice recognition and image recognition.

Advantages of deep learning models that automatically learn the characteristics of data to establish the new features and data classification. The problem of classification when meeting a large number of dimensions is often difficult. The classification model gives good results on the training set but low results on the test set. The common problem is that the data with the number of dimensions is too large up to thousands of dimensions and the data is separated in a large number of dimensions, so finding a good classification model is difficult because there are too many possibilities model selection.

In intelligent fruit recognition and precise agriculture, timely and accurate estimation of the maturity of the orchard will help improve and improve quality, ensure fruit yield and harvest planning. In addition, timely and accurate determination of the ripening period in the fruit maturity stages will allow to reduce farming costs, post-harvest storage costs and increase economic value, to meet the requirement market. These issues are of great concern in the direction of agricultural development to take advantage of high technology in the coming time in Vietnam.

The trained human investigators do the inspection for the quality of fruit by seeing and feeling. This method is significantly inconsistent, fickle and decisions may vary when considered by different investigators. In this type of environment, the analysis of fruit is done for several aspect criterion and it is a task that forms a sequence that includes the repetition of the actions or events. For this purpose, machine vision systems can be best suited for conventional analysis and quality assurance. In precise agriculture, computer vision systems and image processing are already a growing research area which is a significant analyzing technique.

To overcome all the problems mentioned above, the aim is to develop an automated system for recognition of particular fruits using machine learning techniques.

Image processing and computer vision techniques are used on a large scale today and are being applied in many domains or streams. Among these domains, fruit recognition from image is one of them. Fruit recognition is a very challenging task as fruits may contain various similarities between them. Image recognition needs more computation power than most of the text base data classification used widely. However, the fruit recognition model can be beneficial for the people and can be used in a less expensive device.

Object detection is a specific application or process of detecting an object from the specified scene using a certain method or measure. Before the emergence of deep learning technology, the methods of finding the object from a scene were achieved by some prior knowledge and by establishing the mathematical models.

The work is organized as follows: Sect. 2 provides the Related Works; Sect. 3 presents Design and Implementation; Sect. 4 presents Performance Evaluation and Sect. 5 presents our conclusions and suggestions for future work finally.

The paper emphasizes on the use of image processing and computer vision technology that is used in the field of food industry and agriculture. Size, color, shape, texture and defect are the most important quality characteristics of fruits.

2 Related Works

Researcher Shadman [1] and colleagues proposed improved solution developed by Mango Yolo architecture (based on YOLO-v2, YOLO-v3, YOLO-v3 [2]) to develop show mangoes in the garden with an average accuracy of 0.983.

Dang et al. [3] proposed a system which specified the use of image processing and computer vision technology in the field of food industry and agriculture. Size, color, shape, texture and defect are the most important quality characteristics of agricultural products. The manual task of inspection of food done by the investigators is replaced by the computer vision systems which generated more authentic, equitable and non-destructive results. The quality inspection of computer vision included four main steps, namely, acquisition, segmentation, feature extraction and classification.

Li et al. [4] proposed a system that presents object detection based on deep learning for small samples. The proposed system contains three modules: syntactical samples generator, object detection network, and semantic relevant detection. The detection of indoor multiple small objects could be achieved with the use of this system. With considerations of subjective and objective approaches, the experimental results concluded that the proposed system produced better results than the other existing systems.

Tian et al. [5] proposed a system that uses the classical methodologies of object detection. The classic methodologies and the deep learning methodologies used in object detection are studied with respect to its relation and differences between them. Then on the basis study, the limitations and design of the model for object detection is proposed.

Avazjon et al. [6] proposed a system for recognizing fruits in treetops aimed for automatic fruit picking. For this purpose, an expert system must be equipped to an intelligent robotic fruit picker. The robotic fruit picker must analyze the information obtained by vision sensors. The vision sensors used are cameras and laser scanners. The robot must detect physical objects in the observed scene and must decide whether the detected object is fruit or leaf or branch. In uncontrolled situations, such as overlapped fruits or shades, it must make correct decisions. For the detection of fruits on treetop, an approach based on detection of convex surfaces is used. It is assumed that most of the fruits have convex shape. Hence, it is used to distinguish fruits from leaves and branches of the tree.

Rajeshwari et al. [7] proposed a system for accurate apple recognition with robotic fruit picking. A machine learning method is applied to a detection system to improve the accuracy of fruit recognition and achieve a better performance. To help the robot to perform the operation, a three-dimensional (3D) data of a scene is captured directly and accurately by laser radar or depth camera.

Bargoti et al. [8] proposed a system of optimal structure that is applied by design pattern recognition and classification. For solving of pattern recognition and classification tasks, the main problem in artificial networks is the evaluation of the number of hidden neurons. Multi layer perception is the most useful artificial neural network for estimating the functional structure in classification.

Munera et al. [9] proposed a system for green fruit feature extraction. This is done with a combination of the techniques of color model conversion, thresholds, histogram equalization, spatial filtering with Laplace and Sober operators and Gaussian blur. Along with this system, in order to recognize and count the number of green fruits, an algorithm is built and tested.

3 Design and Implementation

The automatic fruit recognition system is developed to make use of computer vision and image processing techniques in the food industry. Segmentation is used to simplify image analysis and image features and descriptors are used for feature extraction. On the basis of color, shape, size and texture, the quality analysis of fruits are done. [2] The architecture of the system is depicted in Fig. 1. The user must give an input image to the system by specifying the path of the image. After that the user will get the output as the name of fruit along with the input image. Fruit prediction is done using the algorithm Deep Convolution Neural Network in Keras. There are two models as one for fresh fruit recognition and second for rotten fruit recognition.

3.1 Fresh Fruit Recognition

There are total 13 classes of fruits used for training and validation purposes. These 13 classes of fruits include apple, blueberry, lemon, mango, orange, pear, pineapple, pomegranate and walnut. The required data set of these fruits were taken from the Fruit-360 data set of Kaggle. The number of fruit images taken for the purpose of training and validation are represented in Fig. 2.

The first step will be to provide an image data set to the system. The data set comprises of 13 classes. The second step will comprise of storing the data set into its memory. This image data set will be then used for pre processing.

Table 1. Fruit images distributed used for training and validation purpose

Number	Name	Train imgs	Test imgs
0	Apple	6404	2134
1	Avocado	918	309
2	Banana	1430	484
3	Blueberry	462	154
4	Lemon	982	330
5	Mango	916	308
6	Orange	479	160
7	Pear	5037	1689
8	Pineapple	983	329
9	Pomegranate	492	164
10	Rambutan	492	164
11	Strawberry	1230	410
12	Walnut	735	249

Pre-processing of the image is done to remove the noise and get clear image. The third step will comprise of feature extraction from it. The features such as edge detection, texture and color are included. Edge detection can be done by detecting the edges of the image. Texture is used to detect the skin disease and defects present on the fruit. Color features are used to identify the color of the fruit including mean and variance of the RGB color model. Based on the features extracted, Convolution Neural Network (CNN) in Keras will be trained.

Due to the high level of performance given by deep learning, a subset of machine learning, it is becoming very popular across many types of data. CNN is built for classifying images which makes use of deep learning. The building of CNN is easier with the use of Keras library in Python. The model uses layers of CNN such as Conv2D and MaxPooling2D. Conv2D is also called as 2D Convolution Layer. This layer produces tensor of outputs by creating convolution kernel that is convoluted with the input layer. MaxPooling2D is used for max pooling operation for spatial data. Spatial data can be defined as representation of information about a physical object by numerical values. Selecting the maximum element from the region of the feature map covered by the filter is the operation performed by max pooling layer. To reduce the dimensions of the feature maps, pooling layers are used so. Thus, it can be concluded that use of pooling layers reduces the number of parameters to learn and the amount of computation performed in the network.

In compiling the model, a loss function, an optimizer and a list of met rices are used. The model includes categorical cross entropy as a loss function and admix as an optimizer [10]. These two were used because after trying with several other loss functions and optimized, amongst them, they gave the best results. A summary of the model is generated using the function after carrying out the first 3 steps. Figure 3 shows the CNN architecture designed for fresh fruit recognition.

The fourth step will be predicting the image based on trained model. The test image will be used in this step. Users need to specify the path of the image that is where the image is stored in the computer system. The image path should be precise. If the image path is incorrect, then prediction will not be done, and error will be thrown. If the path is correct, then users will get the desired output. Trained model will then check for all the possibilities from the given image. The possibilities will include getting the model and weights.

For the purpose of graphical user interface, Flask was used. Flask is a web framework that provides tools, libraries and technologies that allow one to build a web application. This web application is designed and implemented with sample input and output as depicted in Fig. 3.

There are total 3 classes of fruits used for training and validation purposes. These 3 classes of fruits include apple, orange and banana. The required data set of these fruits was taken from the Fruit-360 data set of Kaggle. The number of fruit images taken for the purpose of training and validation are represented in Table 1. The procedure used is the same as used for fresh fruit recognition. In compiling the model, a loss function, an optimizer and a list of met rices are used. The model includes categorical cross entropy as a loss function and ada max as an optimizer. These two were used because after trying with several other loss functions and optimizers, amongst them, they gave the best results. A summary of the model is generated using the function after carrying out the first 3 steps.

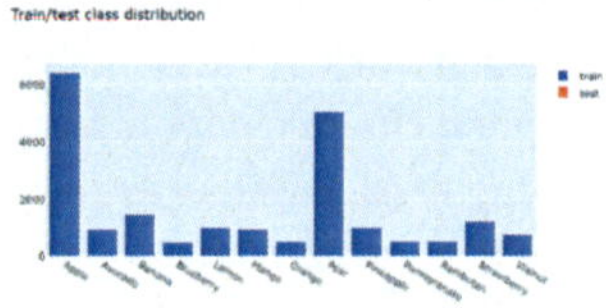

Fig. 1. Summary of training rotten fruit recognition model

No GUI for rotten fruit recognition was implemented because there were only 3 classes of fruits used. But console outputs were generated for this category of recognition. The sample inputs and outputs of the working of system is shown in Fig. 3.

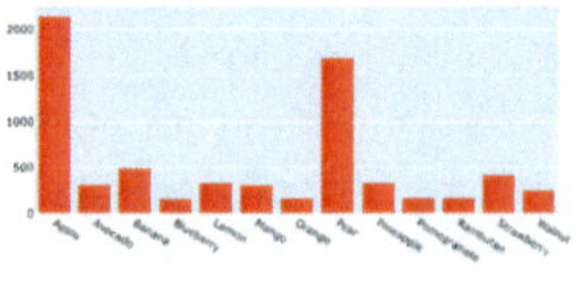

Fig. 2. There is a near identical class distribution in train and test sets

3.2 Improved ResNet50 Fruit Recognition Model

There are total 13 classes of fruits used for training and validation purposes. The required data set of these fruits was taken from the Fruit-360 data set of Kaggle. The number of fruit images taken for the purpose of training and validation are represented in Table 1. This is paper, approach ResNet50 Fruit Recognition Model Architecture from this model, we recommend three more layers: Dropout, Dense and Softmax.

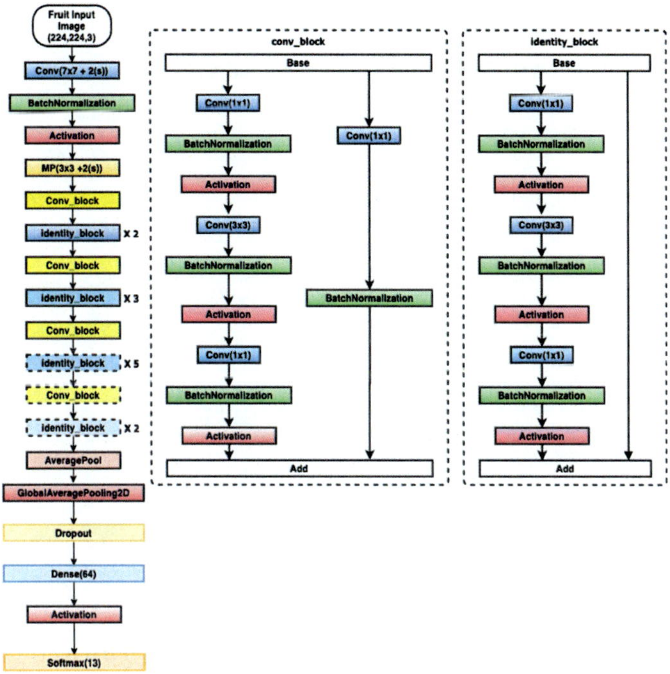

Fig. 3. ResNet50 fruit recognition model architecture

Figure 3. Example network architectures for ImageNet. Based on the above plain network, in this model insert 3 shortcut connections which turn the network into it counterpart residual version. The models are trained on the 5.30 milltion training images, and evaluated on the 50k validation images.

4 Performance Evaluation

The implemented models were checked for classification performance with different optimizers. This is done by representing the values for training accuracy, training loss, validation accuracy and validation loss, respectively. These variations are calculated for fresh fruit recognition with use of different optimizers as follows (Fig. 4).

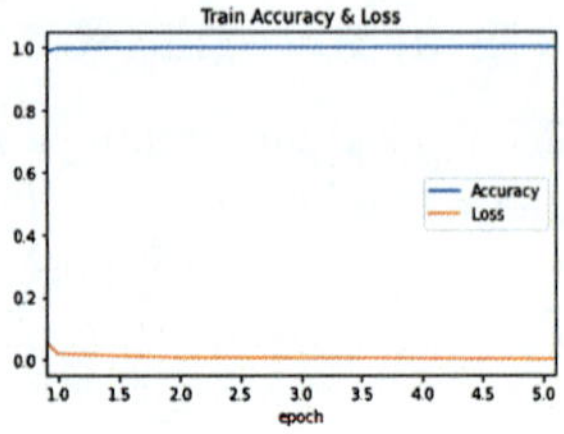

Fig. 4. ResNet50 fruit recognition train and validation

After comparing with optimizers, the best results were obtained from soft max optimizer than others. Hence use of soft max optimizer for fresh fruits recognition was done. Also, false predictions for some fruits were observed. They were as follows.

The image of Pomegranate was inputted as the test image as shown in Fig. 5, and the output was displayed as Apple with 98.88% of accuracy as shown in Fig. 6. Thus, prediction was false because the model considered the color feature of the input image and the prediction for the color of Apple was more in percentage as compared to Pomegranate.

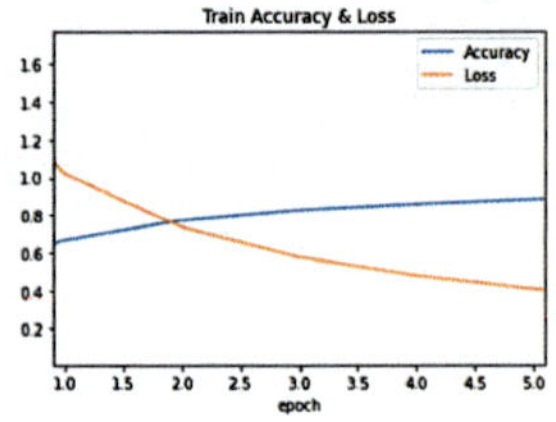

Fig. 5. Convolutional neural network fruit recognition model train

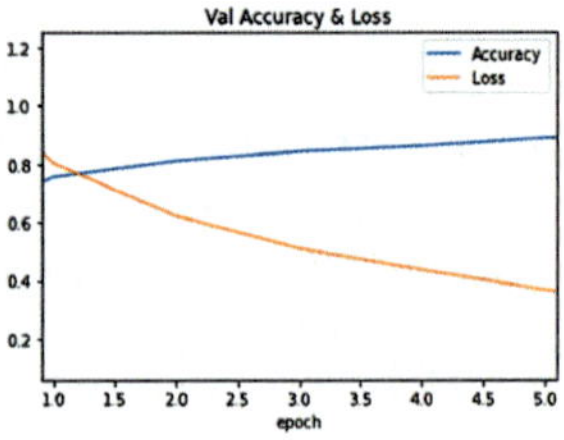

Fig. 6. Convolutional neural network fruit recognition model validation

In the paper, the refining and release techniques were applied to all layers of the input image data set trained from the improved R-CNN model. In order to maintain the calculation results through the trained steps, the set of typical classes of the trained base model. Experiments conducted to extract some of the selected data a subset compared to the initially divided classes. In the experiment script in all the proposed method retains deep learning rate is 0.00005 for the initial class and 0.000055 for the final layer. After 20 min of training, the accuracy achieved is 99.33%.

5 Conclusion

Currently, in precision agriculture, it is very necessary to apply artificial intelligence solutions to create a correct agriculture from the selection of raw materials to the final stage of harvesting. So here is the article we emphasize on the use of computer vision and image processing technology used in the food industry and agriculture sectors. In addition to previously known identifiers: size, color, shape, texture and defects are the most important quality characteristics of the fruit. We fixed the problems encountered during the manual fruit inspection, with the computer vision systems in use. The implemented system provides authentic, fair and non-destructive results for the fruit grading task. In the article, we have conducted research on previously achieved deep learning algorithms from which we discover the strengths and weaknesses of these algorithms. Knowledge is acquired through deep learning and a built-in model is able to recognize fruit from images.

Acknowledgment. The authors wish to express their appreciation to the Ministry of Education and Training for supporting this research project as part of the Ministerial Program of Science and Technology CTB.2021.DNA. "Research on applying deep learning model to recognize ripe pineapple period in Quang Nam - Da Nang".

References

1. Sakib, S., Ashrafi, Z., Siddique, M., Bakr, A.: Implementation of Fruits Recognition Classifier using Convolutional Neural Network Algorithm for Observation of Accuracies for Various Hidden Layers. arXiv e-Journal, pp. 1–4 (2019)
2. Alexey, A.B.: Yolo mark Apple detection during different growth stages in orchards using the improved YOLO-V3 model (2018)
3. Chung, D.T.P., Van Tai, D.: A fruits recognition system based on a modern deep learning technique. In: Journal of Physics: Conference Series, vol. 1327, pp. 1–5 (2019)
4. Hou, L., Wu, Q., Sun, Q., Yang, H., Li, P.: Fruit recognition based on convolution neural network. In: IEEE, 12th International Conference on Natural Computation, Fuzzy Systems and Knowledge Discovery (ICNC-FSKD), pp. 18–22 (2016)
5. Tian, Y., Yang, G., Wang, Z.: Apple detection during different growth stages in orchards using the improved YOLO-V3 model. Comput. Electron. Agric. **157**, 417–426 (2019)
6. Marakhimov, A.R., Khudaybergenov, K.K.: Approach to the synthesis of neural network structure during classification. Int. J. Comput. 20–26 (2020)
7. Rajeshwari, P., Abhishek, P., Srikanth, P., Vinod, T.: Object detection: an overview. Int. J. Trend Sci. Res. Dev. (IJTSRD) **3**, 1663–1665 (2019)

8. Bargoti, S., Underwood, J.: Deep fruit detection in orchards. In: 2017 IEEE International Conference on Robotics and Automation (ICRA), Singapore, pp. 3626–3633 (2017). https://doi.org/10.1109/ICRA.2017.7989417
9. Munera, S., Amigo, J.M., Blasco, J., Cubero, S., Talens, P., Alexios, N.: Ripeness monitoring of two cultivars of nectarine using VIS-NIR hyperspectral reflectance imaging. J. Food Eng. **214**(3), 29–39 (2017)
10. Nguyen, H.H.C., Nguyen, D.H., Nguyen, V.L., Nguyen, T.T.: Smart solution to detect images in limited visibility conditions based convolutional neural networks. In: Advances in Computational Collective Intelligence. ICCCI 2020. Communications in Computer and Information Science, vol. 1287, pp. 641–650. Springer, Cham (2020). https://doi.org/10.1007/978-3-030-63119-2-52

Design and Development of a Dynamic and Efficient PII Data Loss Prevention System

Somchart Fugkeaw(✉), Khanadech Worapaluk, Atiwit Tuekla, and Supakorn Namkeatsakul

School of Information, Computer, and Communication Technology, Sirindhorn International Institute of Technology, Thammasat University, Pathum Thani 12120, Thailand
somchart@siit.tu.ac.th, pete@it-i.net

Abstract. Data Loss Prevention (DLP) tools have been widely employed by many enterprises for their data breach detection and prevention. Currently, there are a number of DLP tools in the market but there are no existing tools tailored for PII (Personal Identifiable Information) sources. In this paper, we present the design and the development of our PII DLP tool called PII Guardian supporting PII data breach detection and prevention. Our PII Guardian system is based on a rule-based approach enabling the accurate and efficient detection of known attacks. Our system classifies the detected data breach activities based on their impact and provides the preventive actions. For the implementation, we developed the prototype system as a proof of concept to demonstrate the accuracy and practicality of our proposed system.

Keywords: DLP · PII · Data Privacy · Rule-based approach · Data breach

1 Introduction

Nowadays personal data protection has been legalized in many countries. For example, General Data Protection Regulation (GDPR) has been enforced for data protection and privacy in the European Union (EU) and the European Economic Area (EEA). It also addresses the transfer of personal data outside the EU and EEA areas [1]. Key data type identified in the regulation is personal identifiable information or PII. PII is any data that can be used to identify a specific person or an individual [2], such as name and surname, home address, email address, IP address, credit card information, etc. Generally, PII that is considered to be protected is sensitive ones since if it is leaked, lost, or compromised, it could result in personal threat, embarrassment, or inconvenience to an individual [3].

As for the privacy concern, PII is thus a target information that needs to be protected from unauthorized access or any unlawful purpose. In Thailand, the Personal Data Protection Act (PDPA) where most contents are adopted from GDPR is scheduled to be enforced in May 2021. As a result, all organizations in Thailand are now actively seeking for the potential PDPA compliance solution. To minimize the risk of data breaches, organizations generally employ DLP systems as a defense mechanism to cope with data leakage and privacy breach of their customers and employees. Therefore,

P. Meesad et al. (Eds.): IC2IT 2021, LNNS 251, pp. 23–33, 2021.
https://doi.org/10.1007/978-3-030-79757-7_3

it is a real challenge to investigate the regulations and propose a DLP tool as a part of PDPA requirements.

Significantly, DLP is one of data privacy solutions to compliment the preventive control for data security and privacy. It is generally exploited to detect potential data breach activities that occur in the system as well as to prevent the undesired incidents. Several organizations have employed DLP to protect their valuable and sensitive information from breach such as leakage, tampering, and loss. The primary function of DLP tools is to scan data breach activities in the platforms where they are installed such as computer, network, or cloud system. Once the suspicious activities are detected, the tool sends alert to system administrator and renders some preventive actions such as block network port, disable user right. Technically, DLP solutions can be categorized either as to detect patterns of known attacks (signature-based or rule-based) or try to find deviations from normal behavior (anomaly-based). Signature-based solution or rule-based solution provides accurate detection for known attacks, while anomaly-based method needs to be trained to find unknown attacks. The latter technique is suitable for complex system environment but it introduces high false positive error. Even though DLP tool provides desirable features for data breach and data loss prevention, its generalization renders superfluous results that are considered as overheads for the specific PII detection. In other words, current off-the-shelf DLP tools are not tailored to work on PII data specifically. The results gained form typical DLP tools are required to be filtered for PII-specific set before they can be used for the PDPA analysis. In addition, the computation and communication overheads of most DLP tools are generally high since the DLP services or system agents need to be available for scanning and examining all file directories in the endpoint systems or throughout the networks.

In this paper, we aim at designing and developing the PII DLP application based on our proposed detection technique called an adaptive signature-based detection approach that specifically supports the detection and prevention of PII data breach to satisfy the PDPA. For the reliability of the tool, it is based on the completeness and accuracy of a set of rules defined.

As for the adaptive feature, the anomaly-based is applied when the legitimate users perform anomalous activities, and they are flagged. We then update the signature rules. Also, we introduce two-phase detection protocol to discover the data breach and detect the type of attacks. The active monitoring feature is exploited to monitor the PII inventory where a number of threads is pointed to the PII sources in the computer.

To the best of our knowledge, we provide the first attempt for the development of the PII DLP system to support the PDPA requirement. The contributions of this paper can be summarized as follows.

1. We design and develop PII Guardian as a novel PII DLP tool based on a rule-based detection approach to support dynamic and computationally cost-effective PII data breach detection.
2. We devise a data breach prevention mechanism based on the impact analysis to classify the detected activities and generate the preventive actions to handle the cases.
3. We present the design and development of our tool to demonstrate that our proposed system can effectively detect the PII data breach occurrences.

The remainders of this paper are organized as follows. Section 2 discusses some works related to ours. Section 3 presents our proposed system. Section 4 describes the evaluation and system implementation. Section 5 gives the conclusion.

2 Related Work

This section describes some existing DLP solutions based on signature-based or rule-based solution, anomaly-based or learning-based solution and hybrid solution.

Signature-Based or Rule-Based Solution
As of the simplicity of the implementation of the signature-based or rule-based approach, there are several tools such as SPIRION [4], CUSpider [5], and CipherCloud [6] using regular expressions, custom keywords to detect sensitive information based on massive dictionary. CipherCloud provides pre-defined rule templates to address multiple global regulations such as PCI, HIPAA, GDPR, GLBA, etc." as a feature highlight.

These tools need to do exhaustive detection and they are not suitable for complex system environment where user behaviors and transactions are unpredictable.

Signature-based solution is applied in Intrusion Detection Systems (IDS) as the common practice for the detection method. For example, SNORT [7], a lightweight IDS uses the signature-based detection as the core component of the system. It captures the network traffic and blocks the anomalous transactions satisfying the predefined rules. NetSTAT [8], a network IDS uses the signature-based detection by analyzing the state of each network connection based on parameters in network table (e.g., routing table, DNS, etc.) against the state transition scenario database.

Signature-based detection is known to give an excellent detection accuracy for the previously known intrusions [9]. However, it has difficulty in detecting unknown intrusions or zero-day attacks. This is because the unknown intrusions are not listed in the predefined rule and the system that uses this approach tends to leave the unknown intrusions as they are considered as normal activities.

Anomaly-Based or Learning-Based Solutions
Basically, the system automatically learns a model of normal behavior by observing past activities and giving flag deviations from the model as anomalies. Different approaches, methods, and sets of features are used to learn the model of underlying anomaly-based solutions. However, it tends to produce high false positive rate and its complexity is high. For example, Caliskan-Islam et al. [10] applied Naive Bayes and Support Vector Machines (SVM) to detect privacy leaks from tweets using a carefully crafted training data. This approach needs a high volume of training data set to make the system reliable and to reduce the false positive rate. Therefore, the system is able to predict malicious and suspicious activities. However, the availability of attack datasets is typically scarce, making it difficult to learn a sophisticated model. In [11], the authors proposed a learning-based detection models that inspect the files in real time allowing the system to immediately respond to any behavior or events violating data sensitivity compliance. In this approach, the system detects sensitive content by leveraging its semantic contextual information through various machine learning and deep learning techniques. Even

though the proposed scheme is able to detect sensitive content at various degrees of granularity within the document, the utilization in supporting several kinds of documents or unstructured data may not be applicable and its complexity is the burden for the integration.

Hybrid Solution

Hybrid solution combines signature-based and anomaly-based solutions. For example, Hwang et al. [12] Proposed a system that integrates anomaly-based solution with SNORT. Basically, signatures were extracted from anomalies detected and they were added into the SNORT signature database for fast and accurate intrusion detection. In [13], the authors proposed a DLP solution based on the combination of signature-based and anomaly-based detection. In their approach, the signature-based detection was applied to detect known anomalous activities against the predefined rules. The anomaly-based detection was proposed in the way that when the suspicious behavior or activities are flagged by the user, the system exploits an operator's feedback on alerts to automatically build and update signatures. However, apart from the flag information done by the user, the proposed technique does not give the details on how the system learns the patterns of suspicious activities and identifies as the new signature pattern.

However, all existing DLP tools have not been designed for PII data breach specifically. Most tools are applied in the target platforms. Furthermore, most DLP tools have not provided the availability feature of its core service when DLP tool keeps running to monitor or poll check the activities exhaustively. In our approach, we use a rule-based solution as for its simplicity and ease of integration. Since PII data is generally deterministic, a rule-based method is suitable. Specifically, we address the drawback of computation cost for searching sensitive words by separating the PII data source and detection tool. For the PII scanning function, we assume that PII data are scanned by the existing scanning tools and the results of PII data represented in regular expression showing the word lists and their source (such as document files, web page) are retained in the PII inventory. Then, our DLP engine (which is the key proposed system of this paper) effectively monitors and detects the activities done over the PII sources.

3 Our Proposed System

Our PII Guardian is an endpoint DLP that runs anonymously in the background of any operating system. Figure 1 illustrates the system architecture of our PII Guardian.

It consists of five major system modules including detection, classification, prevention, alerting, and logging. Overall, the system is designed to connect to the PII Inventory system or any data sources. PII Inventory is the external source storing pertinent PII data tagged with their sources and location. The detected PII data sources are converted to JSON files before they are sent to the detection module. PII data are generally expressed in regular expression as a set of strings. Our system is configured to actively monitor any file directories in a target computer or a server. It enables timely detection and dynamic execution of preventive policy management that are crucial for addressing the PII data compromise cases. After the system detects anomalous transactions, it classifies and provides responses to detected cases. All activities that occur in the system are recorded in the log files for traceability and further analysis.

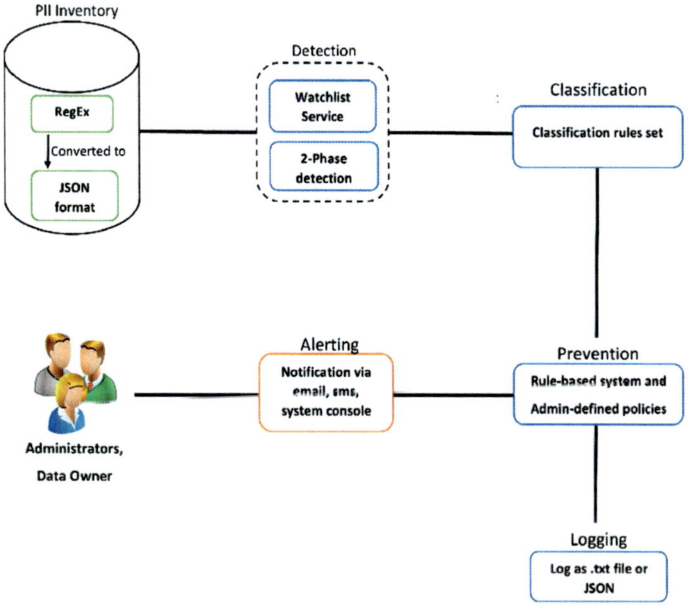

Fig. 1. System architecture of PII guardian

The detail of each system component is described as follows.

Detection

Our PII Guardian is based on a rule-based detection method supporting automatic detection of suspicious activities or PII data breaches. It connects to the external PII Inventory system to monitor PII data sources and their system location. This is to reduce time for exhaustive search of PII data location. Users can also set their own files location that they want the program to monitor. After the system directs to the locations of PII source, they will be automatically added to the watch list of the program. In the watch list, PII sources are monitored and when the suspicious actions (i.e., operations done over the PII) occur, the system will detect automatically. The detection function basically checks suspicious activities, transactions, or PII data breaches, against the predefined rules or policies in the system. Our system will be able to detect any operations in file states such as open, delete, modification, and move. In this system module, we propose two phase detection (2-PD) protocol to dynamically discover and detect the data breach in data at rest. The protocol is detailed as follows.

- **Discovery phase** in which the agent watches the update on PII data in the existing directory path or newly added PII data in the new source.
- **Detection phase** in which the agent realizes anomalous activities, transactions, or PII data breaches, based on a predefined set of rules. In this phase, there are two basic detection rules including:

Detection rule 1: Read the incoming JSON file, to determine file location and priority level of files to be watched and invoke it into the watch list.
Detection rule 2: Recursively recheck the file status of files in the watch list to find anomalous activities, either by logs or by changes.

The following pseudocode describes how the detection phase is executed (Fig. 2).

```
SET pii_inventory folder path
WHILE dlp is running
    READ pii_inventory folder's file status
    IF pii_inventory folder is updated THEN
        SET incoming JSON array as incoming_list
        FOR each JSON in incoming_list
                Append JSON.path to watch list array
        END FOR
    END IF
END WHILE
```

Fig. 2. Pseudo code of detection phase

Classification
This system module classifies and ranks severity levels based on the business or operational impact of the detected activities. Then, the respective action based on the policy rules will be enforced to those cases. The rule used to classify is set in advanced with the fine-grained policies. This helps to reduce the risk of false classification of the activity detected. Our classification encompasses two primary policy rules: priority of the file and types of actions.

***Classification Policy 1*:** Sources of PII files are given the priority based on their content or file type. The watchlist service realizes this information according to the JSON file since the detection phase.
***Classification Policy 2*:** PII data breach activities are monitored based on the operations on file status such as Open, Edit, Delete, and Move.

Table 1 shows classification rules predefined in our classification module. In essence, the priority level of the files in combination with activities or operations that perform over the PII sources are the major analysis factor. Type of PII source determines the sensitivity of the PII data. Significantly, the priority of an incident is determined by both its impact and the time to respond. In our scheme, the frequency or likelihood of the events is not taken into the analysis since the malicious activities are always treated as significant issue. We define impact levels into 3 main levels which are low, medium, and high.

Low level is composed of the impact causing limited adverse effects and the result of data leakage or loss is trivial and comprehensible.
Medium level is composed of the impact causing serious adverse effects to the data owner. Data leakage can cause the data owner to get blackmailing, humiliation, or loss of opportunity.
High level is composed of the impact causing severe or catastrophic adverse effects to data owners. Leakage of the data can cause the data owner to be seriously injured, serious monetary loss, severe company image, or even loss in personal freedom in daily life [14].

Table 1. Sample classification rules

Type of PII source	Sensitivity level	Impact of operation detected
Restricted data –data that are highly sensitive such as financial, Politics opinion, Medical treatment record, etc	Highly Sensitive	High: Open, Edit, Delete, Move
Private data -- data that are relevant to personal attributes such as their date-of-birth, home address and phone number	Sensitive	High: Edit, Delete Medium: Move Low: Open
Public data -- data that can be available in "public media" such as telephone directories, business directories, social media applications, and websites	Non-Sensitive	Medium: Edit, Delete Low: Open, Move

As shown in Table 1, upon receiving the exception from the detection phase. Classification module then checks types of PII source. The system then checks the activities (operations) detected and classifies their impact based on the rules defined. Finally, the classification module generates the impact result to the prevention module to decide on what appropriate preventive action to be enforced.

Prevention
The prevention mechanism provides preventive response based on the impact of detected activities. Table 2 shows preventive rules predefined in our prevention module. For those activities identified as high severity level, they are given high attention and the preventive actions such as disabling connection, encrypting files, revoking user privilege are immediately enforced. In addition to the predefined actions, the administrator can add, update rule anytime for handling new cases.

Table 2. Preventive rules

Impact level	Sensitivity of PII source		
	Non-sensitive	Sensitive	Highly sensitive
High	N/A	(1) Revoke user privilege (for authorized users who perform over privilege) or disable connection (for unauthorized user), and (2) Send notification to system admin	(1) Revoke user privilege (for authorized users who perform over privilege) or disable connection (for unauthorized user), (2) Enforce encryption to compromised files, and (3) Send notification to system admin
Medium	Send notification to system admin	(1) disable connection, and (2) send notification to system admin	N/A
Low	Send notification to system admin	Send notification to system admin	N/A

Alerting
The alerting module is responsible to send alert notification to administrators, personal device owners, or data owners about the suspicious activities detected and appropriate prevention method that has been used in each case. In our system, the alert can be sent via email, sms, and popup screen at the system console.

Logging
Logging system keeps activities details for all events or operations working on the PII sources in.txt file format. It specifically contains suspicious activities detected and the prevention actions generated. It can be also served for further analysis or advanced preventive action generation based on log mining or machine learning technique. The log files can be exported as text files or JSON objects for further use.

4 Evaluation and Implementation

This section provides the comparative analysis of our PII Guardian and existing DLP tools. Then, the details of our system implementation are given.

4.1 System Functionality Comparison

We compare our system with three security incident management schemes including SPIRION [4], CUSpider [5], and CipherCloud [6]. Table 3 presents the comparison of four systems.

Table 3. Functionality comparison

	Deployment model	Target data source	Operations	Advanced preventive control
SPIRION [4]	On-premise and SaaS	Any sensitive files	Any	No
CUSpider [5]	Individual Computer	PII	Scan, Delete	No
CipherCloud [6]	Cloud	Any cloud-based applications	Any	Yes
PII Guardian	Individual computer and remote endpoint	PII	Any	Yes

As can be seen from Table 3, SPIRION [4] and Cipher-Cloud [6] are the commercial software that support cloud-based applications. Both systems provide the detection function executing over any sensitive data files or folders. Hence, the PII is a part of their supporting data. CipherCloud [6] provides the encryption and adaptive control which are crucial for the advanced security enforcement in cloud environment.

CUSpider [5] is an open-source PII DLP tool used to scan PII data stored in personal computers. It scans and presents a list of PII found in the target folders. It also provides a redaction option for a number of file types from within the application. However, it does not support the strong detection function when the PII is viewed or copied. Our PII Guardian has been designed as a comprehensive DLP solution covering both PII data breach detection and prevention. Since our system is dedicated for PII data breach detection, the cost of running the system to detect the PII sources is minimized compared to the compared works [4–6]. Also, our system is only applicable for PII data used for PDPA since some Thai PII data such as citizen ID, car license plate no. etc. have different formats to other countries. In other words, the existing DLP tools do not fully detect certain Thai PII sources.

4.2 System Implementation

We developed the prototype system to demonstrate the efficiency and practicality of our PII Guardian system. For the system development, all system modules were developed by electron is with the desktop application framework, and all nodes were written by JavaScript. We used chokidar [15] library to develop the watching service to monitor the file status. Also, we used *Electron JS* as the main framework of the program to enable operating system independence which is the limitation of most endpoint DLP tools.

In our experiment, we assume that PII data source results are scanned and the list of PII sources are formatted in regular expression and stored in the PII inventory.

Experimentally, we conducted the functional test by simulating 30 test cases of PII data breach. The implementation steps include (1) converting PII instances and their related information in RegEx format into JASON file, (2) locating the watchlist service to monitor JSON files, (3) Detecting anomalous activities, (4) classifying data breach cases, and (5) enforcing preventive actions.

In our experiment, the system was able to detect all activities performing over JSON files. Once the suspicious activities were detected, they will be classified into respective levels based on predefined rules. Obviously, the classification module is able to classify all known cases based on the tagged level. For unknow cases, the system will generate the alert to notify the administrator to flag and update the rule set. In addition, our tool is able to generate alerts for all levels and can enforce actions such as blocking activities, or disabling port for high-level activities. Importantly, all activities are written in a log file for further auditing. The log system is designed to display on the system console screen and it is written in text file format. Figure 3 shows the screenshot of transaction log.

Fig. 3. Transaction log

As for the current stage of our experiment, the system functionality was thoroughly validated and all service functions were proven to be correct. The tool was exploited to run and able to detect anomalous activities, classify data breach cases predefined in the policy rules, and finally generate appropriate prevention action to respond the breach.

5 Conclusion and Future Work

We have proposed the PII Guardian system as a PII DLP tool to support the data privacy protection as well as to serve as a PDPA compliance tool. Our proposed system provides optimized and efficient PII data breach detection and delivers effective preventive functions based on the business impact analysis. Our tool provides ease of use and supports cross OS platforms. The rule sets used by our tool are specifically designed based on PDPA requirement. For future works, we plan to deploy our system in a more complex environment such as cloud or distributed system. The tool will be also calibrated with machine learning technique to enable automatic and adaptive detection of unknown activities. Finally, we will examine the scalable and secure protocols to efficiently handle intensive processing tasks in such environment.

References

1. GDPR. https://en.wikipedia.org/wiki/General_Data_Protection_Regulation. Accessed 29 Aug 2020
2. Roger, A.G.: What is personally identifiable information (PII)? How to protect it under GDPR. https://www.csoonline.com/article/3215864/how-to-protect-personally-identifiable-information-pii-under-gdpr.html. Accessed 29 Aug 2020
3. Al-Fedaghi, S., Al-Azmi, A.A.R.: Experimentation with Personal Identifiable Information, Scientific Research, College of Engineering and Petroleum, Kuwait University, Kuwait City, Kuwait (2021)
4. SPIRION. https://www.spirion.com/. Accessed 30 Aug 2020
5. CUSpider. https://www.columbia.edu/acis/security/spider/. Accessed 30 Aug 2020
6. CipherCloud. https://www.ciphercloud.com/. Accessed 30 Aug 2020
7. Martin, R.: Snort - lightweight intrusion detection for networks. In: Proceedings of the 13th USENIX Conference on System Administration (1999)
8. Vigna, G., Kemmerer, R.A.: NetSTAT: a network-based intrusion detection approach. In: Proceedings 14th Annual Computer Security Applications Conference (Cat. No. 98EX217), pp. 25–34 (1998). https://doi.org/10.1109/CSAC.1998.738566
9. Kreibich, C., Crowcroft, J.: Honeycomb: creating intrusion detection signatures using honeypots. SIGCOMM Comput. Commun. Rev. **34**, 51–56 (2004)
10. Aylin, C., Jonathan, W., Rachel, G.: Privacy detective: detecting private information and collective privacy behavior in a large social network. In: Proceedings of the 13th ACM Workshop on Privacy in the Electronic Society (WPES 2014), Arizona, USA (2014)
11. Yuya, J.O., Mu, Q., Ramani, R., Roger, R.: Context-aware data loss prevention for cloud storage services. In: Proceedings of IEEE International Conference on Cloud Computing, Hawaii, USA (2017)
12. Kai, H., Min, C., Ying, C., Min, Q.: Hybrid intrusion detection with weighted signature generation over anomalous internet episodes. Dependable Secure Comput. **4**(1), 41–55 (2007)
13. Elisa, C., Davide, F., Sandro, E., Jerry, D.H., Nicola, Z.S.: A hybrid framework for data loss prevention and detection. In: Proceedings of IEEE Security and Privacy Workshops, CA, USA (2016)
14. Chulalongkorn University, Law Department: Thailand Data Protection Guidelines 2.0, Chulalongkorn University, Bangkok, Thailand (2019)
15. Paul, M.: Chokidar A neat wrapper around node.js fs.watch/fs.watchfile. https://chromium.googlesource.com/infra/third_party/npm_modules/+/e7396f39cd50de44193622fc2bc48360cb85ce555/node_modules/karma/node_modules/chokidar/README.md. Accessed 25 Nov 2020

Check for updates

Herb Suggestion Model for Treat Disease by Symptoms Using Decision Tree Techniques

Chumpol Mokarat(✉) and Vipa Thananant

Faculty of Business Administration and Information Technology, Rajamangala University of Technology Tawan-Ok, Bangkok 10400, Thailand
{chumpol_mo,vipa_th}@rmutto.ac.th

Abstract. This research paper presents designing and developing model to support a decision making for choosing the traditional herbal medicine to relieve the symptoms in 8 groups of diseases or symptoms and used 38 attributes for filtering the diseases to generate a model. The attributes information related to the diagnosis of disease or symptoms are collected from the knowledge of the herbal medicine books and online medias for use as a learning and testing dataset of 230 records by using the J48 algorithm to create the rule-based from the decision tree model. After that, the model performance is evaluated with k-fold cross validation method and find a suitable accuracy model in the Weka software. From the experimental results, it is found that the data accuracy is 94.40%, a prediction in conjunction with the precision is 0.950, recall is 0.944, and f-measure is 0.938. It is effective to recommend herbs for the symptomatic treatment in alleviating the initial symptoms.

Keywords: Symptoms analysis · Herb suggestion · Decision tree · J48 algorithm

1 Introduction

Nowadays, Thai traditional medicine and alternative medicine are considered as another channel for people to turn their attention to as a health care alternative. They are used in conjunction treatment guidelines with modern medicine. In addition, during the past 20–30 years, although the synthetic chemical drugs were effectively, but also they contained chemical poisoning and they could cause side effects. This is the reason for the social trend towards in the natural products and this trend has grown in popularity [1]. This is because modern medicine only may not be able to cover all health problems. And it is found that there is a problem with the lack of demand for public health services to meet the needs of the people. Therefore, there is a need for an alternative to health care with herbs to assist in treating the initial symptoms.

It is the information that can be used as public health information, especially in the field of Thai traditional medicine. It can be used to study, understand and present as a channel for herbal recommendations for treatment according to the disease group or initial symptoms. It helps users to access the information more conveniently through various applications. Including health care facilities can also be used to provide services

P. Meesad et al. (Eds.): IC2IT 2021, LNNS 251, pp. 34–43, 2021.
https://doi.org/10.1007/978-3-030-79757-7_4

to clients. The data mining [2] technology is a part of Knowledge Discovery in Databases (KDD) [2]. The data mining is a definition to describe the process of modifying large databases for querying areas of interest, including previously unknown data formats. It is also a suitable choice for analysis, design and build decision model of factors that cause diseases or symptoms. And it can be further developed into an application that can suggest herbs for symptomatic treatment with decision tree [2] technique. This technique is a predictive model which can be used to represent both classifiers and regression models. Conducting research with decision tree refers to a hierarchical model of decision and its sequential outcome. It is presented with algorithms such as C4.5 [2] and J48 [2]. Several researches applied decision tree techniques such as Implementation of decision tree using C4.5 algorithm in decision making of loan application by debtor [3], Expert system for rice disease diagnosis [4], and Decision support systems for longan leaf disease diagnosis with decision tree technical [5].

Therefore, the researchers use traditional herb information to analyze, design and develop into a model to be used as a prototype to support application development that can suggest herbs for symptomatic treatment with decision tree technique. The dataset consists of the disease or symptom group data, the training data and test data. It is developed with the J48 algorithm to generate rules and evaluate model performance with k-fold cross validation and find the accuracy model with Weka software [6]. This paper is organized as follows. The introduction is described in Sect. 1. Section 2 reviews the related works and backgrounds. Section 3 describes the research approaches. Section 4 describes the research results. The conclusion is in Sect. 5 and suggestion the future research directions is in Sect. 6.

2 Backgrounds and Related Works

2.1 Herb

Herbs [7] can grow naturally and have meaning for human life, especially in the health dimension both promoting health and treating disease. Most herbs for primary health care are medicinal plants. The use of herbs as medicine must consider the nature of each herb, herbs environment for growing, season and period of harvesting. These are important factors in determining the quality of herbs. Each herb contains many substances such as carbohydrates, fats, essential oils, etc. Herbs for primary health work can be used to treat diseases or illness, such as Gastritis (Turmeric and Banana), constipation (Cassia alata), etc.

2.2 Data Mining

Data mining [2] is a definition to describe the process of modifying large databases for querying areas of interest, including previously unknown data formats. Nowadays, accessing a variety of information is very important and essential in building the data mining. Various tools and techniques can automatically analyze large amounts of data, such as RapidMiner [8], Weka [6], etc. Most of the data mining techniques are based on inductive learning [9]. It is an accuracy model by making forecasts from a sufficient number of training datasets under fundamental assumptions of the inductive learning. The trained model is applicable to future unseen examples.

2.3 Decision Tree

Decision Tree [2] is a predictive model which can be used to represent both classifiers and regression models. Conducting research with decision tree refers to a hierarchical model of decision and its sequential outcome. The decision makers employ decision trees to identify that can best achieve their goals. A classification tree is used to classify an object or an instance. An insurer uses a predefined set of classes such as risky/non-risk and their attributes values such as age or gender. The classification tree is often applied in areas such as finance, marketing, engineering and medicine. The classification tree can also be used as a survey or search technique. The decision tree classifier that is used to facilitate the underwriting process of bank mortgage applications is shown in Fig. 1.

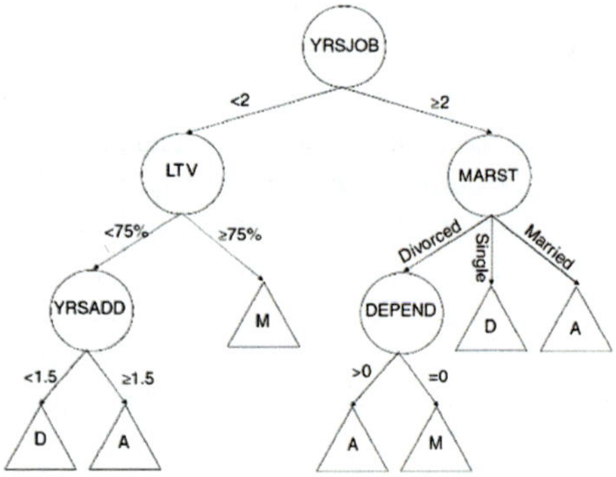

Fig. 1. Decision tree classifier is used to facilitate the underwriting process of mortgage applications of a certain bank [2]

2.4 J48 Algorithm

J48 Algorithm [2] is an open source Java implementation of the C4.5 algorithms in the Weka data mining tool. The J48 algorithm is a reuse of the C4.5 algorithm therefore it performs similar operations to C4.5. The C4.5 algorithm is enhanced from the ID3 algorithm developed by Ross Quinlan [10]. ID3 uses the gain ratio as a criterion for separating pathways. The split path ends when the number of split instances falls below the specified threshold. Error-based pruning is performed after the growing phase. C4.5 can handle numeric attributes.

2.5 K-Fold Cross Validation

k-Fold cross validation [11] is an alternative to Leave-one-out cross-validation (LOOCV). k-Fold Cross Validation randomly divides the set of observations into k groups (folds), which are equal in size. The first fold is treated as a validation set and the first fold is the same size as the remaining k-1 folds. The mean squared error, MSE, is then computed on the observations in the remaining k-1 folds. This procedure is repeated k times and each time, a different group of observations is treated as a validation set. After finishing k times, the MSE results of each time are a set of mean squared error, MSE1, MSE2, …, MSEk. The average of these values represents the estimation of k-fold cross validation value as shown in Fig. 2.

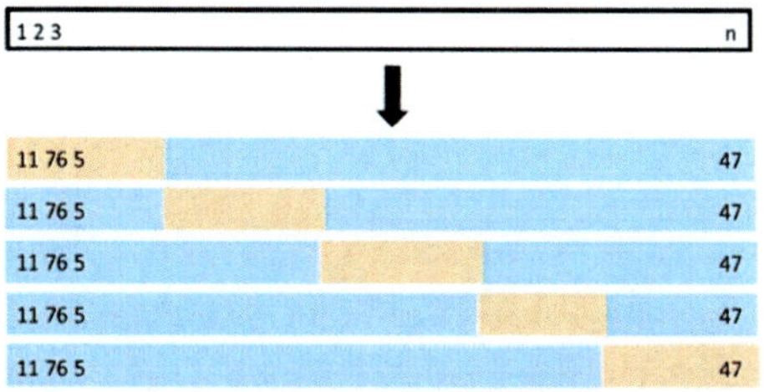

Fig. 2. The k-fold cross validation approach [11]

2.6 Evaluation Model

Confusion matrix [2] is used as an indicator of the classification rule. It contains the number of elements that is correct or incorrect classification for each class. The model performance evaluation is shown in Table 1.

Table 1. Confusion matrix [2]

	Predicted negative	Predicted positive
Negative examples	A	B
Positive examples	C	D

The model evaluation performance is evaluated as follows: Accuracy, Precision, Recall, and F-Measure. They are calculated according to Eqs. 1–4. It is based on the Confusion matrix as shown in Table 1 to evaluate the forecast results compared with the actual results, as the following equations:

- Accuracy

$$Accuracy = ((A + D)/(A + B + C + D)) \times 100 \quad (1)$$

- Precision

$$Precision = (\ D/(B + D)) \times 100 \quad (2)$$

- Recall

$$Recall = (D/(C + D)) \times 100 \quad (3)$$

- F-Measure

$$F - Measure = 2 \times ((Precision \times Recall)/(Precision + Recall)) \quad (4)$$

Where A is the number that the model correctly predicts the negative class. B is the number that the model incorrectly predicts the positive class. C is the number that the model incorrectly predicts the negative class. D is the number that the model correctly predicts the positive class.

2.7 Related Works

In 2011, a decision tree technique was used for medical diagnosis based on information of diabetes, hepatitis and heart disease to develop the decision support system as medical expert. It is based on the C4.5 algorithm, the ID3 algorithm, and the CART algorithm to classify diseases and compare their performances [12]. In 2016, a decision support system was developed to analyze the risk of chronic disease in diabetes and high blood pressure using the decision tree technique from Rapid miner and K-nearest neighbor programs. The model was tested using 10 fold cross validation technique. It was found that the prediction accuracy was 96.45% [13]. In 2017, the medical specialist system was proposed to diagnose common diseases using decision tree technique via mobile application. The J48 algorithm and the Weka program were used to test the model that gives a processing accuracy of 100% [14]. In 2018, a decision tree technique was applied for the diagnosis of disease in cow on mobile phone. The diagnostic models were created using three decision tree algorithms. Cross validation was used to test each model performance to get the best diagnostic model. It was found that the Random Tree algorithm had the most efficient with 99.47% accuracy [15]. In 2019, the research was to examine the relationship of factors affecting pathogenesis and disease risk model of elderly with data mining technique, the result found that the Decision Tree: J48 technique provided the most accurate value with Correctly at 99.796% and generating 10 rules from this model which is an effective and acceptable value [16]. and in 2020, the research aims to analyze coding errors in 43 files database systems for medical records department using data mining techniques, compare performance of decision tree (C4.5) and Naïve Bayes algorithms in analyzing and classifying the errors. The results, when evaluating the classification efficiency with 10-folds cross validation method, showed that the accuracy of the decision tree was 90.16% and the Naïve Bayes algorithm was 89.87% [17]. The decision tree technique was applied in medicine to help support decision making in medicine and other areas such as decision making of loan application by debtor [3], expert system for rice disease diagnosis [4], and decision support systems for longan leaf disease diagnosis with decision tree technical [5].

3 Research Method

The three steps of the research process include the Data Collection and Preparation, Model Construction, and Evaluation Decision Tree Model, as shown in Fig. 3. In the Fig. 3 shows the three steps in the research process as the following:

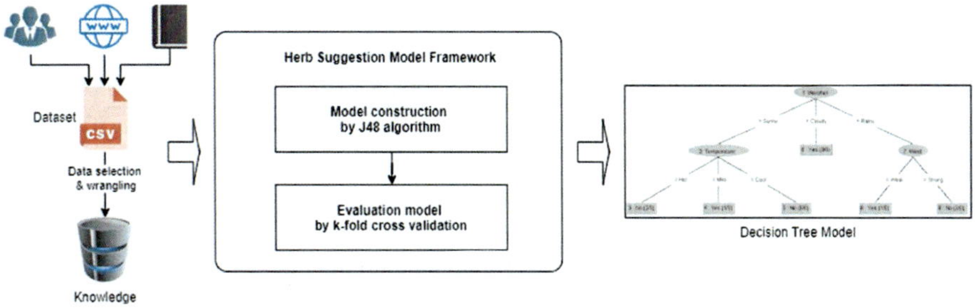

Fig. 3. Three steps in research process: Data Collection and Preparation, Model Construction, and Evaluation Decision Tree Model

3.1 Data Collection and Preparation

In this paper, we collected data set from the book "Herbs for Primary Health" [7] and "Herbs used in Primary Health" website [18]. This will be used as the initial dataset for specifying the data attributes to create a model for the introduction of herbs for the treatment of diseases according to disease and symptoms. It consists of 38 data attributes (including label class). The details of data attributes are described as shown in Table 2.

Table 2. The data attributes used in modeling

No	Code	Attribute	Value	No	Code	Attribute	Value
1	A_0	Colic	yes, no	20	F_0	Dysuria	yes, no
2	A_1	Abdominal wind and other symptoms	yes, no	21	F_1	Mixed blood	yes, no
3	A_2	Stomachache after eating fatty foods	yes, no	22	F_2	Pain when urinating	yes, no
4	A_3	Stomachache before or after meals, cardialgia	yes, no	23	G_0	Toothache	yes, no
5	B_0	Constipation	yes, no	24	H_0	Helminthiasis	yes, no
6	B_Age	Age	<= 50, >50	25	H_1	Malnutrition	yes, no
7	B_1	Symptoms of heart disease	yes, no	26	H_2	Stomachache or vomiting one more	yes, no
8	B_2	How often constipation?	normal, sometime, none	27	H_3	Ascariasis, mouth or anus	yes, no
9	C_0	Diarrhea	yes, no	28	H_4	Stools segmental flat or a long flat line	yes, no

(continued)

Table 2. *(continued)*

No	Code	Attribute	Value	No	Code	Attribute	Value
10	C_1	Liquid transfusions more than 3 times	yes, no	29	H_5	Blister hard under the skin	yes, no
11	C_2	Foamy stool with a strong smell	yes, no	30	H_6	Epilepsy	yes, no
12	D_0	Cough	yes, no	31	H_7	Buttocks or vagina itchy at night	yes, no
13	D_1	Cough and sore throat	yes, no	32	H_8	Bloody diarrhea	yes, no
14	D_2	Dry cough	yes, no	33	H_9	Pale disc	yes, no
15	D_3	Cough and white sputum	yes, no	34	H_10	Fever, diarrhea, and muscle pain	yes, no
16	D_4	Cough and green or yellow sputum	yes, no	35	H_11	Jaundice and enlarged liver	yes, no
17	E_0	Vomit	yes, no	36	H_12	Swollen, itchy on the skin	yes, no
18	E_1	Vomiting once more than and mixed blood	yes, no	37	H_13	Eating a lot but not gaining weight or weight loss (children)	yes, no
19	E_2	Dizziness	yes, no	38	Herb	Herb	class

In the Table 2, there are the following data attributes: no, attribute code (code), attribute, and attribute value (value) respectively. For learning the data attributes, we provide data set for next phase that includes the diseases groups and symptoms, along with the results of herbs for treatments by symptoms of diseases.

3.2 Model Construction

We use the data set that is collected and stored as shown in Table 2. After that we input these data set into model construction process for herb suggestion model using decision tree techniques with J48 algorithm [2] in Weka implementation, we want to predict the data attribute "Herbs to treat disease" based on its measured attributes like abdominal wind and other symptoms or stomachache after eating fatty foods and among others (shown in Table 2). The model result shows in Fig. 4.

Figure 4, shows decision tree model which is constructed from the Weka software. As a result, there are 14 rules form for classifying an herbs to treat diseases by disease groups and symptoms, as shown in Fig. 4.

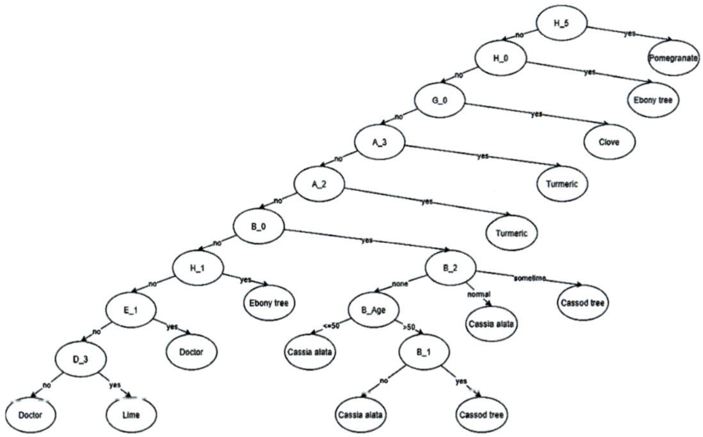

Fig. 4. The herb suggestion decision tree model for treat disease by symptoms using J48 algorithm

3.3 Evaluation Decision Tree Model

This is a step after the model construction step. Evaluation decision tree model step evaluate model performance and find the best performing model. We assess the effectiveness with k-fold cross validation method, splitting the data into 2 to 10 folds. In addition to determining the method of measuring efficiency with accuracy, precision, recall and F-measure values are also taken into consideration of the model's performance. Table 3 shows a comparison of the J48, RandomTree, and REPTree algorithms with fold cross validation, accuracy, precision, recall and F-measure.

Table 3. The performance evaluation results of decision tree model

Algorithm	Fold cross validation	Accuracy	Precision	Recall	F-Measure
J48	9	94.40	0.950	0.944	0.938
RandomTree	10	93.54	0.939	0.935	0.936
REPTree	10	84.05	0.840	0.841	0.835

From Table 3, it is found that J48 algorithm is the best solution with 9 folds (9-Fold Cross Validation). J48 algorithm has accuracy 94.40%, precision 0.950, recall 0.944, and f-measure 0.938.

4 Research Results

The result in Fig. 4 is used to develop as a rule to support the application development for disease analysis by symptoms. The rule condition refers to records in terms of a selected data attributes (disease groups and symptoms). It is presented in the form of conditions

```
IF B_0 = yes AND B_2 = none And B_Age > 50 AND B_1 = yes Then
   Herb = Cassod Tree
End IF
IF B_0 = yes AND B_2 = none And B_Age > 50 AND B_1 = no Then
   Herb = Cassia alata
End IF
```

Fig. 5. The rules for herb suggestion to treat symptoms of "constipation"

IF… THEN…. After that we apply these rules into application. Figure 5 shows rules for the diagnosis of "constipation" symptom.

From Fig. 5, the attribute code B_0, B_2, B_Age, and B_1 represent attributes in table 2. The rules condition for the diagnosis of "constipation" symptom will predicted it based on its measured attributes like having constipation symptom (B_0 = yes), the symptoms are: not frequently constipated (B_2 = none), age over 50 years (B_Age > 50), and heart disease symptoms (B_1 = yes) or no heart disease symptoms (B_1 = no) is the one that contains much more information and for this reason it has been selected as the "Cassod Tree" and "Cassia alata" herbs by "heart disease symptoms" to treat "constipation" symptoms are respectively.

5 Conclusions

When considering the decision tree model and herbal suggestion result by symptoms, it can summarize that the rules from the decision tree classification is 14 rules. Data attribute is related to the disease and the symptoms of the 8 disease groups and symptoms as follows: colic, diarrhea, constipation, cough, vomit, dysuria, toothache, and helminths. Each disease and symptoms group can be identified disease features and symptoms and can be linked to medical herbs for the correct treatment of disease and symptoms. The model can be used to support usage on the application for users to choose to analyze by disease and symptoms. Eventually, this model provides an alternative tool for accessing knowledge in the field of disease treatment by herbal.

6 Suggestion and Future Works

In this research, our main findings are as follows: Firstly, this model should be customized the dataset (as well as attributes) to be used in the design and development of models that are more accurate and applicable in a wider range of applications. Secondly, an application should demonstrate to interact with users to expand other information.

References

1. Food and Drug Administration: The Pathway to Sustainable Thai Herb Development. In: Varavoot, S. (ed.) Med.&Herb Journal 2015, vol. 2, no. 5, pp. 1–16. ACFT, Nontaburi (2015)
2. Rokach, L., Maimon, O.: Data Mining with Decision Trees Theory and Applications, 2nd edn. World Scientific Publishing Co. Pte. Ltd., Singapore (2015)

3. Amin, R.K., Indwiarti, Sibaroni, Y.: Implementation of decision tree using c4.5 algorithm in decision making of loan application by debtor (case study: Bank Pasar of Yogyakarta Special Region). In: 3rd International Conference on Information and Communication Technology (IcoICT), Indonesia, pp. 75–80. IEEE (2015)
4. Patcharaporn, R., Chantana, C.: Expert system for rice disease diagnosis. Veridian E-J. Sci. Technol. Silpakorn Univ. **6**(1), 904–913 (2013)
5. Chidchanok, S., Phaisarn, T., Suradech, B.: Decision support systems for longan leaf disease diagnosis with decision tree technical. Veridian E-J. Sci. Technol. Silpakorn Univ. **1**(6), 1–14 (2014)
6. Weka Software Homepage. https://www.cs.waikato.ac.nz/ml/weka. Accessed 20 Nov 2020
7. Primary Health Care Administrator: Herbal medicine for primary health care. WVO Officer of Printing Mill, Thailand (1997)
8. RapidMiner Homepage. https://rapidminer.com. Accessed 29 Nov 2020
9. Mitchell, T.M.: Machine Learning. McGraw-Hill, New York (1997)
10. Quinlan, J.R.: C4.5: Programs for Machine Learning. Morgan Kaufmann, United States of America (1993)
11. James, G., Witten, D., Hastie, T., Tibshirani, R.: An Introduction to Statistical Learning: With Application in R. Springer, New York (2013)
12. Kumar, D.S., Sathyadevi, G., Sivanesh, S.: Decision support system for medical di-agnosis using data mining. Int. J. Comput. Sci. Issue (IJCSI) **8**(3), 147–153 (2011)
13. Nicha, N.J.: A decision support system for analyzing the risk of two chronic diseases: diabetes mellitus and hypertension. Ind. Technol. Lampang Rajabhat Univ. J. **9**(2), 11–19 (2016)
14. Thanida, S., Pheerasut, B., Sukanchalika, B., Weerachai, K.: The general medical diagnosis expert system by decision tree technique on the smartphone application. J. Innov. Technol. Manag. **4**(2), 84–93 (2017)
15. Nattavadee, H., Phongnarin, S.: Applying decision tree classification techniques for diagnose the disease in cow on mobile phone. J. Sci. Technol. Ubon Ratchathani Univ. **20**(1), 44–58 (2018)
16. Aporn, M., Nithit, S., Bussayamat, H.: A study of disease risk of elderly with data mining technique. Narkbhutparitat J. Nakhon Si Thammarat Rajabhat Univ. **11**(3), 29–34 (2019)
17. Sakarin, H., Nadh, D.: Analysis of ICD-10 coding errors in 43 files database systems for medical record department using data mining techniques. KKU Sci. J. **48**(1), 142–155 (2020)
18. The Herbs in Primary Healthcare Homepage. http://medplant.mahidol.ac.th/pubhealth. Accessed 08 Dec 2020

Enhancing Forex Rates Prediction with Machine Learning on EUR to USD with Moving Average Methods and Financial Factors

Orawan Chantarakasemchit(✉) and Siranee Nuchitprasitchai

King Mongkut's University of Technology North Bangkok, Bangkok, Thailand
orawan.c@email.kmutnb.ac.th, siranee.n@itd.kmutnb.ac.th

Abstract. In this empirical study, we would like to do a performance analysis of EUR/USD FOREX rates by adding three moving averaged methods and four financial factors (Dollar Index, the US Interest rate, Inflation rate, and Real Gross Domestic Product). Four FOREX datasets are FOREX, FOREX with factors, FOREX with MAs, and FOREX with factors and MAs to forecast FOREX rates. In the previous research, we used four factors and Simple Moving Average (SMA) along with FOREX dataset as input data on Linear regression Model (LM) and MultiLayer Perceptron (MLP) and the result showed the Mean Square Error (MSE) on LM is better than MLP. So in this research, we used research efforts on LM, MLP, and Recurrent Neural Networks (RNNs) for predicting EUR/USD FOREX rates. And MSE was used for evaluating the result.

Keywords: FOREX rates · Dollar index · Interest rate · Inflation rate · Gross domestic product · Simple moving average · Weighted moving average · Exponential moving average · Linear regression model · MultiLayer perceptron · Recurrent neural networks · Machine learning

1 Introduction

FOReign EXchange market or FOREX has the primary purpose to support businesses to handle their own business for international trading by using the exchange method. Nowadays, FOREX is used for various purposes including speculation.

Normally, FOREX uses currency pairs in exchanges, for example; We use the EUR/USD currency pair in this research means, we use the Euro/US Dollar rate.

The statistic report titled "Turnover of OTC foreign exchange instruments, by currency" by the Bank for International Settlements (BIS) [1], which updates every three years, showed the daily averaged in billions of US dollars in the years of 2019. OTC stands for Over-Time-Counter, which is the trading done directly between two parties without the supervision of an exchange. It can be summarized that the top five traded currencies in the market (of the total number of transactions) were US Dollar (USD) at 88% next followed by Euro (EUR) 32%, Japanese Yen (JPY) at 17%, Great Britain Pound (GBP) at 13% and the last, Australian dollar (AUD) at 7%. And for this reason, the EUR/USD currencies' pair was chosen as the dataset for predicting in this research.

P. Meesad et al. (Eds.): IC2IT 2021, LNNS 251, pp. 44–54, 2021.
https://doi.org/10.1007/978-3-030-79757-7_5

In our previous research [2], from surveys we knew about the geographical distribution of FOREX turnover (%) from the year 1995 to 2010. It showed the trading of FOREX rates in United Kingdoms is the most, that is more than the United States 9%, and more than the Japan 30% in the years 2010. And it presented the participants in the FOREX that they categorized into two groups; Who needs liquidity and Who provides liquidity. The first group is who needs liquidity or end-customers that contains financial institutions, corporate customers, retail investors, and algorithmic and high-frequency traders. The second group is who provides liquidity or the provider that contains FOREX dealers, global custodians (or administrators) banks, retail aggregators, and asymmetric information and exchange rate determination.

Previously, we used the correlation coefficient values (R) to find the strength of the relationship between the FOREX rates and four factors for proving that four factors can predict the FOREX rates. The correlation coefficient is a statistical measurement of relationship intensity between two variables. The value is in the range from −1 to + 1, a negative value indicates an inversion relationship, a positive value indicates direct variation, and zero value is no relationship between the variables. The four factors are the Dollar Index, Interest Rate, Inflation Rate, and Real Gross Domestic Product. The training data were FOREX rates and the four factors from 2018 and 2019. The results of correlation values showed that Dollar Index and Interest Rate had negative correlation values (−0.96 and −0.79) to the FOREX rates respectively. Inflation Rate and Real Gross Domestic Product to the FOREX rates had positive correlation values (0.52 and 0.73). It can be concluded that if the Dollar Index or Interest Rate is increased; the FOREX rates will be decreased. While Inflation Rate or Real Gross Domestic Product is increased; the FOREX rates will be also increased. Therefore, the same four factors were also used in this research.

We also used the Simple Moving Average (SMA) method with three periods: 3 days, 5 days, and 7 days as input, and they were also computed for proving the relation of predicting by the correlation coefficient values. The results of correlation values showed the most related period is 3 days, next followed by 5 days, and the last is 7 days. There are two model algorithms and two performance matrices were used, they are Linear regression Model (LM), and MultiLayer Perceptron (MLP) for modeling and Mean Square Error (MSE), and Performance Score for evaluating the models. So the same model algorithms were also used in this research for benchmarking the results with Recurrent Neural Networks (RNNs). The results of our previous research showed both values of MSE, and Performance Score from LM is the best on the Forex dataset when enhancing the input with four factors and three series of SMA. While the result values from MLP on the forex dataset without enhancing the input is better than other datasets.

Two graphs of the correlation coefficient values between each factor and FOREX rates, and the correlation coefficient values between each period of Simple Moving Average and FOREX rates are shown in Fig. 1. In graphs, 100% is represented +1 and −100% is represented −1.

In this research, the four financial factors: Dollar Index, Interest Rate, Inflation Rate, and Real Gross Domestic Product were used as input. And the three moving average methods on three periods were used as input, the first method is Simple Moving Average that was used on the periods of 3 days (SMA3), 5 days (SMA5), 7 days (SMA7),

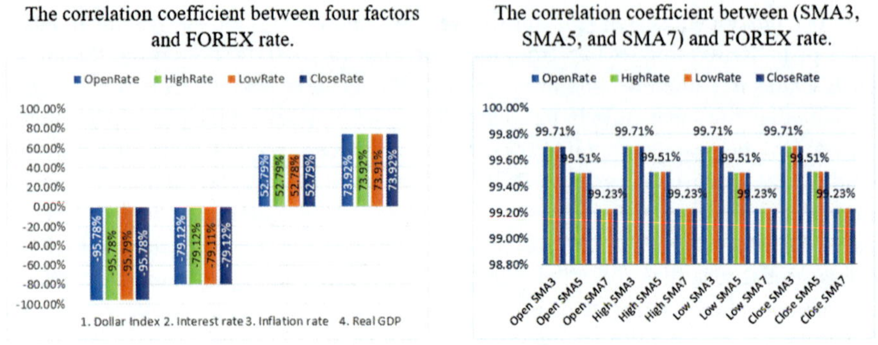

Fig. 1. The correlation coefficient values of our previous research.

the second is Weighted Moving Average on 3 days (WMA3), 5 days (WMA5), 7 days (WMA7) and the last method is Exponential Moving Average on 3 days (EMA3), 5 days (EMA5) and 7 days (EMA7). Moreover, the machine learning models were used for modeling and evaluating the result in the modeling process of this research. Three Models that were used, are the Linear regression Model (LM), MultiLayer Perceptron (MLP), and Recurrent Neural Networks (RNNs). And the performance metric for evaluating results is the Mean Square Error (MSE).

2 Literature Survey

In this section, we describe the previous survey research related to our works.

The research by Ming, Yang, Qing-Guo, and Tong [3] predicted Forex with Artificial Neural Networks (ANN) that is MultiLayer Perceptron (MLP) using Fundamental Data in the year 2008. The economic factors data: Interest rates, Gross Domestic Products (GDP), quarterly trade balance numbers, and the Consumer Price Index (CPI), was used as input and the result of the research showed the Root Mean Square Error (RMSE) and the Directional Accuracy (DA) for GBP/USD on the different number of hidden neurons 4 to 6 in each fundamental data and all. The conclusion of the research was shown the economic fundamentals did not improve the performance when they are used as an input because the indicators were updated quarterly. So the economic fundamentals which are more frequently updated may be a greater contribution to the predictive performance of the ANN.

Khurshid and Terry [4] showed FOREX rates prediction for Great Britain Pound (GBP), Canadian Dollar (CAD), and Japanese Yen (JPY) against US Dollar (USD). They used Linear Auto Regressive Moving Average (ARMA) models, Gaussian State Space models (SS), feedforward Artificial Neural Networks (ANN) and three different structure of Recurrent Neural Networks (RNN1, RNN2, RNN3) models with two categories for evaluating, the first was traditional loss functions such as Mean Percentage Errors (MPE), Root Mean Square Errors (RMSE) and the second was using parametric and non-parametric approaches for example; the AGS Test is the parametric test that provides a test for statistical significance of the difference between RMSE, and the SIGN Test is the non-parametric test for equal forecast accuracy between two models. The conclusion of this research indicated that the RNN3 model is the best for forecasting GBP and both RNN1 and RNN2 for forecasting JPY while CAD on the ANN model was leading for forecasting.

José, Germán, and Núria [5] showed the experiment of FOREX rates prediction on EUR/USD with Simple Moving Average (SMA) and Exponential Moving Average (EMA) on 12, 26, 34 and 50-days periods along with Economic Sentiment Indicator (ESI), which summarizes all positive and negative developments in the surveyed sectors of each EU member country in each month, by using Support Vector Machine (SVM) as a model algorithm. The data for training in their research was between December 15, 1999, and November 22, 2000. The results of the research showed that when used the ESI indicators as input increased the performance. And the Normalized Mean Squared Error (NMSE) of EMA without using the ESI indicators was better than SMA without using the ESI (4.47 and 4.55). While the NMSE of SMA with ESI is better than EMA with the ESI (4.10 and 4.12).

Raudys, Lenčiauskas, and Malčius [6] showed the experiment of 19 moving average methods and using smoothness and lag for benchmarking with the simple moving average on the real-world time series including Forex data in the year 2013. They used four-period smoothness levels which are 5 days (a week), 10days (two weeks), 21 days (a month), and 63 days (a quarter). Moreover, they studied and presented the trend of comparison between the short and long period that showed a shorter period followed the trend better than a longer period. The result of their research showed Exponential Hull Moving Average (EHMA) is the best method on Forex data of all 4 periods and the second best is Triple Exponential Moving Average (TRIX).

Dautel, Härdle, Lessmann, and Seow [7] showed the experiment of Forex rates prediction by using four deep learning frameworks: Feedforward neural networks (FNN), Simple Recurrent Neural Network (SRNN), Long Short-Term Memory (LSTM) and Gated Recurrent Units (GRU). They used four currency pairs as input datasets: EUR/USD, GBP/USD, USD/JPY, and USD/CHF. The period of all datasets was between January 4, 1971, and August 25, 2017, on daily transaction so the total records were 12,710. They concluded that the training loss (Log Loss) on LSTM and GRU were better that FNN and SRNN.

In conclusion, in 2008, there are two papers presented about FOREX rates prediction. The first papers [3], enhanced the GBP/USD with four factors by using MLP for modeling and using RMSE, and DA for evaluating. They concluded that enhancing input dataset with those factors did not improve the performance. In the second [4], the researchers used three currency pairs by using six model algorithms and using MPE, RMSE, AGS Test, and SIGN Test for evaluating. The results showed that RNN is the best algorithms on GBP and JPY, and ANN is leading on CAD. In 2009, the paper [5] showed enhancing EUR/USD with two moving average methods and ESI by using SVM for modeling and using NMSE for evaluating. The result showed that enhancing with ESI increased performance. Moreover, enhancing with SMA and ESI was better than EMA and ESI. And only enhancing with EMA was better than SMA. In 2013, the paper [6] showed experiments about 19 moving average methods on three level of periods on the financial data including FOREX rates. And they concluded that the EHMA is the best on FOREX rates. In 2020, the paper [7] showed four currency pairs by using FNN, SRNN, LSTM, and GRU for modeling and using Log loss for evaluating. And the result showed that LSTM and GRU were better than FNN and SRNN.

3 Background

3.1 Financial Factors

Dollar Index, the U.S. Interest Rate, Inflation Rate, and Real Gross Domestic Product are four selected factors as input. The data are from January 2018 to December 2019.

Dollar Index is a measure of the value of the US Dollar against the value of six currencies. Six currencies are Euro, Swiss Franc, Japanese Yen, Canadian dollar, British pound, and Swedish Krona. The U.S. Interest Rate or Federal Funds rate, Inflation Rate, and Real Gross Domestic Product are usually used for analyzing the national economy as the financial indicators, and sometimes, they are also used as the factors for machine learning.

3.2 Simple Moving Average

The Simple Moving average (SMA) is a technique of time series technical analysis and the definition of SMA is the average values over the defined time, for example; SMA on period 7 days that means the average values of current and six previous days. The formula as shown below,

$$SMA = \frac{Price_1 + Price_2 + \ldots + Price_n}{n} \quad (1)$$

Where n is the number of total periods, and $Price_n$ is the value at period n.

3.3 Weighted Moving Average

The Weighted Moving Average (WMA) is a moving average technique with assigned weighting. The formula as shown below,

$$WMA = \frac{Price_1 \times n + Price_2 \times (n-1) + \ldots + Price_n}{\frac{n \times (n+1)}{2}} \quad (2)$$

Where n is the number of total periods, and $Price_n$ is the value at period n.

So the weights that were used in this research on the periods of 3, 5, and 7 days as shown in Fig. 2.

3.4 Exponential Moving Average

The Exponential Moving Averages (EMA) is also known as an Exponentially Weighted Moving Average (EWMA). The EMA has also applied weights that decrease exponentially. The formula as shown below,

$$EMA = Price_t \times k + SMA_y \times (1-k) \quad (3)$$

Where t is today (current), y is yesterday (current - 1), k is 2/(number of days in period + 1) and SMA is the Simple Moving Average value of closing price for the number of days in the period.

Period 3 days

(P) Period : 3	(S) Sum of Period	(P) / (S) = (W)	(W) Weight
Current = 3	3+2+1 = **6**	3/6 = 0.5	**0.500**
Current-1 = 2	3+2+1 = 6	2/6 = 0.3333	**0.333**
Current-2 = 1	3+2+1 = 6	1/6 = 0.1667	**0.167**

Period 5 days

(P) Period : 5	(S) Sum of Period	(P) / (S) = (W)	(W) Weight
Current = 5	5+4+3+2+1 = **15**	5/15 = 0.3333	**0.333**
Current-1 = 4	5+4+3+2+1 = 15	4/15 = 0.2667	**0.267**
Current-2 = 3	5+4+3+2+1 = 15	3/15 = 0.2	**0.200**
Current-3 = 2	5+4+3+2+1 = 15	2/15 = 0.1333	**0.133**
Current-4 = 1	5+4+3+2+1 = 15	1 / 15 = 0.0667	**0.067**

Period 7 days

(P) Period : 7	(S) Sum of Period	(P) / (S) = (W)	(W) Weight
Current = 7	7+6+5+4+3+2+1 = **28**	7/28 = 0.25	**0.250**
Current-1 = 6	7+6+5+4+3+2+1 = 28	6/28 = 0.2143	**0.214**
Current-2 = 5	7+6+5+4+3+2+1 = 28	5/28 = 0.1786	**0.178**
Current-3 = 4	7+6+5+4+3+2+1 = 28	4/28 = 0.1429	**0.143**
Current-4 = 3	7+6+5+4+3+2+1 = 28	3/28 = 0.1071	**0.107**
Current-5 = 2	7+6+5+4+3+2+1 = 28	2/28 = 0.0714	**0.071**
Current-6 = 1	7+6+5+4+3+2+1 = 28	1/28 = 0.0357	**0.036**

Fig. 2. The weights of three periods.

4 Data Description

This research used Forex historical data and four financial factors based on the location of the United States of America for matching the pair of US Dollar currency.

The FOREX dataset contained five important features that are Date-time Stamp, Open Bid Quote, High Bid Quote, Low Bid Quote, and Close Bid Quote. The Date-time Stamp was used as an ID key. The FOREX data is a minute per transaction so total transactions in the years 2018 were 372,607 and 354,773 transactions in years 2019.

The four financial factors; Dollar Index, Interest Rate, Inflating Rate, and Real Gross Domestic Product (GDP) were used to enhance the input dataset. The total number of transactions of the Dollar Index, which was updated daily, was 261 transactions in 2018 and 269 transactions in 2019. Next followed by the total transactions of the Interest Rate, which was updated by the economic policy, which was 16 transactions in 2018 and 13 transactions in 2019. The Inflation Rate was updated monthly, which was 12 transactions in 2018 and 12 transactions in 2019. And the last, Real GDP was updated quarterly but the source website provides the raw data monthly so the transaction in 2018 was 12 transactions and in the years 2019 was also 12 transactions as shown in Table 1.

Table 1. Data description.

Name	Source	Updated transition	2018	2019
FOREX	www.HistData.com	Minutes	372,607	354,773
Factor^{1st} - Dollar Index	www.investing.com	Daily	261	269
Factor^{2nd} - Interest Rate	www.federalreserve.gov	On announce	16	13
Factor^{3rd} - Inflation Rate	https://inflationdata.com	Monthly	12	12
Factor^{4th} - Real GDP	https://apps.bea.gov	Quarterly	12	12

Besides, the three methods of moving average were calculated from features of the FOREX dataset except the Date-time Stamp, which is the ID key, in the Data Manipulating and was used as the enhancing input dataset.

5 Algorithms Used

The survey titled "A survey of the applications of text mining in the financial domain" by Kumar and Ravi [8] in the year 2016 showed 58 surveyed papers from journals and 31 papers from conferences. The papers reviewed were between 2000 and 2016. They showed methods distribution and it can be summarized that the top three methods were Support Vector Machine (SVM), Naive Bayes (NB), and Neural Networks (NN). The SVM and NB are suitable for the target of classification and in this research the target that we used is regression so the NN, which is Recurrent Neural Networks (RNNs).

Moreover, the Linear regression Model (LM) and MultiLayer Perceptron (MLP) were the models used in the previous work so we used their results for comparing the results with RNNs. Therefore, the Machine Learning (ML) algorithms used in this research are LM, MLP, and RNNs.

5.1 Linear Regression Model

The Linear regression Model (LM) is a supervised learning technique that is normal used on regression targets. It performs the regression task to predict a dependent variable value (output) based on a given independent variable (input). So this technique finds out a linear relationship between input and output. The 10-fold cross-validation is performed to eliminate the bias and to increase accuracy.

5.2 MultiLayer Perceptron

The MultiLayer Perceptron (MLP) is a class of feedforward artificial neural network (ANN). It has three types of layers; input, hidden, and output. Each layer is fully connected to the next layer. And Every layer contains nodes and connections, which weight to multiply and pass through it. The backpropagation is used as a learning part of MLP by using gradient descent of weight with an error of output or target.

5.3 Recurrent Neural Networks

The Recurrent Neural Networks (RNNs) are a class of feedforward neural networks that are powerful for modeling time series or natural language. The RNNs perform the same function for every input of the dataset while the output was produced and sent back into the recurrent network as input.

6 Research Methodology and Design

6.1 Machine Learning Libraries

The experiment is coded in Python programming language in Jupyter notebook using the libraries: pandas – a library providing data analysis tools for the Python programming language, NumPy – a library that facilitates high-level mathematical functions. The machine learning algorithms are taken from sklearn, which is a machine learning library in the Python programming language containing various classification, regression, clustering, dimensionality reduction, and preprocessing algorithms, and TensorFlow that provides the library for performing deep learning models.

6.2 Design and Experiment

The research was performed into two processes. The first process is Data Manipulation that contains four sub-processes. They are data collection, data preparation, data cleansing, and data processing. And the second process is Modeling.

Data Manipulation: The process is for performing the datasets for training in the model algorithms. The research's conceptual model of the Data Manipulation Process is shown in Fig. 3. The detail in each process as follows,

Data Collection: The FOREX raw data and four financial factors were collected into excel format and imported into the SQL database server.

Data Preparation: Three moving average methods in three periods were calculated from FOREX historical data in each feature. The first is the Simple Moving Average for 3 days (SMA3), 5 days (SMA5), and 7 days (SMA7). Next followed by Weighted Moving Average for 3 days (WMA3), 5 days (WMA5), and 7 days (WMA7). And the last is Exponential Moving Average for 3 days (EMA3), 5 days (EMA5), and 7 days (EMA7). For example; the computed value of Exponential Moving Average for 3 days (EMA3) of Open rates, would be set to the Open_EMA3. The format of the datatype and header name of features also were set in this step.

Data Cleansing: The 'Volume' of FOREX was found that is the NULL column so it was not used in this research. After that, the duplicate records, NULL values, and missing values were checked and removed.

Data Processing The dataset was separated into four datasets. They are the FOREX dataset, the FOREX dataset with factors, the FOREX dataset with series of MAs, and the FOREX dataset with factors and series of MAs. And they were used as input datasets for training in the model algorithms.

Modeling: After the Data Manipulation Process is completed, the output that we got is four datasets that were performed for training in the Modeling process. And before training the model, each dataset was split into 70% as training datasets and 30% as testing datasets. The research's conceptual model of the Modeling Process is shown in Fig. 4.

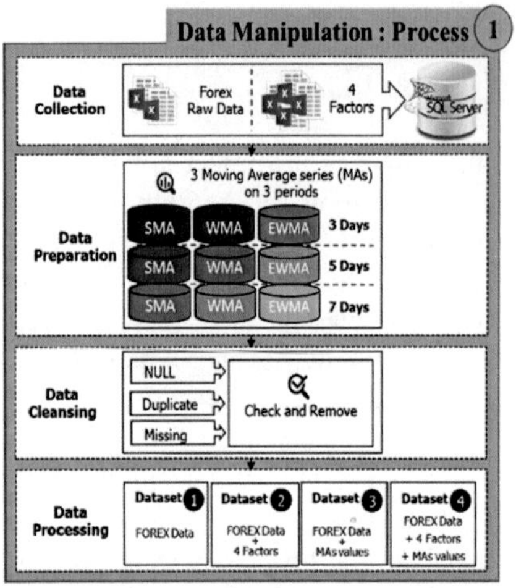

Fig. 3. The research's conceptual model of the Data Manipulation Process.

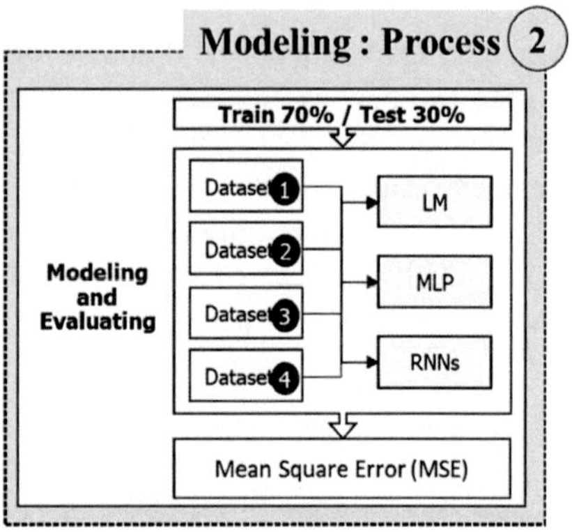

Fig. 4. The research's conceptual model of the Modeling Process

Three models were used to train all datasets and each model was run four times (four output: Open, High, Low, Close Rates) in each dataset. So the total number of times is sixteen times for sixteen outputs. For example; the Dataset❶ on LM predicted to Open rates will be counted as one output.

Table 2. The number of neurons of MLP.

MultiLayer Perceptron (MLP)	Neurons	MultiLayer Perceptron (MLP)	Neurons
Input Layer (1 neuron per 1 feature)		**Hidden Layer** 50 layers with 'relu' activation	10
FOREX Dataset (4 features)	4		
FOREX with Factors	8	**Output Layer** (1 neuron per 1 output layer) Run 16 times (4 outputs × 4 datasets)	1:1
FOREX with MAs	40		
FOREX with Factors and MAs	44		

In this research, we also used MLP parameters set in the same as our previous settings. In the input layer, one neuron per one feature so the maximum number of neurons in the input layer was 44 based on the FOREX with factors and series of MAs dataset, for example; SMA3 on Open Rates that is one feature so there is one neuron. The hidden layer was set on 10 neurons and 50 layers with 'relu' activation, which is the activation function for the hidden layer is the rectified linear unit function. For the output layers, the number of neurons was one output. So the number of neurons of MLP presented in 0 as shown in Table 2. For RNNs, we used "hyperbolic tangent (tanh)" function as activation, "sigmoid (sigmoid)" as recurrent activation, one batch size, and one time-step.

The Mean Square Error (MSE) is the performance metric that was used to evaluate and compare each output with three algorithms; LM, MLP, and RNNs. The MSE would be represented the prediction error values so if the result shows that is nearest zero (0), it will be the best result.

7 Result

The effectiveness of four datasets were measured by the Mean Square Error (MSE) as shown in Table 3.

Table 3. Mean square error of LM, MLP, and RNNs

Dataset	Mean Square Error (MSE)		
	MLP	LM	RNNs
Open rates			
FOREX Dataset	0.000000020634	0.000000003534	0.00002156
FOREX with Factors	0.000041300426	0.000000003532	0.00014897
FOREX with MAs	0.000004087383	0.000000003323	0.00239382
FOREX with Factors and MAs	0.000053266943	0.000000003327	0.02348068
High rates			
FOREX Dataset	0.000002476364	0.000000003165	0.00002701
FOREX with Factors	0.000591857656	0.000000003163	0.00047661
FOREX with MAs	0.000009129903	0.000000002927	0.01132709
FOREX with Factors and MAs	0.000009235991	0.000000002939	0.00085023
Low rates			
FOREX Dataset	0.000000210315	0.000000003136	0.00001814
FOREX with Factors	0.000408355624	0.000000003126	0.00016787
FOREX with MAs	0.000009314231	0.000000002932	0.00014974
FOREX with Factors and MAs	0.000050706673	0.000000002953	0.00235375
Close rates			
FOREX Dataset	0.000000177391	0.000000002856	0.00001156
FOREX with Factors	0.000037111233	0.000000002854	0.00021991
FOREX with MAs	0.000003444577	0.000000002706	0.00089064
FOREX with Factors and MAs	0.000273166881	0.000000002708	0.00202506

It can be concluded that the performances of the LM show significant improvements as shown in Table 4. The average value of MSE from LM on both the FOREX data with MAs and the FOREX data with factors and MAs is the best, they are 0.000000002972 and 0.000000002982. While the average value of MSE from MLP and RNNs on FOREX data is better than other datasets that MLP is 0.000006494024 and RNNs is 0.000019567500.

Table 4. Average of mean square error of MLP, LM, and RNNs

Dataset	Average of Mean Square Error (MSE)		
	MLP	*LM*	*RNNs*
FOREX Dataset	**0.000000721176**	0.000000003173	**0.000019567500**
FOREX with Factors	0.000269656235	0.000000003169	0.000253340000
FOREX with MAs	0.000006494024	**0.000000002972**	0.003690322500
FOREX with Factors and MAs	0.000096594122	**0.000000002982**	0.007177430000

8 Conclusion

The purpose of this study is to determine the performance of FOREX rates prediction adding financial factors and three moving average methods trained by Linear regression Model (LM), MultiLayer Perceptron (MLP), and Recurrent Neural Networks (RNNs). This paper has compared the Mean Square Error (MSE) of FOREX, FOREX with factors, FOREX with MAs, and FOREX with factors and MAs to forecast FOREX rates.

The result showed LM is the best model algorithm in this research. While the result of MLP and RNNs showed that it did not get high performance with the enhanced datasets in this experiment.

References

1. Turnover of OTC foreign exchange instruments, by currency. https://stats.bis.org/statx/srs/table/D11.3. Accessed 01 Sept 2020
2. Chantarakasemchit, O., Nuchitprasitchai, S., Nilsiam, Y.: Forex rates prediction on EUR/USD with simple moving average technique and financial factors. In: 17th International Conference on Electrical Engineering/Electronics, Computer, Telecommunications and Information Technology (ECTI-CON), Phuket, Thailand, pp. 771–774. IEEE (2020)
3. Eng, M.H., Li, Y., Wang, Q.G., Lee, T.H.: Forecast forex with ANN using fundamental data. In: 2008 International Conference on Information Management, Innovation Management and Industrial Engineering, Taipei, Taiwan, pp. 279–282. IEEE (2008)
4. Khurshid, M.K., Terry, L.K.: Testing forecast accuracy of foreign exchange rates: predictions from feed forward and various recurrent neural network architectures. Comput. Econ. **32**(4), 383–406 (2008)
5. Sanabria, J.A., Sánchez, G., Agell, N.: A boost in exchange rate forecasting: qualitative variables, technical indicators and parameters selection. In: 12th International Conference of the Catalan Association for Artificial Intelligence, pp. 310–317 (2009)
6. Raudys, A., Lenčiauskas, V., Malčius, E.: Moving averages for financial data smoothing. In: Skersys, T., Butleris, R., Butkiene, R. (eds.) ICIST 2013. CCIS, vol. 403, pp. 34–45. Springer, Heidelberg (2013). https://doi.org/10.1007/978-3-642-41947-8_4
7. Dautel, A.J., Härdle, W.K., Lessmann, S., Seow, H.-V.: Forex exchange rate forecasting using deep recurrent neural networks. Digital Finance **2**(1–2), 69–96 (2020). https://doi.org/10.1007/s42521-020-00019-x
8. Kumar, B.S., Ravi, V.: A survey of the applications of text mining in financial domain. Knowl.-Based Syst. **114**, 128–147 (2016)

A Recommendation Model Based on User Behaviors on Commercial Websites Using TF-IDF, KMeans, and Apriori Algorithms

Piyanuch Chaipornkaew(✉) and Thepparit Banditwattanawong

Department of Computer Science, Kasetsart University, Bangkok, Thailand
{piyanuch.chai,thepparit.b}@ku.th

Abstract. In recent years, recommendation systems have been widely introduced in various domains. The three main types of recommendation system are content-based, collaborative and hybrid filtering. Nowadays, much research applies machine learning techniques to construct recommendation models. This research also implemented a recommendation model using three machine learning techniques: TF-IDF, KMeans, and Apriori algorithms. TF-IDF was applied to form word vectorization from webpage headings. KMeans was utilized for clustering webpage headings while the Apriori algorithm was employed to find the association of webpage clusters. The elbow method was utilized to obtain the optimal number of clusters. KNN, Decision Tree, and Multi-Layer Perceptron were employed to evaluate the prediction accuracy. The dataset analyzed in the research was collected from a specific commercial website. User behaviors on the website were considered as the dataset in the research. The recommendation lists were retrieved from webpages in the same cluster and associated clusters. The prediction accuracy of the proposed model was approximately 88.62%.

Keywords: Recommendation model · Webpage heading · TF-IDF · KMeans · Apriori algorithm · Elbow method · KNN · Decision tree · Multi Layer Perceptron

1 Introduction

Steve Jobs said "get closer than ever to your customers. So close that you tell them what they need well before they realize it themselves." [1]. One of the effective ways to get closer to customers is providing them "an intelligent website". The definition of an intelligent website is a website that can offer what customers need at an early stage without any request from them. Most organizations aim to have their own intelligent website. In order to implement an intelligent website, machine learning techniques are required.

In the case of commercial websites, machine learning techniques are utilized in many ways. Data extracted from back-end websites are analyzed using machine learning and turned into useful information. Such information could be illustrated by specific patterns, which depend on audience purpose. Results from machine learning in the form of

P. Meesad et al. (Eds.): IC2IT 2021, LNNS 251, pp. 55–65, 2021.
https://doi.org/10.1007/978-3-030-79757-7_6

graphical representation could help administrators in decision making. Machine learning is also applied to construct a prediction model which can assist in the planning of business campaigns. The recommendation model is also generated from machine learning techniques. The definition of a recommendation system is a system that is used to find new items or services which are related to the interests of users [2]. Such systems could help administrators to run businesses more effectively. The recommendation system is applied in many ways, such as decision-making, maximizing profits, and minimizing risk. In accordance with the method used, there are three main categories of recommendation system, namely collaborative filtering, content-based filtering, and hybrid filtering.

The collaborative filtering method considers a group of similar users and then predicts the results based on the interests of those users. The content-based filtering method considers the items which users are interested in, and then formulates a set of items that correlate with the specified items. The hybrid filtering method is a combination of collaborative and content-based filtering. In some cases, collaborative filtering is applied to provide first-stage results, and then such results are used as input data for content-based filtering to generate the final output. In some cases, content-based filtering is applied in the first phase while collaborative filtering is utilized in the second phase after the result has been extracted from the content-based filtering.

There are many widely-used machine learning algorithms to implement a recommendation system. Examples of machine learning techniques are clustering, kernel methods, ensemble, matrix factorization, decision tree, graphical model, and regression. The top 5 application domains of the recommendation system are movie, social, academic, news, and e-commerce domains [3]. This research also falls under the last domain, which is a recommendation system for e-commerce.

This paper proposes a recommendation model based on user behaviors on commercial websites using TF-IDF, KMeans, and the Apriori algorithm. Each technique was applied in each stage of the proposed model to provide input for the next stage. The proposed model was divided into three phases, namely data preprocessing, clustering, and association rule mining. The recommendation lists derived from the proposed model were generated from webpages in the same cluster and associated clusters.

The remainder of the paper is as follows. Section 2 is the literature review. Section 3 presents the methodology. Section 4 presents the experimental results. Section 5 provides the conclusion and suggestions, and Sect. 6 presents the acknowledgements.

2 Literature Review

Recommendation models are related to various machine learning techniques, such as clustering, classification, and association. The research of Erra et al. focused on topic modelling. The paper presented approximate TF–IDF in order to extract the topic from a massive message stream using GPU [4]. There were two main contributions in the research. Firstly, an approximate TF–IDF measure was devised. Secondly, parallel implementations of the calculation of the approximate TF–IDF, based on GPUs were provided. The research claimed that the parallel GPU architecture met fast response requirements and overcame storage constraints when processing a continuous flow of data stream in real time. The experiment also illustrated that the GPU implementation was stable and performed well even with limited memory. Furthermore, the time to compute the approximate TF–IDF measure on the GPU was not affected by the data source.

Another research on clustering was entitled "An evaluation of document clustering and topic modelling in two online social networks: Twitter and Reddit" [5]. The research focused on document clustering and topic modelling methods for social media text data. The research presented document and word embedding representations of online social network data. Furthermore, doc2vec and tf-idf weighted meant word embedding representations delivered better results than simple averages of word embedding vectors in document clustering tasks. The results also demonstrated that k-means clustering provided the best performance with doc2vec embeddings. The top term analysis was conducted based on combining TF-IDF scores and word vector similarities. This method provided a representative set of keywords for a topic cluster. The doc2vec embedding with k-means clustering might successfully recover the latent hashtag structure in Twitter data.

The research of Zhang et al. [6] also implemented a recommendation system which provided a user rating score for unseen items. It would then recommend the items which gave the highest predicted score to that user. The paper mentioned that collaborative filtering algorithms were one of the main algorithms used in recommendation systems. Although such algorithms were simple and efficient, the sparsity of the data and the scalability of the method limited the performance of these algorithms. Therefore, the paper proposed a recommendation model combining a collaborative filtering recommendation algorithm with deep learning technology. The research was divided into two parts. First, the model used a feature representation method based on a quadric polynomial regression (QPR) model. The QPR model obtained the latent features more accurately by improving upon the traditional matrix factorization algorithm. These latent features were then regarded as the input data for the deep neural network (DNN) model, which was the second part of the proposed model. DNN was applied to predict the rating scores. The experimental results illustrated that the recommendation performance could be effectively improved by the proposed model.

Research on clustering and association mining by Djenouri et al. [7] mentioned that the task of association rule mining was very time consuming. Although evolutionary and swarm-based algorithms were designed to find approximate solutions, these approaches still had long execution times, especially when applied on dense and big databases, or when low minimum support and minimum confidence threshold values were used. Moreover, these approaches suffered from the lack of diversity in the rules presented to the user. To address these drawbacks, this paper proposed an efficient parallel algorithm named CGPUGA. It is a genetic algorithm that runs on clusters of GPUs to efficiently discover diversified association rules. Then, to evaluate the rules, which was the most time-consuming task, the designed algorithm relied on the massively parallel GPU threads. Furthermore, to deal with the issue of rule quality, the search space of rules was partitioned into several regions assigned to different workers, and rules found by each worker were then merged to ensure diversification. The results revealed that CGPUGA was 600 times faster than the sequential version of the algorithm for big datasets. In terms of rule quality, the results showed that the CGPUGA algorithm provided rules of higher quality compared to the state-of- the-art NIGGAR, MSP-MPSO and MPGA algorithms.

The research on word embeddings, namely Unsupervised sentence representations as word information series: Revisiting TF-IDF, was constructed in the research of Arroyo-Fernandez et al. [8]. The paper presented an embedding method that was aimed at learning unsupervised sentence representations from unlabeled text. The paper proposed an unsupervised method that modeled a sentence as a weighted series of word embeddings. The weights of the series were fitted by using Shannon's Mutual Information (MI) between words, sentences and the corpus. The proposed method offered advantages over existing ones: identifiable modules, short-term training, online inference of (unseen) sentence representations, as well as independence from the domain, external knowledge and linguistic annotation resources. The paper represented sentences by using the link between the contexts learned by word embeddings, and the amount of information in the words within a sentence. The research exploited the mentioned link to learn, without supervision, the weights of a series of word embeddings representing a sentence. The proposed model had a high level of concreteness and low computational cost.

3 Methodology

The research proposes a recommendation model based on user behaviors on commercial websites. The proposed model is divided into three phases as shown in Fig. 1. The first phase is data pre-processing. The second phase is data analysis, and the third phase is Association Rule Mining. There are two main tasks to complete in the first phase. The first task is excluding some data, which are missing data, invalid data, and dependent attributes. Whenever data could not be described, such data would be automatically defined as invalid data. Therefore, instances and features that included invalid data were also ignored. In the dataset, some attributes depended on other attributes; therefore, it was necessary to remove dependent attributes from the analysis to reduce computational time. The second task in the first phase was to merge all related data tables to form a suitable dataset and provide convenience for the analysis.

The second phase of a proposed model was data analysis. The dataset which was derived from the first phase had 19,034 instances with 27 features. These data were ready for analysis. The first step was feature representation of webpage headings in order to classify webpages. All webpage headings were separated into words using the word_tokenize method from pythainlp.tokenize library. Next, the TF-IDF algorithm was applied to generate vectors from such words. Clustering was the last task in phase 2. In order to cluster these generated vectors, KMeans was adopted to complete this task. The elbow method was utilized to obtain the optimal number of clusters. The prediction results provided target classes of webpages. In order to discover the suitable number of clusters, KNN, Decision Tree, and Multi-Layer Perceptron were applied. There were two main relationships from phase 2. The first result was the relationship between the ID of the webpage and its cluster label while the second result was the relationship between the ID of the visitor and his/her favorite cluster labels.

The third phase of a proposed model was association rule mining. The objective of this phase was to generate association rules for webpage clusters. The two main inputs were processed in this phase. The first input was the relationship between the ID of the visitor and his/her favorite cluster labels. Since each visitor visited more than one type of cluster, such data were rearranged to provide the variation of clusters. The Apriori algorithm was applied to these prepared data to mine the association rules for webpage clusters. When the association rules were generated, the association of each webpage cluster was also discovered. The last task in this phase is the recommendation process, which was performed by aggregating data, which were the IDs of the webpages in each cluster, the association cluster labels, and the IDs of visitors and their favor cluster labels. The final results were recommendation lists, which were the top ten most possible webpages. The top five webpages were extracted from the same cluster while another five webpages were brought from associate clusters.

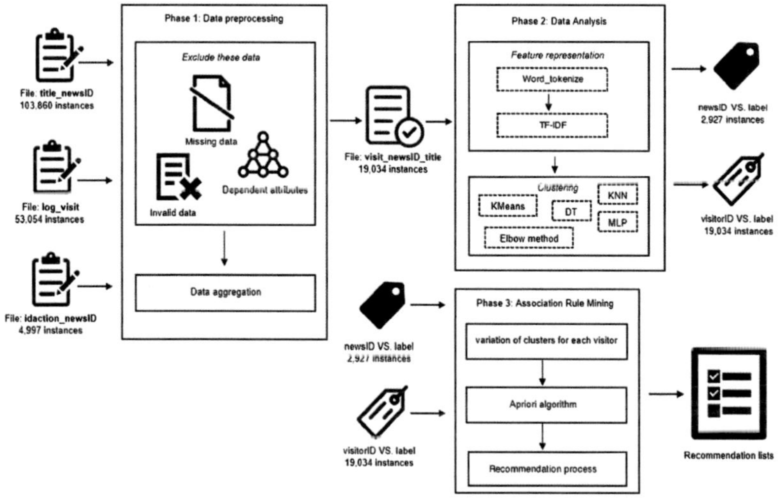

Fig. 1. The proposed model

4 Experimental Results

Three main data files were collected from back-end websites. Data from the first file were the headings of webpages and their IDs. The number of instances for the first file was 103,860. The second file were data of visits for nine months, such as the ID of the visitor, the ID of action, and the visitor location. There were 53,054 instances with 25 features. The third file was composed of the ID of action and the ID of the webpage heading, and there were 4,997 instances. The number of instances was reduced to 19,034 instances after missing values, invalid data, and dependent attributes were removed. Machine learning techniques, namely Word-tokenize, TF-IDF, KMeans, and the Apriori algorithm, were applied to provide recommendation lists of webpages. The elbow method was adopted to obtain the optimal k (no. of clusters). The optimal k would be 3, 7, and 10 when Fig. 2 was considered. There was no obvious optimal k when k was more than 10. To conclude from Fig. 2, the experiment for clustering would specify a value for k between 3 and 13.

Another three machine learning techniques, namely KNN, Decision Tree, and Multi-Layer Perceptron (MLP), were utilized to evaluate a suitable number of clusters. The prediction accuracy of each machine learning technique is shown in Tables 1–3. The experiment of KNN was conducted with the variation of k; however, the optimal value of k was 300. When Decision Tree was applied, the variation of depth was evaluated and it was revealed that the optimal depth was 27. MLP was conducted with various numbers of hidden layers and hidden nodes. The optimal number of hidden layers was 2 with 50 hidden neurons.

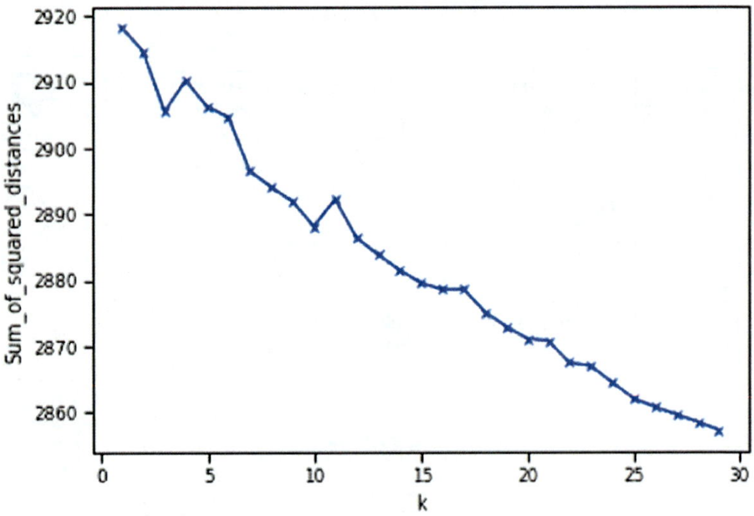

Fig. 2. The elbow method for optimal k between 1 and 29

The results from Tables 1, 2 and 3 were plotted as shown in Fig. 3. There were three main conclusions. Firstly, there was an inverse correlation between k and accuracy. Secondly, Decision Tree and MLP yielded similar results when compared to KNN. Lastly, the optimal value of k was between 3 and 7 when at least 80% accuracy was required. The results of the experiments when k was between 3 and 7 are presented in Table 4. Decision Tree was adopted and the depth was 27. Table 4 reveals that Decision Tree yielded high performance results in every case of k. Therefore, any value of k between 3 and 7 could be adopted. Whenever more clusters are needed, a high number of k would be selected. Similarly, when less clusters are required, a low number of k should be considered. The research selected six clusters which provided 88.62% accuracy.

Table 1. The prediction accuracy of KNN (K = 300)

No. of clusters	Accuracy
3	77.21
4	71.83
5	66.18
6	66.73
7	75.81
8	72.94
9	62.56
10	60.27
11	65.20
12	64.09
13	65.15

Table 2. The prediction accuracy of Decision Tree (depth = 27)

No. of clusters	Accuracy
3	95.49
4	92.72
5	87.87
6	88.62
7	82.54
8	77.72
9	72.92
10	70.58
11	71.16
12	62.51
13	61.37

When data were aggregated from six clusters, the relationships between visitors and clusters appeared. Since some visitors explored more than one cluster, association rule mining of clusters could be extracted. The total number of instances for association rule mining was 15,711. The experimental results are illustrated in Table 5.

Table 3. The prediction accuracy of MLP (2 hidden layers, 50 hidden neurons)

No. of clusters	Accuracy
3	94.29
4	90.47
5	84.93
6	81.00
7	80.94
8	73.83
9	68.48
10	70.89
11	64.50
12	59.67
13	62.29

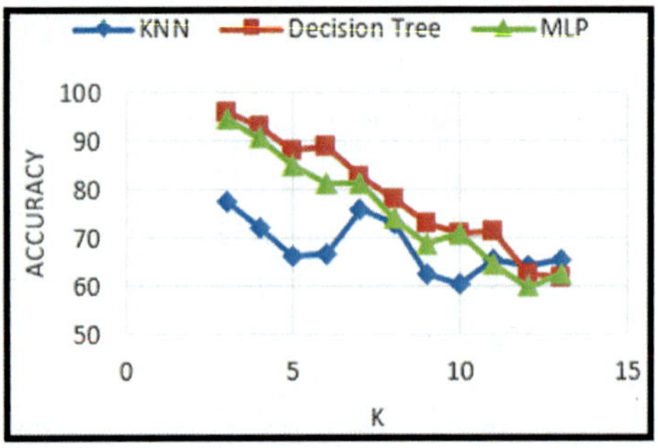

Fig. 3. The prediction accuracy of KNN, Decision Tree, and MLP

Table 4. Results of Decision tree

No. of k	Precision	Recall	F1 score
3	95.31	95.22	95.24
4	92.40	92.21	92.17
5	89.70	89.48	89.29
6	89.23	88.93	88.98
7	81.24	81.01	81.00

The association rules could be generated when minimum support and minimum confidence were low because there were only a few visitors who explored more than one

Table 5. No. of visited clusters vs. no of visitors

No. of clusters	No. of visitors
1	15,711
2	767
3	241
4	104
5	43
6	16

cluster. For example, only 241 visitors explored three clusters. The Apriori algorithm was conducted with the minimum support, minimum confidence and minimum lift set to 0.0005, 0.4, and 70, respectively. Six association rules were generated with different values for support, confidence, and lift. The experimental results are presented in Table 6.

Table 6. Association rules of clusters

Rule No.	Description	Support	Confidence	Lift
1	2 → 1	0.00146	0.40350	71.23
2	**2 → 0**	**0.00115**	**0.58065**	**84.47**
3	**3 → 2**	**0.00108**	**0.54839**	**87.03**
4	2 → 0	0.00134	0.43750	87.01
5	3 → 2	0.00134	0.43750	70.14
6	**5 → 2**	**0.00102**	**0.64000**	**102.60**

According to the results in Table 6, there were three rules that started with cluster no. 2. Rule number 2 was accepted because it provided more confidence than other rules. Similarly, rule no. 3 was adopted instead of rule no. 5 because it yielded a higher level of confidence. In conclusion, there were three main association rules, which were rules nos. 2, 3, and 6. The meaning of rule no. 6 is that visitors who explored webpages in cluster 5 may possibly prefer webpages in cluster 2 with 64.00% confidence. An example of a recommendation list for one visitor is shown in Table 7. The top five webpages were retrieved from webpages in the same cluster and were sorted by date. The latest one would be in the first order. The next five webpages were selected from the associated clusters. Similarly, the latest webpages would be higher in the list.

Table 7. Recommendation lists for one visitor

Webpage	Rank	Cluster No.
????.com/sport/news/102619	1	5
????.com/news/ประเด็นร้อน/10368 4	2	5
????.com/news/ประเด็นร้อน/10296 2	3	5
????.com/sport/news/92507	4	5
????.com/news/ประเด็นร้อน/74082	5	5
????.com/news/ประเด็นร้อน/96364	6	2
????.com/news/ประเด็นร้อน/96406	7	2
????.com/news/ประเด็นร้อน/99515	8	2
????.com/news/ประเด็นร้อน/95733	9	2
????.com/sport/news/100689	10	2

5 Conclusion and Suggestions

The research proposed a recommendation model based on user behaviors in commercial websites using three main machine learning techniques, namely TF-IDF, KMeans, and the Apriori algorithm. Decision Tree, KNN, and MLP were also applied to evaluate the algorithms. The experimental results revealed that the optimal number of clusters was between 3 and 7. However, the research conducted six clusters of webpages defined as 0 to 5. The prediction accuracy was 88.62%. Three association rules of webpage clusters were generated by using the Apriori algorithm. The values for support and confidence were low because only a few visitors explored more than one webpage. Other adaptations and experiments could be employed in future research. Future work could also concern deeper analysis of particular mechanisms to cluster and generate stronger association rules.

Acknowledgments. The authors would like to express their gratitude to an anonymous company which cannot be mentioned because of confidentiality. It is appreciated that the business data provided by this selected company is sensitive and will not be disclosed or used for any purpose other than for research purposes.

References

1. Amazing Steve jobs quotes (that will motivate you). https://wisdomquotes.com/steve-jobs-quotes/. Accessed 23 Nov 2019
2. Adomavicius, G., Tuzhilin, A.: Toward the next generation of recommender systems: a survey of the state-of-the-art and possible extensions. IEEE Trans. Knowl. Data Eng., 734–739 (2005)
3. Portugal, I., Alencar, P., Cowan, D.: The use of machine learning algorithms in recommender systems: a systematic review. Exp. Syst. Appl., 205–227 (2018)
4. Erra, U., Senatore, S., Minnella, F., Caggianese, G.: Approximate TF–IDF based on topic extraction from massive message stream using the GPU. Inf. Sci., 143–161 (2015)
5. Curiskis, S., Drake, B., Osborn, T., Kennedy, P.: An evaluation of document clustering and topic modelling in two online social networks: Twitter and Reddit. Inf. Process. Manag. (2020)
6. Zhang, L., Luo, T., Zhang, F., Wu, Y.: A Recommendation model based on deep neural nctwork. IEEE Access **6**, 9454–9463 (2018)
7. Djenouri, Y., Belhadi, A., Fournier-Viger, P., Fujita, H.: Mining diversified association rules in big datasets: a cluster/GPU/genetic approach. Inf. Sci., 117–134 (2018)
8. Arroyo-Fernandez, I., Mendez-Cruz, C., Sierra, G., Torres-Moreno, J., Sidorov, G.: Unsupervised sentence representations as word information series: revisiting TF-IDF. Comput. Speech Lang., 107–129 (2019)

On the Utility of Combining Topic Models and Recurrent Neural Networks

Justin Wood(✉), Bohan Li, Jae Lee, Corey Arnold, and Wei Wang

University of California, Los Angeles, USA
{juwood03,cwarnold,weiwang}@ucla.edu

Abstract. Learning under deep neural network models has shown interesting abilities to generate semi-structured sequences. Such generated fragments can appear to be remarkably like real data, yet applications of these generative techniques are generally lacking. This is partly due to the inability of recurrent neural network sequences to be semantically connected amongst long related fragments. One way to increase the semantics of these fragments is to combine features learned from topic modeling with those of recurrent neural networks. This combination ensures a higher correlation between generated sequences, much like real data. In this paper we investigate existing approaches to adding topics into the recurrent neural network. We then develop a novel approach called Topic-RNN, which represents the state-of-the-art technique of integrating topics into the RNN. Our results show this method outperforms existing approaches substantially. Additionally, we show that in such cases where topic information is available, Topic-RNN surpasses word level RNNs for the task of word prediction.

Keywords: Topic modeling · Topic-RNN · Recurrent neural network · Latent Dirichlet allocation

1 Introduction

Deep neural networks have been established as an effective technique to pattern recognition and machine learning [1–3]. A subclass of these models is the recurrent neural network (RNN), which utilizes previous states in training of the current state. An interesting effect to training these models is the ability to generate sequences of data on a diverse range of input. The RNN is particularly interesting in its ability to capture and generate the appropriate context. This is due to the model in which it is trained. For each time-step the hidden layer is influenced by the input axons as well as axons from previous hidden layers. This allows the model to learn not only the current pattern, but previous occurrences as well. In the generation phase, it is easy to predict the next output by sampling from the model's predictive distribution. A series of predictions can then be chained together giving an entire sequence of predicted output. When put all together, the stochastic sampling produces output which is often referred to as a product of dreaming.

P. Meesad et al. (Eds.): IC2IT 2021, LNNS 251, pp. 66–76, 2021.
https://doi.org/10.1007/978-3-030-79757-7_7

Although there are applications that utilize these generated sequences [4–7] they often rely on isolated sequences. The ability to connect sequences would have tremendous utility in areas such as anonymization of private data where a foolproof way to guarantee anonymization is to generate a representation of the data that does not share any of the original data.

This however is not a trivial task. In this paper, we work towards the goal of incorporating context around generated segments by integrating into the RNN model information acquired from topic modeling. Because topic modeling is most naturally used in text, we restrict the methods to those applied in text; and because we are interested in applications where a new word is generated, we focus on character level generation.

Probabilistic topic models have long been established as an effective means of discovering underlying semantic themes in a set of input [8–10]. The current state of the art topic models are derived from latent Dirichlet allocation (LDA) [8]. LDA makes a simplifying assumption about how a corpus is generated. Since it is too hard to mathematically model the precise steps a human generates a corpus, LDA assumes that words and topics are selected from sampling of distributions and repeated for the length of the corpus.

Given topic models are used at the word level it may be a natural inclination to seek a combination with a word-based RNN. However we are interested in utilizing the RNN for words that do not belong to a given input vocabulary (such as anonymization of patient data). A word-based RNN would simply not work. Additionally, word-based RNN's can be problematic due to a large input size. For input layers such as one-hot encoding the size of each input equals the size of the vocabulary. For large corpora this can degrade computation time and predictive power [4].

While there continues to be progress in other approaches to text mining, such as the Transformer [11], or convolutional neural network (CNN) based approaches [12], the RNN is still incredibly important. The authors estimate that RNNs comprise 25% of current deep learning publications (based on a Google Scholar search for papers published in 2019). Our approach presented in this paper is not only limited to text analysis. We envision our technique of combining topic modeling to RNNs to be useful in data sets used with RNNs where topic models form meaningful clusters.

2 Preliminaries

2.1 Recurrent Neural Networks

Recurrent neural networks (RNNs) are a class of neural networks well suited for sequential input. They are straightforward extensions of the standard feed-forward multilayer perceptron networks [13], which add a cycle to the hidden layer. At each time step, the RNN updates its hidden state per the new input. The RNN's hidden state is thus maintained and passed through time, which can distill information from an infinite-length context window. Such mechanisms can endow RNNs with long-term learning, which allows them to model long-distance dependencies effectively.

2.2 Topic Models

Topic modeling is an unsupervised learning algorithm that groups the words of corpus into meaningful clusters called topics. Techniques are usually based off of Latent Dirichlet allocation (LDA) [8]. The input to LDA is a set of documents and the output is two sets of distributions θ and ϕ. θ represents a mixture of topics for each document and ϕ becomes the mixture of words for each topic. Hidden distributions are inferred using Gibbs sampling where each token is assigned a topic sampled from a posterior distribution derived from LDA's Bayesian model.

3 Related Work

3.1 Deep Learning Sequence Generation

Recent research has successfully injected side information into RNN models on a word-level basis to increase performance [14]. Hoang et al. analyzed different methods of subjecting information into the input, hidden, and the output layer of recurrent neural network language models (RNNLM). The research shows and effective and consistent performance boost to the vanilla RNNLM.

One drawback to the RNNLM is that long-range information cannot be forwarded between phrases. To improve on the basic RNN structure, recent research has looked into combining the long short-term memory (LSTM) with the RNN to generate long sequences [4]. This work demonstrates the possibility of generating complex long-range sequences. And also successfully extended the approach to a handwriting synthesis, where they could condition the text sequence to a highly realistic handwriting style.

Kim et al. demonstrated an approach to predicting next characters in text streams by using a large RNN-powered by a Hessian-Free optimizer [15]. Similarly, character-level generation has shown to be attractive to achieving the asymptotic limit in text compression [5]. While word-level generation is limited to the vocabulary set of the training set, character-level generation can generate an "understanding" using the intelligence of the model.

Character-level generation has found success in other areas as well such as named entity recognition. Nogueira dos Santos et al. introduced a deep neural network named CharWNN that can attain state-of-the-art results for a language-independent model [6]. Additional research by Zhang et al. suggests that character-level generation can be done without structure if several conditions are satisfied [16]. These are: the size of the dataset, if the texts are curated, and the right selection of the character sets. The results show that on these conditions, a deep CNN network performs measurably well on text classification problems.

3.2 Topic and Document Modeling RNNs

Utilizing deep neural networks has alleviated several document classification challenges in research today. Tang et al. developed a neural network model for sentiment classification [17]. The results over various datasets have achieved performance improvements over state-of-art algorithms and the current sentiment classification model using a neural network. Their results show the LSTM achieved a higher performance compared to a multi-filtered CNN.

Wan et al. demonstrates the use for topic modeling in neural networks on image classification [18]. The image data contains scenic images such as an office, bedroom, living room, etc., a total of 15 scenic categories. The experiments were designed in three different settings: hierarchical topic modeling, a neural network, and a hybrid model that combined both the hierarchical model and the neural network. Throughout the experiment, the hybrid model outperforms the neural network by a wide margin and achieves comparable results against the supervised LDA model.

4 Proposed Approach

The ability to combine RNNs and topic models can result in a beneficial outcome for prediction, but the best approach is not so intuitive while maintaining the restriction of using a character level RNN. For consistency in measurement, we choose the same parameters for all models. For all RNNs we choose maximum likelihood estimation for the loss function and backpropagation through time (BPTT) [19] as the technique to minimize loss. We explore different methods of adding the topic modeling information into the RNN.

4.1 RNNs with Topic Models

Topic Input Vocabulary. In the simplest approach, we employ a character-level RNN to generate character sequences (Π-RNN). This is done by appending topic modeling information into the existing input. The change to the existing RNN model is at the input layer. Before training the RNN on the data, we generate a mapping of tokens to topics using an existing topic model. We then take the input set to be the Cartesian product of the character input set and the topic set.

Since the change is only made in the input itself, no differences are made to the corresponding forward and back propagation equations.

Two-Hot Encoding. To reduce the input size of the topic input vocabulary approach it is possible to give as input the character to train as well the topic as a two-hot encoded vector (2-RNN). This serves to shrink the input from size $C \times K$ down to size $C+K$. With C being size of the character vocabulary (Π) and K the total number of topics. In this model the forward and back propagation equations remain unchanged. The execution time and memory are thus reduced from $C \times K$ to $C + K$ in the big O notation for Π-RNN.

Context Dependent. If we interpret the K input layer in the two-hot encoding approach to be independent of the C input layer and allow a direct influence on the output layer from the K input layer then we form the basis of the context dependent RNN [20] (CD-RNN). This model includes the influence from the K input layer in both the hidden layer calculations as well as the output layer. Given a sequence of input vectors $(x_1, ..., x_T)$, the RNN predicts the output sequence $(\hat{y}_1, ..., \hat{y}_T)$ using the following equations:

$$\mathbf{h_t} = tanh(W^{hx} \times \mathbf{x_t} + W^{hh} \times \mathbf{h_{t-1}} + W^{hk} \times \mathbf{k_t} + \mathbf{b^h}) \tag{1}$$

$$\mathbf{\hat{y}_t} = softmax(W^{yh} \times \mathbf{h_t} + W^{yk} \times \mathbf{k_t} + \mathbf{b^o}) \tag{2}$$

Where h_t is the high-dimensional hidden state at the time-step t, W^{hx}, W^{hh}, and W^{yh} are the weight matrices connecting the input to hidden layer, hidden to hidden layer, and hidden to output layer respectively, and the vectors b^h and b^o are the biases. The new weights must also be factored into BPTT. The additional updates are:

$$\Delta W^{yk}_{pz} = \eta \times \sum_{t=1}^{T} \delta_{t,p} \times k_{t,z} \tag{3}$$

$$\Delta W^{hk}_{jz} = \eta \times \sum_{t=1}^{T} \delta_{t,j} \times k_{t,z} \tag{4}$$

With z corresponding to a binary topic input at time t, η is the learning rate and δ are the respective gradients from BPTT. Since we are adding an additional edge set to the output, this increases the time and memory for prediction as $\mathcal{O}(C \times H + K \times H + H^2 + K \times C)$ with H defined as the size of the hidden layer. The training execution and memory will be multiplied by the total number of characters in the corpus (D) and the number of time steps to unroll (U) respectively.

Topic-RNN. While the context dependent RNN effectively uses the topic to help the predictive power of the RNN, it is our hypothesis that better results can be achieved by separating topics and their predictions. That is, each topic is given its own RNN and the model overall encompasses these K RNNs. Then the current topic can dictate the entire model's output. Given an additional input of topic assignments, the change needed to the loss function, forward and backward propagations are to add an extra dimension for each topic in the set of topics as follows:

$$\mathbf{h_{t,s}} = tanh(W^{hxs} \times \mathbf{x_t} + W^{hhs} \times \mathbf{h_{t-1,s}} + \mathbf{b^{hs}}) \tag{5}$$

$$\mathbf{\hat{y}_{t,s}} = softmax(W^{yhs} \times \mathbf{h_{t,s}} + \mathbf{b^{os}}) \tag{6}$$

$$E = -\frac{1}{T} \times \sum_{t=1}^{T} \ln(\hat{y}_{t,s,p}) \tag{7}$$

$$\delta_{t,s,p} = -\frac{\partial E}{\partial \hat{y}_{t,s,p}} \times \frac{\partial \hat{y}_{t,s,p}}{net_{t,s,p}} = -\frac{1}{N} \times (1 - \hat{y}_{t,s,p}) \tag{8}$$

$$\begin{aligned}\delta_{t,s,j} &= \sum_{p} \delta_{t,s,p} \times W_{pj}^{yhs} \times \frac{\partial h_{t,s,j}}{\partial net_{t,s,j}} \\ &= \sum_{p} \delta_{t,s,p} \times W_{pj}^{yhs} \times (1 - h_{t,s,j}^2)\end{aligned} \tag{9}$$

$$\begin{aligned}\delta_{t-1,s,m} &= \sum_{j} \delta_{t,s,j} \times W_{jm}^{hhs} \times \frac{\partial h_{t-1,s,m}}{\partial net_{t-1,s,m}} \\ &= \sum_{j} \delta_{t,s,j} \times W_{jm}^{hhs} \times (1 - h_{t-1,s,m}^2)\end{aligned} \tag{10}$$

$$\Delta W_{pj}^{yhs} = \eta \times \sum_{t=1}^{T} \delta_{t,s,p} \times h_{t,s,j} \tag{11}$$

$$\Delta W_{jm}^{hhs} = \eta \times \sum_{t=1}^{T} \delta_{t,s,j} \times h_{t-1,s,m} \tag{12}$$

$$\Delta W_{ji}^{hxs} = \eta \times \sum_{t=1}^{T} \delta_{t,s,j} \times x_{t,s,i} \tag{13}$$

With s being the topic at time t (also time $t-1$).

The intuition behind Topic-RNN is a combination of specialization and the addition of learning from a different type of context. A parallel example of why specialization works would be if one patient is suffering from a skin condition and another patient from a heart condition, then there may be better outcomes if the first patient is seen by a dermatologist and the second by a cardiologist rather than if both are seen by the same specialist in family medicine. RNNs also only consider one type of context, which is the previous input. Topic-RNN also considers the topic of the word which adds a new type of context—only the makeup of words of the topic.

To evaluate the effectiveness of prediction we consider the case when each word (v) in the non-trained data is reflected in ϕ and the opposite case; where ϕ is the topic to word distributions from topic modeling. In the case where v is reflected in ϕ then we simply take the output of the current s-dimension RNN (Topic-RNN-ϕ). However, when v is not already considered in ϕ it may not always be the best decision to use the s-dimension topic model as output. For example, if v does not exist in ϕ_s then this is similar to asking an untrained RNN the prediction. In such cases it may be better to use a vanilla character RNN (Char-RNN) which contains context over every word in the corpus.

The difficulty in this approach is knowing when to use the Topic-RNN-ϕ and when to use the Char-RNN. Since we are predicting at the character level, we do not know the word of the current character. We do however know the previous characters. In this way we can ask each Topic-RNN-ϕ how well it can predict the next character given the existing input characters for all prefix matching words after training. With a word vocabulary of Υ, and $\mathbf{l}$ being the prefix $(c_1, ..., c_{t-1})$ this formalizes as:

$$A^{k}_{t,c_t,1} = \overline{E^{k}_{t,c_t,1}} : \forall (c_1, ..., c_t) \sqsubseteq v \in \Upsilon \tag{14}$$

Additionally, we can ask the Topic-RNN-ϕ how certain it is of its prediction. In cases where the Topic-RNN-ϕ is very certain then this may be reflective of a "good" guess. For certainty we measure entropy, defined as:

$$N_t = -\sum_{i} \hat{y}_{t,i} \times \log(\hat{y}_{t,i}) \tag{15}$$

We can also measure how similar the prediction is to the Char-RNN. Since the Char-RNN is a "good" baseline guess, then something that agrees with that may also be good and may be even better. We choose Jensen-Shannon distance as our similarity metric:

$$\hat{J}_t(\hat{\mathbf{y}}'_{\mathbf{t}}, \hat{\mathbf{y}}''_{\mathbf{t}}) = \frac{1}{2} \times \sum_{i} \hat{y}'_{t,i} \times \left(\ln \hat{y}'_{t,i} - \ln \frac{\hat{y}'_{t,i} + \hat{y}''_{t,i}}{2} \right) \tag{16}$$

$$J_t = \sqrt{\hat{J}_t(\hat{\mathbf{y}}'_{\mathbf{t}}, \hat{\mathbf{y}}''_{\mathbf{t}}) + \hat{J}_t(\hat{\mathbf{y}}''_{\mathbf{t}}, \hat{\mathbf{y}}'_{\mathbf{t}})} \tag{17}$$

J_t, N_t, and A_t become the features that we use to decide when to choose between Char-RNN and Topic-RNN-ϕ. But we must also consider the location of the current input. In cases where the input is the first character of a word it may always be preferable to use Char-RNN, which has more context than Topic-RNN-ϕ. Because context and location are important, we choose to feed the input features into an RNN (E-RNN) to help decide which model to use. If we assume the hidden layer size to be equal between Char-RNN, Topic-RNN-ϕ, and E-RNN, V to be the size of Υ, and ζ to be the max length word in Υ then the memory needed would be $\mathcal{O}(K \times [U \times (C \times H + H^2) + C \times \zeta \times V])$ and a training execution of $\mathcal{O}(K \times [S \times (C \times H + H^2) + C \times \zeta \times V])$. In this model the extra cost of memory and training execution time are somewhat alleviated by the prediction execution time which is $\mathcal{O}(C \times H + H^2)$.

5 Evaluation

5.1 Perplexity

To test the predictive power of the various topic model based RNNs we create an experiment to determine the ability of each RNN to correctly guess the next character given a set of input characters from a corpus.

Experimental Setup. Each corpus is split into a training and test set in a ratio of 80/20. All models are then built off the training set. After 50 epochs of training, we feed characters from the test set into the models and compare the error (E) of each model. We choose 50 epochs because the error rate in the training set appears to converge, which also matches a standard used in other

Table 1. Topic-RNN compared against baseline methods for the prediction of characters.

	MedlinePlus	Reuters-21578	20-Newsgroups	Sent-Web
Topic-RNN	**0.98**	**1.99**	**2.31**	**1.5**
Topic-RNN-ϕ	**0.276**	**0.284**	**0.34**	**0.24**
CD-RNN	1.72	2.32	2.42	1.72
2-RNN	1.32	2.31	2.4	1.7
Π-RNN	8.42	8.65	8.84	8.52
Char-RNN	1.6	2.28	2.4	1.65

research [21,22]. To show the extensibility with more complicated cells we repeat the experiment using only Char-RNN as a baseline for GRU and LSTM cells. We choose only Char-RNN because it is the most competitive among the baseline methods.

Experimental Results. The average error for each model and dataset is shown in Table 1. The best results come from Topic-RNN when each test word is already apart of ϕ. However, even when the test word is not known Topic-RNN outperforms the other models. It is our contention that the vocabulary-based model is constrained by having too many outputs, while the two-hot encoding and context dependent models do not specialize. Topic-RNN can specialize while not overcomplicating the predictions and therefore leads to the best results.

The effect of input and output size seems to be substantial. Π-RNN far underperforms the others models and also has largest input and output size—while the other models have sizes more or less the same. Topic-RNN also demonstrates that specialization can improve the predictive power or RNN's as opposed to asking a single model to learn the entire input set. The results are also promising

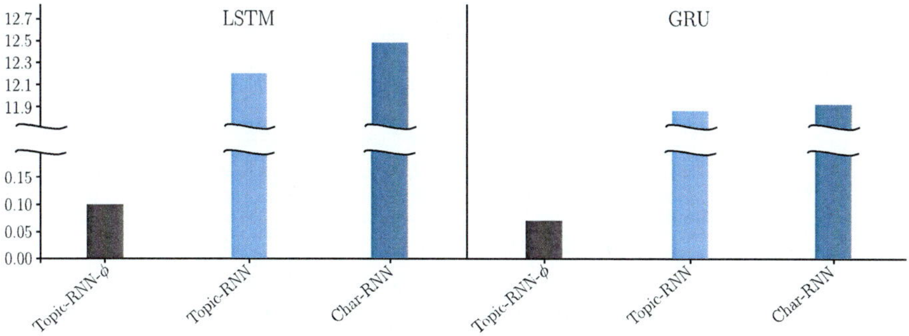

Fig. 1. LSTM and GRU cells used in Topic-RNN compared against a vanilla Char-RNN.

for LSTM and GRU cells. Figure 1 shows the total error for all datasets when Topic-RNN and Char-RNN use LSTM and GRU cells. The results are consistent with the basic cell results and show our technique is likely applicable across different cells.

5.2 Word Prediction

Given the ability of Topic-RNN to accurately predict the next character, we aim to discover whether it can predict words as well. We compare this against a word-level RNN (Word-RNN) to highlight the effectiveness of Topic-RNN.

Experimental Setup. As with the previous experiment, the corpora are split 80/20 into a train and test set. Word-RNN is trained against the words and Topic-RNN against the characters and topics. After training we determine the proportional probability of a word under the Topic-RNN model by using a simple average probability for each character. Given the two distributions, we can determine the error for both models.

Experimental Results. The results of all datasets are given by Table 2. Surprisingly, Topic-RNN gives better predictive error in most datasets than Word-RNN. In the other sets, such as the MedlinePlus data set, Word-RNN only does marginally better. We hypothesize Word-RNN is outperformed by Topic-RNN mainly because it is hindered by large input and output sizes; additionally, the inclusion of topics to specialize give more gains. This experiment underscores the effectiveness of adding topic modeling into RNN predictions.

Table 2. Topic-RNN compared against baseline methods for the prediction of words.

	MedlinePlus	Reuters-21578	20-Newsgroups	Sent-Web
Topic-RNN	7.867	**9.97**	**10.64**	**8.09**
Topic-RNN-ϕ	**7.175**	**8.366**	**8.85**	**6.786**
Word-RNN	**7.53**	15.53	17.42	9.318

6 Discussion

This paper investigates the utility of combining RNN and topic models. We show that this combination results in improved performance for the recurrent neural network. Training input from the RNN with the addition of topic modeling input can improve prediction of held out data. Although the evaluation focused on perplexity, there remains other areas worthy of investigation. We hypothesize our approach of adding topic modeling into the RNN would be useful to other tasks such as machine translation. Empirical evaluation is left as an aim in future work.

One potential area of use is in semi-supervised topic modeling [23,24]. For the task of assigning meaningful labels to topics, Topic-RNN could potentially improve upon the performance of existing methods, especially when there exist a large portion of words not found in the semi-supervised input set. Another open area of interest would be the addition of existing work which focuses on non-labeled topics [25,26] to see if these changes can propagate to labeled topics.

7 Conclusion

Topic-RNN represents a novel and outperforming method of combining topics with the recurrent neural network. Our analysis shows the advantage of learning from both previous data as well as data partitioned by topic. Topic-RNN's improvements over existing topic-based methods have been validated experimentally. Under these experiments, and likely others the combination of topic models and RNNs leads to improvements in the recurrent neural network.

References

1. Hinton, G., et al.: Deep neural networks for acoustic modeling in speech recognition: the shared views of four research groups. IEEE Sig. Process. Mag. **29**(6), 82–97 (2012)
2. Krizhevsky, A., et al.: ImageNet classification with deep convolutional neural networks. In: Advances in Neural Information Processing Systems 25: 26th Annual Conference on Neural Information Processing Systems 2012, pp. 1106–1114 (2012)
3. Collobert, R., Weston, J.: A unified architecture for natural language processing: deep neural networks with multitask learning. In: Machine Learning, Proceedings of the Twenty-Fifth International Conference (ICML 2008), Helsinki, Finland, 5-9 June 2008, pp. 160–167 (2008)
4. Graves, A.: Generating sequences with recurrent neural networks. arXiv preprint arXiv:1308.0850 (2013)
5. Ward, D.J., et al.: Dasher - a data entry interface using continuous gestures and language models. In: Proceedings of the 13th Annual ACM Symposium on User Interface Software and Technology, UIST 2000, San Diego, California, USA, 6-8 November, 2000, pp. 129–137 (2000)
6. Nogueira dos Santos, C., Zadrozny, B.: Learning character-level representations for part-of-speech tagging. In: Proceedings of the 31th International Conference on Machine Learning, pp. 1818–1826 (2014)
7. Gregor, K., et al.: DRAW: a recurrent neural network for image generation. In: Proceedings of the 32nd International Conference on Machine Learning, ICML 2015, Lille, France, 6–11 July 2015, pp. 1462–1471 (2015)
8. Blei, D.M., Ng, A.Y., Jordan, M.I.: Latent Dirichlet allocation. J. Mach. Learn. Res. **3**, 993–1022 (2003)
9. Griffiths, T.L., Steyvers, M.: Finding scientific topics. Proc. Natl. Acad. Sci. **101**(Suppl. 1), 5228–5235 (2004)

10. Rosen-Zvi, M., et al.: The author-topic model for authors and documents. In: UAI 2004, Proceedings of the 20th Conference in Uncertainty in Artificial Intelligence, Banff, Canada, 7–11 July 2004, pp. 487–494 (2004)
11. Devlin, J., et al.: BERT: pre-training of deep bidirectional transformers for language understanding. In: Proceedings of the 2019 Conference of the North American Chapter of the Association for Computational Linguistics: Human Language Technologies, NAACL-HLT 2019, vol. 1 (Long and Short Papers), Minneapolis, MN, USA, 2–7 June 2019, pp. 4171–4186 (2019)
12. Luo, L.: Network text sentiment analysis method combining LDA text representation and GRU-CNN. Pers. Ubiquitous Comput. **23**(3–4), 405–412 (2019)
13. Rosenblatt, F.: Principles of neurodynamics. Perceptrons and the theory of brain mechanisms. Technical Report, DTIC Document (1961)
14. Hoang, C.D.V., et al.: Incorporating side information into recurrent neural network language models, pp. 1250–1255 (2016)
15. Sutskever, I., Martens, J., Hinton, G.E.: Generating text with recurrent neural networks. In: Proceedings of the 28th International Conference on Machine Learning, ICML 2011, Bellevue, Washington, USA, 28 June–2 July 2011
16. Zhang, X., Zhao, J.J., LeCun, Y.: Character-level convolutional networks for text classification. In: Advances in Neural Information Processing Systems 28: Annual Conference on Neural Information Processing Systems 2015, 7–12 December 2015, Montreal, Quebec, Canada, pp. 649–657 (2015)
17. Tang, D., Qin, B., Liu, T.: Document modeling with gated recurrent neural network for sentiment classification. In: Proceedings of the 2015 Conference on Empirical Methods in Natural Language Processing, EMNLP 2015, Lisbon, Portugal, 17–21 September 2015, pp. 1422–1432 (2015)
18. Wan, L., Zhu, L., Fergus, R.: A hybrid neural network-latent topic model. In: Proceedings of the Fifteenth International Conference on Artificial Intelligence and Statistics, pp. 1287–1294 (2012)
19. Werbos, P.J.: Backpropagation through time: what it does and how to do it. Proc. IEEE **78**, 1550–1560 (1990)
20. Mikolov, T., Zweig, G.: Context dependent recurrent neural network language model. In: 2012 IEEE Spoken Language Technology Workshop (SLT), Miami, FL, USA, 2–5 December 2012, pp. 234–239. IEEE (2012)
21. Sainath, T., et al.: Making deep belief networks effective for large vocabulary continuous speech recognition. In: 2011 IEEE Workshop on Automatic Speech Recognition & Understanding, pp. 30–35. IEEE (2011)
22. Tampuu, A., et al.: Multiagent cooperation and competition with deep reinforcement learning. PLoS ONE **12**(4), e0172395 (2017)
23. Wood, J., et al.: Source-LDA: enhancing probabilistic topic models using prior knowledge sources. In: 33rd IEEE International Conference on Data Engineering, ICDE 2017, San Diego, CA, USA, 19–22 April 2017, pp. 411–422 (2017)
24. Hansen, J.A., Ringger, E.K., Seppi, K.D.: Probabilistic explicit topic modeling using Wikipedia. In: Proceedings of the Language Processing and Knowledge in the Web - 25th International Conference, GSCL 2013, Darmstadt, Germany, 25–27 September 2013, pp. 69–82 (2013)
25. Dieng, A., et al.: TopicRNN: a recurrent neural network with long-range semantic dependency. In: 5th International Conference on Learning Representations, ICLR 2017, Toulon, France, 24–26 April 2017, Conference Track Proceedings (2017)
26. Nguyen, D.Q., et al.: Improving topic models with latent feature word representations. Trans. Assoc. Comput. Linguist. **3**, 299–313 (2015)

Detection of Depression-Positive Thai Facebook Users Using Posts and Their Usage Behavior

Siranuch Hemtanon[1], Saifon Aekwarangkoon[2], and Nichnan Kittiphattanabawon[1](✉)

[1] School of Informatics, Walailak University, Nakhon Si Thammarat, Thailand
knichcha@mail.wu.ac.th

[2] School of Nursing, Walailak University, Nakhon Si Thammarat, Thailand

Abstract. Detecting clinical depression is an important task to find affected patients for effective treatment, especially in an early state with a higher effective treatment. This work proposes a method for automated detecting the possible depression-positive person from Facebook data, which refers to the user's textual posts and usage behavior. A machine-learning classification then uses the data to create a model to determine features signifying depression-positive users. We consider used words and statistical data of actions made on Facebook platforms, such as the number of posts, comments, and replies a user made daily, along with time and frequency information of these actions. An experiment was conducted to examine the potential and capability of the proposed method. A model from Neural Networks' behavior data yielded the best result, a 1.0 F1 score. In contrast, the model of text data from Neural Networks acquired the results as 0.88 F1 scores for classification results. From the models, we also obtain a list of significant features indicating a depression-positive state of users as keywords from text data and notable behavior from action data based on the calculated weight from machine learning.

Keywords: Depression · Mental disorder detection · Behavior features · Social network activity · Facebook data

1 Background

Clinical depression or Major depressive disorder is one of the most globally prevalent mental disorders [1]. Over 322 million people, equivalent to 4.4% of the world's population, are estimated to suffer from clinical depression [2]. Clinical depression can be chronic or episodic, leading to long-lasting or recurrent feelings of sadness, loss of interest, guilt, and low self-worth. As symptoms progress, depression can lead to have suicidal thoughts and commit suicide in a severe case [3]. In Thailand, the Department of Mental Health reported that there are suicidal attempters in a ratio of 4.94 and 5.33 per 100,000 population for Thai teenagers in 2018 and 2019, respectively [4], with the estimation of a continuously increasing number. Studies [5, 6] signify that recognizing the disorder in the early state brings more benefits, including more effective treatment and lessening negative impacts from affected patients. However, services for the detection

P. Meesad et al. (Eds.): IC2IT 2021, LNNS 251, pp. 77–87, 2021.
https://doi.org/10.1007/978-3-030-79757-7_8

of clinical depression are limited to self-report or family-reports such as Patient Health Questionnaire-9 (PHQ9) [7] and Hamilton Rating Scale for Depression (HAM-D) [8] for a professional report.

Recent researches are examining the potential of social network media such as Twitter [9, 10], Reddit [11], and Facebook [12] as a medium to detect signs of clinical depression as manifested in user expressions, including posts and comments. Studies [13] found a relation between expressed language and clinical disorders; thus, depression-affected patients can be detected from their posts and comments in social network media. These social media networks have been accepted and used in daily lives as tools to express thoughts, opinions, and feelings. These social media networks have become a large data bank of people's thoughts and activities for data-analyzing. In Thailand, Facebook has been the most used social media, as over 50 million users accounted for 76.7% of the entire Thai population in April 2020.

For text-based analysis of health problems and metal disorders, supervised machine learning is one of the most used methods. The labelled text data of health states from Social Medias are trained to create a prediction model able to classify users. Prieto et al. proposed a method to use textual information from Twitter to measure a set of illness conditions including flu, depression, pregnancy and eating disorders. They used a method based on regular expressions from calculating the log-likelihood of the words and supervised classification (Support Vector Machines (SVM), Naïve Bayes (NB), Decision Trees (DT) and K-Nearest Neighbor (KNN)) to identify a set of tweets indicating diseases or health states. Birnbaum et al. [9] developed a collaborative approach of using supervised machine learning and predefined clinical appraisals to identify schizophrenia disorder from Twitter data. Nadeem [10] developed a method to recognize depression disorder through analysis of large-scale records of user's linguistic history in Tweeter with supervised classifiers. Cacheda et al. [11] proposed a text-analysis using random forest (RF) technique on Reddit public data to determine users' depression states based on features from textual, semantic, and writing similarities. They proposed two different approaches as RF classifier with two threshold functions to detect depression-positive persons and independent RF classifiers to identify depression-negative persons. Eichstaedt et al. [12] studies status posts from depression-diagnosed Facebook users to create a topic model of language markers for depression using latent dirichlet allocation (LDA), an unsupervised method, for finding relation between depression and used terms.

Aside from textual expression in posts and comments, social media usage behavior can be another clue for depression detection since clinical depression has manifested how one expresses and how one acts. Depression-positive users should also have a different pattern of social media usage. However, there are very few studies on detecting mental disorders based on users' social media usage. Hence, this study aims to develop a framework to detect Thai depression-positive Facebook users using text data and usage data.

2 Method

This research proposes a framework of depression detection from Facebook data in both content and usage aspects. Thus, data from Facebook are separated into two parts as content and user behavior. The core method for achieving a detection model is a supervised classification using machine learning. An overview of the proposed framework is illustrated in Fig. 1. Subsequent paragraphs; however, are indented.

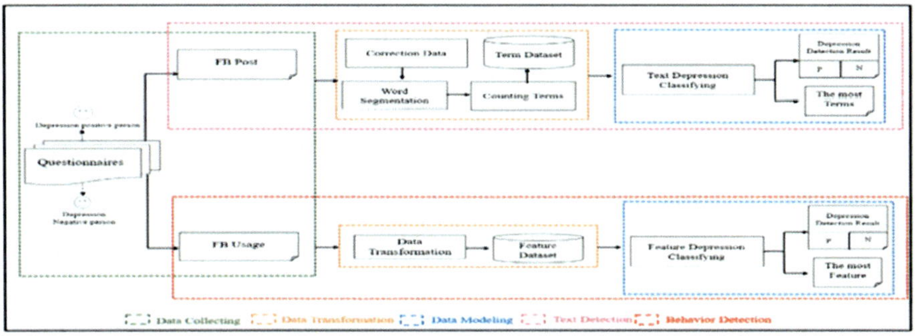

Fig. 1. Proposed framework of depression detection with Facebook data

From Fig. 1, there are three main processes. The first one is data preparation, which includes data collection, data transformation, and data labeling. The second part is to generate a text-based classification model for content-based depression detection. In this part, statistical information at the language level is calculated. Words as features are used to generate a classification model, and the found keywords indicating a depression-positive state based on statistic calculation. The third part is the development of the classification model based on Facebook usage data.

2.1 Data Preparation

Data in this work are Facebook data from actual users. Facebook is an online social media required registration for an individual user. Its users can post text messages, photos, and multimedia, with attention to share with other users. We split the Facebook data into two types. The first data type is posted content in a textual kind. The text posted by a user is intended for others to read. Thus, the posted texts have meaning and intention in them. The second data type is the data of the post and other reactions allowed in Facebook media. These data are about time and frequency of user actions as one's behavior in Facebook media. We expect that these two data types should contain a good set of significant features in determining depression-affected persons on the Facebook platform. Data in 14 consecutive-day periods is required from a Facebook user for training classification models in terms of collection.

For posts, all text-based contents made for the public are collected along with user ID. Text-based content is the only content owner makes; thus, content shared from another account is not collected. Since social media is a place for casual expression, typos, and misspellings always present among words in a post and require manual editing to reduce noises and prevent scattered term statistics. As there is no word boundary in Thai, a word segmentation or tokenizer is required to indicate a word boundary. Specifically, several Thai tokenizers are available online; however, we expect the word to be on a concept level, and Lexto+ [14] provided by NECTEC is suitable for the task. Segmented words in a text are transformed into a numerical representation in the form of a vector using a bag of word approach [15]. A bag of word approach is to represent each unique word with a frequency of the word in a post. Each post is treated as a separate document and made into a list of all words from all documents, excluding the punctuation. Then, a vector of posts and unique words is created. The word-vector will then be used for finding features significant for classifying specified classes. For determining the significance of terms, term frequency (TF) and inverse document frequency (IDF) are exploited. The TF-IDF score is a multiplication of TF and IDF, and the higher the score shows, the higher the significance.

We observe users' newsfeed and count their actions related to Facebook function with a timestamp as features for user behavior. A list of collected behavior features is given in Table 1.

Table 1. Behavior features designed to capture signs of depression from Facebook usage.

Aspect	Definition	Features and their details
Facebook functions	Posts made by an account owner	- Post_N: Average number of daily posts - Post_SD: Standard deviation of daily posts
	Share made by sharing other account's post	- Share_N: Average number of daily shares - Share_SD: Standard deviation of daily shares
	Comment as replying in another person's post and share	- Cmnt_N: Average number of daily comments - Cmnt_SD: Standard deviation of daily comments
	Reaction to another person's comments in own post and share	- React_P: Average percentage of daily replies - React_SD: Standard deviation of daily replies
Time	Weekday - Weekend	- Weekd_N: Number of actions done in Mon-Fri - Weeknd_N: Number of actions done in Sat-Sun

(*continued*)

Table 1. *(continued)*

Aspect	Definition	Features and their details
	Hour of a day	- Morn_N: Number of actions done in 06:01–12:00 - Aftn_N: Number of actions done in 12:01–18:00 - Evng_N: Number of actions done in 18:01–24:00 - Laten_N: Number of actions done in 00:01–06:00
Consecution	Consequent action	- Gap0_N: Number of times that actions are done consecutively within 5-s gap - Gap1_N: Number of times that actions are done consecutively within 1-min gap - Gap5_N: Number of times that actions are done consecutively within 5-min gap - Gap10_N: Number of times that actions are done consecutively within 10-min gap - Gap15_N: Number of times that actions are done consecutively within 15-min gap

The features involve the main functions of Facebook, including posting, commenting, replying to a comment, and sharing a post. The activity regarding the features is counted and calculated to daily statistics. This work excludes the chatting function provided with Facebook messenger because we can monitor only the activities that users set as public or friend only.

From the obtained numerical data representing either depression-state Facebook usage behavior, we then normalize the data to scale the numerical data into a range between 0 and 1 using the min-max Normalization method, which is the most common method normalize data [16]. The minimum value of that feature is transformed as ‘0’, and the maximum value is transformed into ‘1’ while the other value is calculated into a decimal between 0 and 1. The formula for a min-max of [0, 1] is given in (1).

$$x' = \frac{x - \min(x)}{\max(x) - \min(x)} \tag{1}$$

Where x refers to an original value and x′ is a normalized value. Furthermore, we separate the normalized value into five categories regarding the range as very low (0–0.19), low (0.2–0.39), moderate (0.4–0.59), high (0.6–0.79), and very high (0.8–1). These five categories should widen the model to be more versatile.

To assign a class for supervised machine learning, we rely on a questionnaire to determine Facebook users of their depression status. This work's chosen depression test is the PHQ9 questionnaire in the Thai version, widely used and accepted by the Department of Mental Health to detect signs of depression in Thailand. That version of PHQ9 consists of nine questions aiming to capture symptoms of depression that affect ones' mind, emotional state, cognitive state, physical state, and behaviors by asking the respondent for his/her current feeling. The respondent is asked to answer the question in a 0–3 scale answer where 0 means not at all and 3 refers to often in an increasingly frequent manner. The depression state from PHQ9 is defined into five classes: no depression, mild depression, moderate depression, moderately severe depression, and severe depression. Then, we group mild depression, moderate depression, and severe depression as 'depression-positive' and no depression as 'depression-negative'. The depression result of a respondent is used as an assigned class to his/her post for textual post-classification and his/her behavior.

2.2 Depression Prediction

The depression prediction in this work can be formalized as a binary classification problem using the label from the PHQ9. A supervised machine learning of a features-based approach with designed features is expected to capture correlations between different aspects of the individual user's Facebook data and their depression state.

With the different nature between text data and behavior data, separated classification models are developed. The text-based classification takes a TF-IDF score of words from a collection of posts for features. The posts' label is inherited from the user's depression state; thus, if the posts belong to a depression-positive user, posts will be labeled as positive, accordingly, and vice versa for a depression-negative user. To extend the words for more conceptual representation, an n-gram is also applied to group consecutive words up to 3 words; hence, text-based features become a collection of unigram, bigram, and trigram. With bigram and trigram, the negation word "ไม่" (mai) in Thai has a function to negate a meaning of a word or a phrase, which will play a more crucial role to opposite the meaning of the following semantic concept. In terms of depression prediction from usage behavior, users represent instances in training with PHQ9 results as their label. Table 1, mentioned specified features, and their value represent correlations between users' depression state and Face-book usage behavior.

2.3 Key Feature Detection

The detection model that applies information gain (IG) calculation, such as decision tree algorithm ID3, can rank features following their obtained IG score. IG score is calculated as a measure of the difference in entropy values from before to after the set S is partitioned regarding a feature A by (2).

$$IG(A) = H(S) - \sum\nolimits_{t \in T} p(t)H(t) = H(S) - H(S|A) \tag{2}$$

Where $H(S)$ is entropy of set S. T refers to subsets from separating set S by a feature A, and $P(t)$ is a proportion of the number of items in subset t to the number of items in set S. Last, $H(S)$ refers to entropy of the subset t.

The obtain IG score can represent how much impact a feature contains for depression-positive users; thus, the higher IG score a feature gets, the more significant it is to identify depression state. Therefore, exploiting the IG score's rank can answer the key feature representing the depression state in both models.

3 Experiments

To test the potentials of the proposed method, demonstrative experiments are designed and conducted. First, a selection of participants for getting Facebook data is described. Then, experiments are split into two main parts based on data type as text and behavior-based data. The experiments are to examine the performance of the proposed method. Last, the result of the found key factors indicating the positive-depressed persons is provided.

3.1 Participants and Experimental Data

Participants in the experiments were voluntary Thais who willingly applied to our experiment with the following criteria: 1) Age over 18-year-olds, 2) Owning a personal Facebook account for more than 1 year, 3) Willing to take a PHQ9 test, and 4) Allowing and opening to be monitored for their Facebook activity for 14 consecutive days.

By regulation, we need a certificate of ethical approval for using human personal data, and the certificate grants us a maximum of 160 participants. Hence, we obtained the participant with details as given in Table 2.

Table 2. Participant information

Participant information	Amount (160 participants)
Gender	Male: **42**, Female: **114**, LGBT: **4**
Age group	18–22: **56**, 22–35: **43**, 36–50: **39**, < 50: **2**
Depression state from PHQ9	Negative: **40**, Positive: **120**

The participants' Facebook data are collected once they consent to monitor their account and finish the PHQ9 for their depression status. For both experiments, five machine-learning techniques as NB, NN, DT, SVM, and KNN, are applied to generate a depression prediction model. The experiment is performed in Rapid Miner software (version 9.6). All the parameters for the applied techniques are set as default. Data are equally separated into five folds to be tested in 5 fold cross-validation fashion. Each of the separated folds is made sure to contain the same number of positive and negative depression instances. The evaluation measurement in both experiments is precision (A), (P), recall (R), and F-measure (F1) score.

3.2 Evaluation Results of Depression Detection

This experiment involves two data types as text-based data from posts and behavior based-data from usage. The posts are preprocessed, including removing non-words and correct misspellings; then, they are transformed, as mentioned in Sect. 2.2. The post-collection has characteristics, as shown in Table 3.

Table 3. Characteristic of text data

Post	Depression-positive users		Depression-negative users	
	4,353		1,889	
Word	Unigram	Bigram	Trigram	Combination
	7,342	16,382	14,418	38,142

We then process to remove specific stop-words, including preposition, conjunction, and interjection. Examples of words in gram level are as follows:

- Unigram: Unigram: รู้สึก, เครียด, ชีวิต, เศร้า, เบื่อ, มาก
- Bigram: Bigram: รู้สึก|เครียด, รู้สึก|เบื่อ, ชีวิต|เศร้า, เศร้า|มาก, เบื่อ|มาก
- Trigram: Trigram: รู้สึก|เบื่อ|มาก, ชีวิต|เศร้า|มาก

Another detection in this experiment is to test depression detection from Facebook usage behavior. With the designed behavior features, the raw data are in the form of numerical value. The performance results in comparison of depression detection are given in Table 4.

Table 4. Comparison of performance results from depression detection

	Text-based data				Behavior-based data			
	A	P	R	F1	A	P	R	F1
NB	0.73	0.98	0.75	0.84	0.98	0.99	**1**	0.99
NN	0.77	0.98	0.79	**0.88**	0.98	0.99	**1**	0.99
SVM	0.76	0.96	0.78	0.86	0.94	0.92	**1**	0.96
DT	**0.8**	0.88	**0.86**	0.87	0.91	0.96	0.91	0.94
KNN	0.76	**1**	0.76	0.87	**1**	**1**	**1**	**1**

The experimental results indicate that behavior data is superior to post data in precision and recall. The evaluation results of behavior-based depressing detection are very high, regardless of applied machine learning techniques. From the analysis, we found the following points to discuss experimental results.

The designed features of behavior data involve many aspects that can be formative in discriminating between depressive-positive users and depressive-negative users. According to data distribution, it can clearly see that some aspects are different by a wide margin, such as stand derivation of the daily post (Post_SD) and average daily comments (Cmnt_avg). Their information gain score is also high and can comfortably determine the users' depression state.

On the other hand, posts from depressive-positive users can be either a normal post or a depression-informative post since depressive disorder patients do not always depress, especially mildly depressed individuals. Moreover, depressive-negative users can also be emotional and make a post using depressive words occasionally. Thus, the distribution of words can be mixed up and lower the discrimination power of classification. This can be concluded that the text-based detection of depression is sensitive to training data. It needs a prepared or selective dataset consisting of only depressive post content to improve their performance. Unfortunately, to obtain such a dataset, considerable efforts and budget are needed to ascertain dataset quality. Furthermore, the prepared dataset should be large and contain variations of words to be more effective in classification.

3.3 Key Features Indicating Depression from Models

The classification model learned from Facebook data can indicate the highly impacted features to determine positive-depression users. This information reveal yet to be known facts or prove the existing theory of depression disorder.

In the process of classification model generation, especially Decision Tree model creation, given data is calculated into statistical information called 'Information Gain' (IG). IG refers to a measurement of how much information about the class that one feature gives. IG value lies within the range between 0 and 1, while 1 is the highest informative value indicating a highly impacted feature, and 0 is the lowest. We can determine the impact of the features for positive-depression Facebook users by sorting the IG value.

For text-based classification, features are words in the posts; thus, the top-ranked features should indicate words that depressive-positive users often use. The 10-top-ranked words are as follows:

<table>
<tr><td>ไร้|ค่า (0.97)</td><td>ตาย (0.92)</td><td>เครียด (0.87)</td><td>เบื่อ (0.87)</td><td>เกลียด (0.86)</td></tr>
<tr><td>เหงา (0.86)</td><td>แคร์ (0.86)</td><td>ไม่|ไหว|แล้ว (0.85)</td><td>ไม่|สน (0.83)</td><td>โดดเดี่ยว (0.83)</td></tr>
</table>

In terms of behavior, a feature is an action in a specific duration or usage style. The top-10-ranked behaviors are given below.

React_P(0.94)	Resp_SD(0.93)	Gap20_N(0.93)	Post-SD (0.87)
Laten_N (0.82)	Gap15_N (0.81)	Gap10_N(0.76)	During_Ln (0.72)
Gap5_N (0.74)	Avg_During_Ln (0.71)		

4 Conclusions and Future Works

Detecting clinical depression is an important task to find affected patients for effective treatment, especially in an early state with a higher treatment rate. This work proposes a method for automated detection of the possible depression-positive person from Facebook data. The Facebook data used in this work as features for the classification model are split into two types. The first type is a text from a post. Words in posts are processed into a vector for their term-frequency and inverse-post frequency. A machine-learning classification then uses the score to create a model to determine a set of terms that depression-positive persons are usually used. The second type of Facebook data is statistical data of actions made Facebook platform, such as the number of posts, comments, and replies a user made daily and time and frequency information of these actions. The features are also trained to generate a classification model from the supervised machine learning approach. The label for depression-positive and negative Facebook users in this work comes from the PHQ9 as a label for a post and user behavior profile. An experiment was conducted to examine the potential and capability of the proposed method. In finding participants, the criteria of the 160 volunteers in this experiment include having a Facebook account for more than a year and having at least 7 Facebook actions weekly. For classification results, the model from behavior data from KNN yielded the best result as 1 F1 score, while the model of text data from NN acquired the best classification for 0.88 F1 scores. We found a list of significant features as keywords from text data and notable behavior from action data based on the calculated weight from machine learning from the models. This work applied a limited set of behavior features and texts from a post that can be numerically processed from users' actions in a Facebook service; thus, it can be expanded to cover another behavior set such as a mood and color from an image post and behavior data from the chatbox. Moreover, we plan to improve the model's capability by applying more data to classify a severity degree of depression, including mild depression, moderate depression, and severe depression. Then, another mood disorder such as bipolar and schizoaffective disorder will be studied and applied to the extent of the method's scope. Last, an assistant system to help a detected depression-positive user such as a Chabot to console the sad person should be implemented and included in the system.

References

1. Charlton, B.G.: The malaise theory of depression: major depressive disorder is sickness behavior and antidepressants are analgesic. Med. Hypotheses (2000)
2. World Health Organization: Depression and other common mental disorders: global health estimates (No. WHO/MSD/MER/2017.2). World Health Organization (2017)
3. Mackenzie, S., Wiegel, J.R., Mundt, M., Brown, D., Saewyc, E., Heiligenstein, E., Fleming, M.: Depression and suicide ideation among students accessing campus health care. Am. J. Orthopsychiatry **81**(1), 101 (2011)
4. THE ASEAN POST Homepage. https://theaseanpost.com/article/suicides-rise-thailand. Accessed 20 Nov 2020
5. Halfin, A.: Depression: the benefits of early and appropriate treatment. Am. J. Manag. Care **13**(4), 92–97 (2007)

6. World Health Organization: Promoting mental health: concepts, emerging evidence, practice: a report of the World Health Organization, Department of Mental Health and Substance Abuse in collaboration with the Victorian Health Promotion Foundation and the University of Melbourne. World Health Organization, Melbourne (2005)
7. Manea, L., Gilbody, S., McMillan, D.: A diagnostic meta-analysis of the Patient Health Questionnaire-9 (PHQ-9) algorithm scoring method as a screen for depression. General hospital psychiatry (2015)
8. Hamilton, M.: The Hamilton rating scale for depression, pp. 143–152. Springer, Heidelberg (1986)
9. Birnbaum, M.L., Ernala, S.K., Rizvi, A.F., De. Choudhury, M., Kane, J.M.: A collaborative approach to identifying social media markers of schizophrenia by employing machine learning and clinical appraisals. J. Med. Internet Res. **19**(8), 289 (2017)
10. Nadeem, M.: Identifying depression on Twitter. arXiV preprint arXiv:1607.07384 (2016)
11. Cacheda, F., Fernandez, D., Novoa, F.J., Carneiro, V.: Early detection of depression: social network analysis and random forest techniques. J. Med. Internet Res. **21**(6), 12554 (2019)
12. Eichstaedt, J.C., et al.: Facebook language predicts depression in medical records. Proc. Natl. Acad. Sci. **115**(44), 11203–11208 (2018)
13. Kapoor, K.K., Tamilmani, K., Rana, N.P., Patil, P., Dwivedi, Y.K., Nerur, S.: Advances in social media research: past, present and future. Inf. Syst. Front. **20**(3), 531–558 (2018)
14. Haruechaiyasak, C., Kongthon, A.: LexToPlus: a Thai lexeme tokenization and normalization tool. In: Proceedings of the 4th Workshop on South and Southeast Asian Natural Language Processing, pp. 9–16 (2013)
15. Sethy, A., Ramabhadran, B.: Bag-of-word normalized n-gram models. In: Ninth Annual Conference of the International Speech Communication Association (2008)
16. Al Shalabi, L., Shaaban, Z.: Normalization as a preprocessing engine for data mining and the approach of preference matrix. In: 2006 International Conference on Dependability of Computer Systems, pp. 207–214. IEEE (2006)

Abusive Language and Hate Speech Detection for Indonesian-Local Language in Social Media Text

Shofianina Dwi Ananda Putri(✉), Muhammad Okky Ibrohim, and Indra Budi

Faculty of Computer Science, Universitas Indonesia, Depok, Indonesia
shofianina.dwi91@ui.ac.id, {okkyibrohim,indra}@cs.ui.ac.id

Abstract. In social media, people are free to express their feelings and thoughts. However, people can also use abusive language and hate speech to insult or humiliate individuals or groups on social media, such as Twitter. Various detection methods have been developed to control the spread of abusive language and hate speech in Indonesia, but the detection process is still focused on monolingual. As a country with various ethnicities and cultures, Indonesia also has a variety of local languages. This study examines abusive language and hate speech detection on Twitter, which also contains five local languages, including Javanese, Sundanese, Madurese, Minangkabau, and Musi. In this work, we present a preliminary evaluation to find the best performance of machine learning methods in detecting abusive language and hate speech on Twitter as preliminary study for each local language. We use several machine learning algorithms, such as Naïve Bayes (NB), Support Vector Machine (SVM), and Random Forest Decision Tree (RFDT) as classifiers and TF-IDF weighted word n-gram and character-n gram as feature extraction. The experiments use the 5-Fold cross-validation approach and evaluated by measuring the F-1-Score. After the experiment, we have obtained the SVM classifier with word n-gram features show the best F-1-Score for each dataset.

Keywords: Abusive · Hate speech · Machine learning · Indonesian local languages · Twitter

1 Introduction

The development and use of social media change rapidly, along with the ease of access to the internet today. As per January 2020, the data shows that social media users, especially in Indonesia, reached 160 million users, where this value increased to around 12 million users (+8.1%) between April 2019 and January 2020[1]. In social media, people are used to express their though, or sharing their ideas and creations. Furthermore, people also use social media to communicate with others and reach information. However, the ease and freedom of expression in social media can increase the potential of spreading abusive language and hate speech. Abusive language is an expression that contains

[1] https://datareportal.com/reports/digital-2020-indonesia.

P. Meesad et al. (Eds.): IC2IT 2021, LNNS 251, pp. 88–98, 2021.
https://doi.org/10.1007/978-3-030-79757-7_9

harsh/dirty words addressed to interlocutor both verbally and in writing, while hate speech as an act, either directly or not based on hatred towards religion, belief, race, skin color, ethnicity, gender, people with disabilities, and sexual orientation, that contains incitement to individual or groups through various media [1]. These are very dangerous because these utterances can trigger conflict and potentially cause material loss and victims [1].

Indonesia is a nation with vast diversity of tribes, races, cultures. With a lot of dis-similarities between one's region and the other, Indonesians tend to use their own local language to communicate [2]. Among the 706 local languages in Indonesia, there are at least 20 languages that are still spoken by millions of people in Indonesia [3]. The five largest local spoken languages, which are Javanese, Sundanese, Madurese, Minangkabau, and Musi [2, 3]. The massive use of local languages gives positive impact on expanding language theories, as well as nurture and flourishing regional sense of belonging [2]. However, it also elevates some communication issues. For example, social media enables someone to use hate and abusive speech using local language and make the detection process more difficult.

Abusive language and hate speech are serious problems worldwide and many countries and organizations have agreed to resist [4]. Abusive language and hate speech detection process on social media are important to filter hateful tweets and analyze public sentiment from one group to another and avoid potential incorrectly linked activity [5]. The problem of detecting abusive language and hate speech have been raised as a topic in several previous studies. The hate speech classification by using NB was done by [6]. They used NB to classify hate speech in three categories, which are non-political hate speech, hate speech with political motive, and non-hate speech. It achieves 93.22% of accuracy.

In [7], the hate speech detection in Indonesian tweets has been done. In their research, the dataset was generated from tweets related to the governor election of DKI Jakarta. The classification process is then carried out using the NB, SVM, and Bayesian Logistic Regression and RFDT methods as classifiers and uses the word n-gram and character n-gram as features. It achieves 93.5% of $F_1 - Score$ with word n-gram feature and the RFDT algorithm. A study for abusive language detection has been done by [8]. They use NB, SVM, and RFDT to identify not abusive language, abusive but not offensive and offensive language in Indonesian tweets. The result shows that the NB classifier with unigram + bigrams features has the highest performance with 70.06% of $F_1 - Score$. In [9], the hate speech detection was done using ensemble methods. They used TF-IDF as features and five independent classification algorithms, namely NB, k-Nearest Neighbors, Maximum Entropy, Random Forest, and SVM. Then, they use hard voting and soft voting as ensemble methods. It obtained the soft voting has better performance with 79.8% of $F_1 - Score$ on imbalance dataset and 84.7% of $F_1 - Score$ on balanced dataset. The dataset was obtained from [7].

According to [10] and [11], these methods are considered effective and computationally efficient because they do not require extensive data for the training process, simple to implement and produce high accuracy. Some previous researches support that NB, SVM, and RFDT show good performance in the abusive and hate speech detection process. The various features and classifiers that have been obtained will be used for text classification in this study. This research will focus on detecting abusive language and hate speech on Indonesian Twitter data, which contains local languages from five regions in Indonesia, namely Javanese, Sundanese, Madurese, Minangkabau, and Musi. A machine learning approach will be implemented to solve this problem. The classification algorithms used are NB, SVM, and RFDT. Meanwhile, the features used are word n-gram and character n-gram with TF-IDF as term weighting scheme.

2 Method

The workflow in this research can be seen in Fig. 1. The first stage is crawling the data. The dataset were collected using Twitter API[2] and Tweepy Library[3]. Then the data is filtered to obtain tweets containing each local language characteristic.

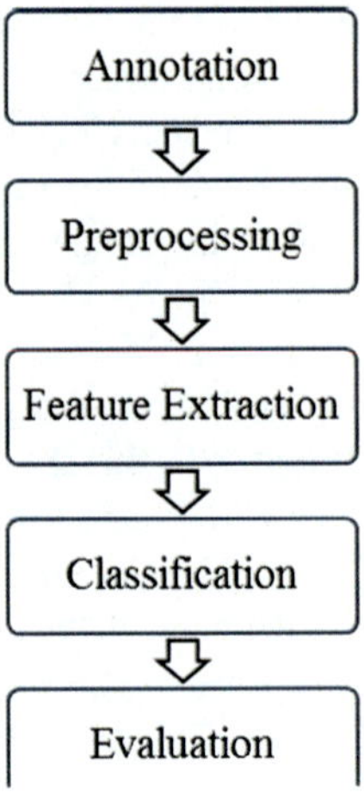

Fig. 1. Research flow

2.1 Annotation

The dataset five dataset are annotated into two labels that are abusive language and hate speech. We use ten annotators to annotate the dataset, where each tweet from each language is annotated by two annotators that native in each local language dataset will be

[2] https://developer.twitter.com/en/docs/twitter-api.

[3] http://docs.tweepy.org/en/latest/.

annotated by them. The examples of tweets in each local language dataset are shown in Table 1, while the examples of annotated tweets for each language can be seen in Table 2. In this paper, the final label was decided using the 100% agreement technique, where the annotation results can be seen in Table 3. Furthermore, to validate the annotation result, we also calculate the agreement level using Cohens Kappa [12] as shown in Table 3. The table shows that the obtained Cohen's Kappa coefficients for each language have a moderate agreement. It indicates this research use valid dataset. For each annotated dataset, the top words can be seen in Table 4.

2.2 Pre-processing

The next step is pre-processing the data text. Pre-processing is done to change the word representation in the data, so that it can be easier to be processed in next step. Pre-processing is an important step in conducting text analysis, this is because most of the data on social media contains words and sentences that are informal, unstructured and have a large amount of noise [13]. The pre-processing steps that have been carried out include case folding to change all characters in the dataset into lower cases, then data cleansing to remove usernames, URLs, punctuation, hashtags, and detected emotes.

Table 1. Examples of dataset for each local language

Language	Tweet
Javanese	*@USER Konco asu yo ngunu kui.. ehh astaghfirullah'*
	Translation: @USER This is how a friend who resembles a dog looks like. ehh astagfirullah (asu is Javanese which means dog)
Sundanese	*@USER @USER Anjir ari ngomong teh sok bener'*
	Translation: @USER @USER Anjir, you talk like it's true (anjir is Sundanese which means dog or word to express shock)
Madurese	*@USER tim london taun iye ampas kabeh'*
	Translation: "@USER the london team this year is trash again" (ampas is the Madurese means trash and use as abusive word)
Minangkabau	*@USER kampuang nan jauh di mato* **Translation:** @USER hometown far in sight
Musi	*@USER Botak pitak banyak lagak,bengak* Translation: @USER bald, partially bald, ostentatious, stupid

Table 2. Examples of labelled dataset

Tweets	Abusive	Hate speech
@USER @USER gemes anjir aing masih gakuat'	Yes	No
Translation: @USER @USER exasperated anjir I'm still not strong		
@USER @USER Henceut sia beureum	Yes	No
Translation: @USER @USER you black ass		
@USER asu buzzer2an ki yo nggo ngopo to? pemerintah ra nyenengke, dikritik sambi guyon we ra ditompo!	Yes	Yes
Translation: @USER dog, What is buzzer for? The government did not like it, it was criticized for making jokes		
@USER Kamu cari masalah sama saya? kamu tarung sama bapak saya aja gimana.? bapak saya juara 1 gelud sekelurahan lhoo	No	Yes
Translation: @USER you jackass trying to start something? How about you take on my pops instead? He is the undisputed champ in our district		
@USER @USER Gambar na buren anjir panon aing nyeuri	No	Yes
Translation: @USER @USER the picture is blurred anjir, my eyes hurts		

Table 3. Cohen's kappa for each dataset

Language	Number of Annotated Tweets	Number of Tweets with 100% Agreement	Cohen's Kappa	Cohen's Kappa Level Interpretation
Javanese	5000	3449	0.437	Moderate
Sundanese	5000	2207	0.462	Moderate
Madurese	5000	2773	0.439	Moderate
Minangkabau	5000	3125	0.508	Moderate
Musi	5000	2564	0.525	Moderate

2.3 Feature Extraction

The next step is feature extraction. In dataset, the features on are converted from text to vector representations. We use character n-gram features with n = 3 (Trigrams), n = 4 (Quad grams) and a combination of both [7, 8, 14]. Then, we use word n-gram feature with n = 1 (unigram), n = 2 (bigrams), n = 3 (trigrams), and combination of unigram + bigrams, bigrams + trigrams and unigram + bigrams + trigrams. Term Frequency-Inverse Document Frequency (TF-IDF) were also used as term weighting.

In this research, we use Synthetic Minority Over-Sampling Technique, namely SMOTE to tackle imbalance data issue, in order to reduce probability of overfit [15]. In

Table 4. Top 10 used abusive word for each dataset

Javanese	Sundanese	Madurese	Minangkabau	Musi
asu	*gelo*	*aseng*	*Anjiang*	*Ngoceh*
bejad	*lebok*	*laknat*	*Asiang*	*Anjeng*
goblog	*tolol*	*ampas*	*Ateis*	*Cino*
banci	*perek*	*burik*	*Autis*	*gilo*
bangsat	*anjir*	*gendeng*	*Babi*	*bengak*
kirik	*bagong*	*sinting*	*bangkai*	*buyan*
bokong	*anying*	*ancok*	*bencong*	*kapir*
keparat	*kejem*	*buduh*	*Bilih*	*najes*
picek	*geuleuh*	*ateis*	*Bisu*	*bokep*
edan	*kanyut*	*toyol*	*bongak*	*jembot*

creating the synthetic data, SMOTE use K-nearest neighbour and choosing the random data from minority class. This process will be continuously repeated until the minority and majority class has the same synthetic data proportion. We use Scikit-Learn library[4] to implement this technique.

2.4 Classifiers

In this research, several classifiers used are Naïve Bayes (NB) [16], Support Vector Machine (SVM) [17], and Random Forest Decision Tree (RFDT) [18]. Naïve Bayes classification is done by calculating the data's probability of entering a specific label category based on its features. In classification, NB has a robust naive assumption, where each feature used is assumed to be mutually independent [16]. The SVM classification method is based on the principle of Structural Risk Minimization (SRM). Vapnik first proposed this algorithm in 1995. SVM will map the input into a high-dimensional vector in the classifying process then look for the best hyperplane to separate data input that has different classes [17]. Meanwhile, RFDT is an ensemble learning method, which is a method that combines several classification algorithms. The combination is including either different algorithms or the same algorithm but repeatedly does. This classification is carried out by combining several decision trees, which are then determined based on the majority voting of the formed decision trees [18]. These methods are the most popular classifier and show promising results in solving classification tasks [19]. We use the Scikit-Learn library in implementing these classifiers[4].

For the training and testing phase, we use a 5-Fold cross-validation strategy. In this strategy, the data is split into five equal folds, the 80% of the data act as train data and test data for the rest [20]. All the classifier performances were evaluated using $F_1 - Score$ measurement [19]. To classify the data, we use character n-gram, word n-gram and combination of them all as features and NB, SVM and RFDT as classifiers. Table 5–9

[4] https://scikit-learn.org/.

present the evaluation performance using $F_1 - Score$ measurement. This evaluation will see how each feature affects the classification results. The imbalance issue was tackled for all dataset using an imbalanced learn API library[5].

Based on Table 5 for the Javanese dataset, we can see that the SVM classifier shows the best performance for abusive and hate speech classification. It achieves 87.5% of $F_1 - Score$ with the combination of bigrams and trigrams features. From the table, we also obtained that the RFDT also perform better than Naive Bayes.

Table 5. $F_1 - Score$ for Javanese dataset

Feature	NB	SVM	RFDT
Char Trigrams	0.483	0.483	0.489
Char Quadgrams	0.445	0.765	0.478
Char Trigrams + Quadgrams	0.448	0.723	0.508
Word Unigram	0.460	0.504	0.492
Word Bigrams	0.476	0.495	0.495
Word Trigrams	0.481	0.508	0.493
Word Unigram + Bigrams	0.461	0.780	0.463
Word Bigrams + Trigrams	0.469	0.875	0.497
Word Unigram + Bigrams + Trigrams	0.465	0.704	0.489

Table 6. $F_1 - Score$ for Sundanese dataset

Feature	NB	SVM	RFDT
Char Trigrams	0.527	0.776	0.559
Char Quadgrams	0.503	0.791	0.549
Char Trigrams + Quadgrams	0.533	0.775	0.596
Word Unigram	0.532	0.733	0.683
Word Bigrams	0.281	0.627	0.556
Word Trigrams	0.264	0.769	0.435
Word Unigram + Bigrams	0.502	0.795	0.536
Word Bigrams + Trigrams	0.298	0.732	0.527
Word Unigram + Bigrams + Trigrams	0.510	0.769	0.566

For the Sundanese dataset, the SVM classifier also works optimally by producing a higher $F_1 - Score$ value than other classifiers. Moreover, the SVM classifier returns similar values of $F_1 - Score$ for all feature variations, it indicates that the word n-gram

[5] https://pypi.org/project/imbalanced-learn/.

Table 7. $F_1 - Score$ for Madurese dataset

Feature	NB	SVM	RFDT
Char Trigrams	0.486	0.521	0.399
Char Quadgrams	0.455	0.668	0.383
Char Trigrams + Quadgrams	0.456	0.592	0.391
Word Unigram	0.446	0.739	0.382
Word Bigrams	0.252	0.640	0.270
Word Trigrams	0.250	0.652	0.196
Word Unigram + Bigrams	0.349	0.727	0.399
Word Bigrams + Trigrams	0.259	0.582	0.259
Word Unigram + Bigrams + Trigrams	0.338	0.705	0.403

Table 8. $F_1 - Score$ for Minangkabau dataset

Feature	NB	SVM	RFDT
Char Trigrams	0.570	0.548	0.481
Char Quadgrams	0.480	0.583	0.465
Char Trigrams + Quadgrams	0.527	0.560	0.486
Word Unigram	0.446	0.690	0.402
Word Bigrams	0.207	0.592	0.350
Word Trigrams	0.195	0.670	0.156
Word Unigram + Bigrams	0.402	0.605	0.469
Word Bigrams + Trigrams	0.227	0.541	0.395
Word Unigram + Bigrams + Trigrams	0.409	0.609	0.496

and character n-gram features have a good effect on the classifier. The highest $F_1 - Score$ is achieved by combining unigram and bigram features and followed by the character quad gram feature.

The word n-gram feature, which is unigram feature with SVM classifier achieved the best performance for the Madurese and the Minangkabau dataset. Variations of word n-gram features still show good performance compared to character n-gram features, where only below 0.600 of $F_1 - Score$ are obtained with these features.

On Musi dataset, SVM classifier with trigrams feature shows the best performance with 0.802 of $F_1 - Score$. This is similar to other datasets, where the highest value is obtained with the SVM classifier with word n-gram and its combination features. On this dataset, we can see that the NB and RFDT classifier constantly show lower performances in classification process.

Table 9. $F_1 - Score$ for Musi dataset

Feature	NB	SVM	RFDT
Char Trigrams	0.560	0.618	0.337
Char Quadgrams	0.456	0.646	0.424
Char Trigrams + Quadgrams	0.464	0.629	0.419
Word Unigram	0.382	0.693	0.330
Word Bigrams	0.232	0.730	0.381
Word Trigrams	0.287	0.802	0.170
Word Unigram + Bigrams	0.309	0.641	0.344
Word Bigrams + Trigrams	0.282	0.652	0.416
Word Unigram + Bigrams + Trigrams	0.333	0.614	0.344

The entire dataset shows good performance with the SVM classifier. This is in line with research conducted by [21–23] in detecting abusive language and hate speech in social media text. The best performance of SVM is obtained by using variations of the word n-gram feature, that are the Javanese dataset obtained by combination of bigrams and trigrams features, on Sundanese dataset is obtained with a combination of unigram and bigrams, in Madura and Minang dataset are obtained with the unigram feature, and the trigrams feature show the best performance on the Musi dataset.

3 Conclusion

This paper aims to find the best method for detecting abusive language and hate speech on Twitter. The used data is Twitter text in Indonesian which contains five local languages, including Javanese, Sundanese, Madurese, Minang, and Musi. We use a machine learning approach using the Naïve Bayes (NB), Support Vector Machine (SVM), and Random Forest Decision Tree (RFDT) algorithm as classifiers. We use TF-IDF weighted word n-gram and character-n gram as feature extraction. Feature variations, balancing data and pre-processing are applied to the entire dataset.

The results show that the SVM classifier can classify abusive language and hate speech in the text compared to NB and RFDT. For the Javanese dataset, the $F_1 - Score$ is 0.875, for the Sundanese dataset, the $F_1 - Score$ is 0.795, 0.739 of $F_1 - Score$ is for the Madura dataset, then the 0.690 and 0.802 of $F_1 - Score$ are obtained for Minangkabau and Musi dataset, respectively. The best results are achieved by applying a data balancing process. Therefore, it is necessary to increase the number of tested data in the future, due to the high volume of data circulating on social media.

Acknowledgement. The authors acknowledge the PUTI Doktor research grant NKB-3226/UN2.RST/HKP.05.00/2020 from Directorate Research and Community Services, Universitas Indonesia.

References

1. Komnas, H.A.M.: Buku Saku Ujaran Kebencian (Hate Speech). Komisi Nasional Hak Asasi Manusia, Republik Indonesia
2. Saputri, M.S., Adriani, M.: Identifying Indonesian local languages on spontaneous speech data. In: 2019 International Conference on Advanced Computer Science and Information Systems, ICACSIS 2019, pp. 247–254 (2019). https://doi.org/10.1109/ICACSIS47736.2019.8979939
3. Cohn, A.C., Ravindranath, M.: Local languges in Indonesia: language maintenance or language shift? Linguist. Indones. **32**, 131–148 (2014)
4. Davidson, T., Warmsley, D., Macy, M., Weber, I.: Automated hate speech detection and the problem of offensive language. In: Proceedings 11th International Conference Web Social Media, ICWSM 2017, pp. 512–515 (2017)
5. Joulin, A., Grave, E., Bojanowski, P., Mikolov, T.: Bag of tricks for efficient text classification. In: 15th Conference European Chapter of the Association for Computational Linguistics EACL 2017 - Proceedings Conference, vol. 2, pp. 427–431 (2017). https://doi.org/10.18653/v1/e17-2068
6. Akbar, R.R., Shofa, R.N., Paripurna, M.I.S.: The implementation of Naïve Bayes algorithm for classifying tweets containing hate speech with political motive. In: 2019 International Conference on Sustainable Engineering and Creative Computing (ICSECC) (2019)
7. Alfina, I., Mulia, R., Fanany, M.I., Ekanata, Y.: Hate speech detection in the Indonesian language: a dataset and preliminary study. In: 2017 International Conference Advanced Computer Science and Information Systems, ICACSIS 2017, 2018-January, pp. 233–237 (2018). https://doi.org/10.1109/ICACSIS.2017.8355039
8. Ibrohim, M.O., Budi, I.: A dataset and preliminaries study for abusive language detection in indonesian social media. Procedia Comput. Sci. **135**, 222–229 (2018). https://doi.org/10.1016/j.procs.2018.08.169
9. Fauzi, M.A., Yuniarti, A.: Ensemble method for indonesian twitter hate speech detection. Indones. J. Electr. Eng. Comput. Sci. **11**, 294–299 (2018). https://doi.org/10.11591/ijeecs.v11.i1.pp294-299
10. Nikam, S.S.: A comparative study of classification techniques in data mining algorithms. Int. J. Mod. Trends Eng. Res. **4**, 58–63 (2017). https://doi.org/10.21884/ijmter.2017.4211.vxayk
11. Pranckevičius, T., Marcinkevičius, V.: Comparison of Naive Bayes, random forest, decision tree, support vector machines, and logistic regression classifiers for text reviews classification. Balt. J. Mod. Comput. **5** (2017). https://doi.org/10.22364/bjmc.2017.5.2.05
12. Bendavid, A.: Comparison of classification accuracy using Cohen's Weighted Kappa. Expert Syst. Appl. **34**, 825–832 (2008)
13. Hakiem, M., Fauzi, M.A., Indriati: Klasifikasi Ujaran Kebencian pada Twitter Menggunakan Metode Naïve Bayes Berbasis N-Gram Dengan Seleksi Fitur Information Gain. J. Pengemb. Teknol. Inf. dan Ilmu Komput. **3**, 2443–2451 (2019)
14. Ibrohim, M.O., Budi, I.: Multi-label hate speech and abusive language detection in Indonesian Twitter, 46–57 (2019). https://doi.org/10.18653/v1/w19-3506
15. Chawla, N.V., Bowyer, K.W., Hall, L.O., Kegelmeyer, W.P.: SMOTE: synthetic minority over-sampling technique. J. Artif. Intell. Res. **16**, 321–357 (2002)
16. Lewis, D..: Naive (bayes) at forty: the independence assumption in information retrieval. In: European Conference on Machine Learning (EMCL), pp. 4–15 (1998)
17. Srivastava, D.K., Bhambhu, L.: Data classification using support vector machine. J. Theor. Appl. Inf. Technol. **12**, 1–7 (2010)
18. Ali, J., Khan, R., Ahmad, N., Maqsood, I.: Random forests and decision trees. Int. J. Comput. Sci. Issues. **9**, 272–278 (2012)

19. Kovács, B., Tinya, F., Németh, C., Ódor, P.: Unfolding the effects of different forestry treatments on microclimate in oak forests: results of a 4-yr experiment. Ecol. Appl. **30**, 321–357 (2020). https://doi.org/10.1002/eap.2043
20. Yadav, S., Shukla, S.: Analysis of k-fold cross-validation over hold-out validation on colossal datasets for quality classification. In: 2016 IEEE 6th International Conference on Advanced Computing (IACC) (2016)
21. Ibrohim, M.O., Budi, I.: Translated vs non-translated method for multilingual hate speech identification in Twitter. Int. J. Adv. Sci. Eng. Inf. Technol. **9**, 1116–1123 (2019). https://doi.org/10.18517/ijaseit.9.4.8123
22. Almatarneh, S., Gamallo, P., Pena, F.J.R., Alexeev, A.: Supervised classifiers to identify hate speech on English and Spanish Tweets. In: Jatowt, A., Maeda, A., Syn, S.Y. (eds.) ICADL. LNCS, vol. 11853, pp. 23–30. Springer, Cham (2019). https://doi.org/10.1007/978-3-030-34058-2_3
23. Desrul, D.R.K., Romadhony, A.: Abusive language detection on Indonesian online news comments. In: 2019 2nd International Seminar on Research of Information Technology and Intelligent Systems, ISRITI 2019, pp. 320–325 (2019). https://doi.org/10.1109/ISRITI48646.2019.9034620

English-Vietnamese Machine Translation Using Deep Learning

Tuan Nguyen Minh[1], Phayung Meesad[2](✉), and Huy Cuong Nguyen Ha[3]

[1] Department of Information Technology, King Mongkut's University of Technology North Bangkok, Bangkok, Thailand
minh.tuan@itd.kmutnb.ac.th

[2] Department of Information Technology Management, Faculty of Information Technology and Digital Innovation, King Mongkut's University of Technology North Bangkok, Bangkok, Thailand
phayung.m@itd.kmutnb.ac.th

[3] The University of Da Nang, Da Nang, Vietnam
nhhcuong@sdc.udn.vn

Abstract. Recently, artificial intelligence-based machine translation has been much improved over the traditional methods. A machine translator is very useful for translating text or speech from one language to another. Machine translators have replaced the word mechanism in one language for words in another with verbatim translations. However, a good translation should be employed a both a sentence and a word that have complete meaning in accordance with the context of relevant sentence. In this paper, we studied on English – Vietnamese translation using deep learning methods including Recurrent Neural Network (RNN), Long Short-Term Memory (LSTM), Gated Recurrent Units (GRU), Attention, and Transformer. The deep learning-based machine translators were compared based on the test accuracy of results translation. It was found that best deep learning-based machine translator model was the Attention mechanism, achieving 98.8% accuracy. The Transformer yielded second rank or 98.5% accuracy.

Keywords: Machine translator · Deep learning · RNN · LSTM · GRU · Attention · Transformer

1 Introduction

With increasing many kinds of data in the world, getting useful information from those is especially important to make a precise decision in learning and management. Some of data could develop strongly in the future especially image, text and speech, and we need to solve the data with more modern technologies or algorithms [1]. The computers increase doubled or over every year to handle the data and process big user data and document like images, text, speech, or audio every minute. Billions of applications also develop every day to solve, organize and analyze the suitable data for users. In the past, hardware of the computer is the most important device to save information, solve the problem of a procedure. After that, development of the computers helps users to solve

P. Meesad et al. (Eds.): IC2IT 2021, LNNS 251, pp. 99–107, 2021.
https://doi.org/10.1007/978-3-030-79757-7_10

the big problem base on connecting many computers in a noticeably short time. Beside that they can save much memory base on the cloud computing. The prediction is data growth could increase largely over what technology is now. Most of data can be saving or processed in hardware or software design but set of data dwarf could become too large and fit memory [2]. To get the information of data to understand could help people to find a suitable choosing and reduce cost in time and money. Analyzing the text by learning deep neural network models to appreciate or translate is a new approach where we could know and have more information for expressing appropriate manner in a conversion.

There have been some researchers did experiments for machine learning translators for translating from English to Vietnamese and Vietnamese to English. For example, reference [3] built two models: ConvS2S model base on RNN, and Transformer model base on RNN. They reported that Transformer model was the best results when given a suitable batch size and learning rate. However, this model did not show the full comparing to get the best way for translation. With the new trend development of technology and algorithms, analyzing text also becomes one of main points to understand the nature of a language.

In this paper, we study on deep learning machine translations models for language translation from English to Vietnamese. Five deep learning models were experimented based on RNN, LSTM, GRU, Attention mechanism, and Transformer model. The main purpose was to compare them to get the best solutions in English-to-Vietnamese translation. Another purpose was to seek for a new approach in analyzing text in general and translation English-to-Vietnamese.

2 Literature Reviews

2.1 Machine Translator

A Machine Translator (MT) is a type of computational linguistics basing on application of software for translating text or speech from one language to another. MT shows the replacement the words by the words; however, sometimes it is not good enough because the words are not corresponding meaning in some situation. Current MT is improved every day and allows users to translate more and more accurate by awareness of machine and by the methods or algorithms. Good MT based on encoder-decoder structure. In this paper, we apply the function of machine to build the models and MT will analyze by algorithms to recognize the outputs of the language from inputs. MT has played an important role in applying recurrent neural network for model encoder-decoder model especially in translating language model (as shown in Fig. 1).

2.2 Deep Learning

Deep learning is a subset of machine learning methods based on artificial neural network with representation. Deep learning uses many functional layers to extract information from raw input data. Deep learning could be applied to supervised, semi-supervised and unsupervised learning tasks. This is particularly important because some unlabeled data is good to label data. For example, deep structure could be trained in an unsupervised manner are neural history compressors and deep belief networks.

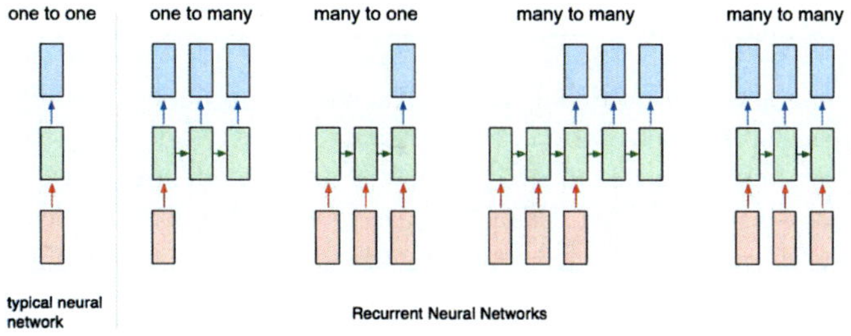

Fig. 1. Diagram of different RNN sequence types

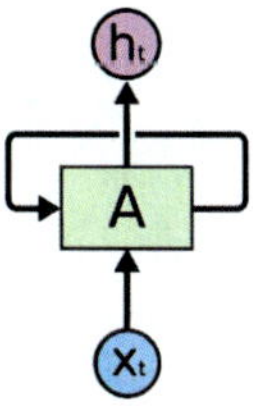

Fig. 2. Recurrent Neural Network model

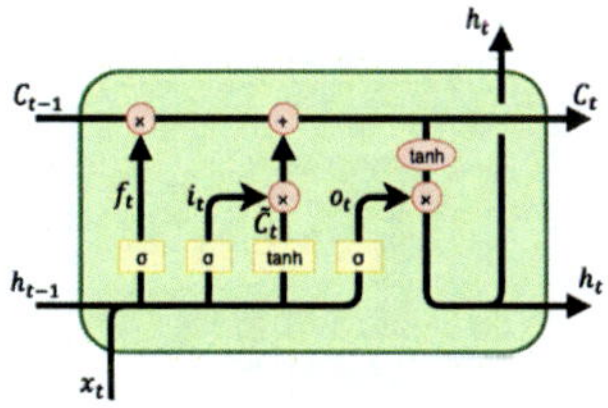

Fig. 3. LSTM model

2.3 Recurrent Neural Network (RNN)

RNN is a type of artificial neural networks that have input nodes connected to hidden node and then hidden node connected to output nodes as feedforward fashion. In addition, the hidden state or outputs from hidden nodes are fed back as additional inputs allowing temporal dynamic behavior to be captured. The RNN can take up the state within to process a sequence of variable length inputs. This makes them useful tasks such as unsegmented, connected handwriting or speech recognition. The architecture of RNN is shown in Fig. 2. [4].

2.4 Long Short-Term Memory (LSTM)

LSTM shown in Fig. 3. [5] is deep learning that is a kind of recurrent neutral. Not only it could process single data points, but also entire sequences of data. For example, LSTM is applied to unsegmented, connected handwriting or speech recognition. Moreover, LSTM is a good tool for classification, processing and prediction based on time series data. LSTM is applied to deal with the vanishing gradient and exploding problems.

2.5 Gated Recurrent Unit (GRU)

GRU [6] is also a kind of artificial recurrent neutral networks used in deep learning. GRU combined information form a new source annotation vector to generate context vector. GRU could perform on tasks of polyphonic music modeling, speech, and natural language processing like LSTM. Using less parameters, GRU is faster and sometimes shows better performance on certain smaller and less frequent datasets. Figure 4 shows the architecture of GRU (Fig. 5).

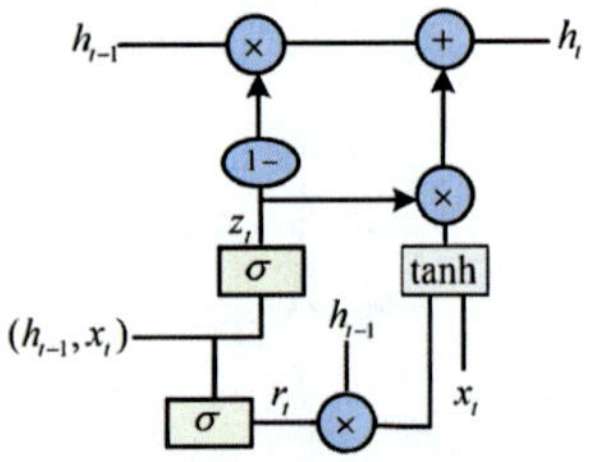

Fig. 4. GRU model

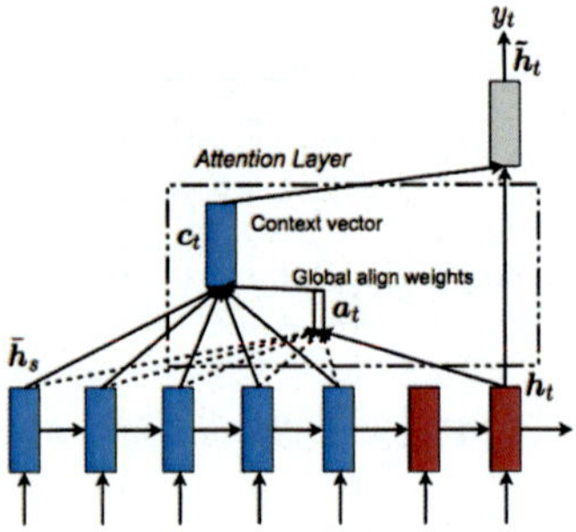

Fig. 5. Attention model

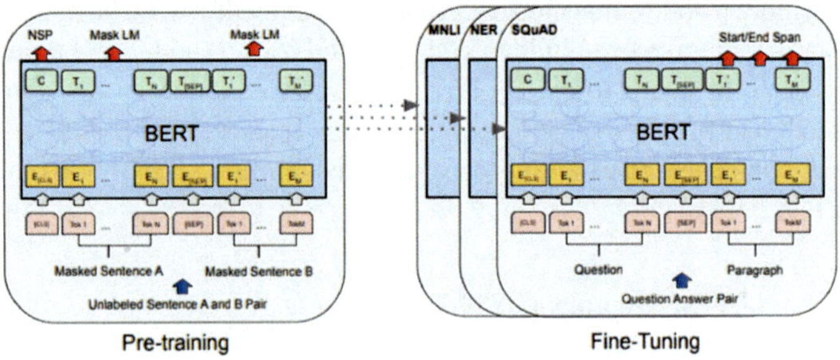

Fig. 6. BERT pre-training and fine-tuning phases

2.6 Attention Mechanism

Attention [7] is a mechanism that is developed to improve the performance of Encoder-Decoder RNN on machine translation. Attention in deep learning is to confirm the factors when processing data. Attention is one component of network architecture and is oversees managing and quantifying the relation between inputs and outputs. Attention has been applied in natural language processing tasks especially in problem of long sequence in machine translation.

2.7 Pre-training of Deep Bidirectional Transformers (BERT)

Bert [8] is really a powerful language representation model in natural language programming. BERT could produce vector representation for each word to make as the output. The main content of Bert is to generate language model. The input of Bert needs to be encoded as a sequence of tokens which are converted to vectors and predicted to decoder output. Sharing the same architecture, BERT has two phases: pre-training and fine-tuning phases. The same pre-train model parameters are used to initialize models for different tasks that can be fine-tuned. Figure 6 shows BERT in Pre-training and Fine-tuning phases.

3 Research Methodology

To achieve the objective, the research methodology in this research as 3 major steps: data preparation, model settings, and model training and evaluation.

3.1 Data Preparation

In this paper, we used a language data set retrieved from the website: http://www.manythings.org/anki/vie-eng.zip. This data set contains three columns of sentences in English, Vietnamese, and descriptions. The data have totally 4,083 records divided into 50% training set, 25% validation set, and 25% test set.

3.2 Model Settings

Five deep learning types were used including RNN, LSTM, GRU, Attention and Transformer. Each model had their setting as shown in Table 1.

Recurrent Neural Network (RNN)

Firstly, we prepared the data separated into 3 columns with English, Vietnamese, and description. We also dropped out the description column, not related to the translation. To receive the best model, we built the encoder, decoder input and decoder target based on tokenizing words. Secondly, we built five layers with two inputs encoder and decoder consisting two layers SimpleRNN. After that, we combine two inputs by the Dense and Time Distributed layers. In this model, the vectors through SimpleRNN also turned into two dimensions. The combination of two layers: Dense and Time Distributed turned the

Table 1. Model setting.

Models	Type	#Layers#Cells	#Parameters
RNN	Simple RNN	5, (73,163)	260,623
LSTM	LSTM classic	5,(73,163)	836,623
GRU	Gated recurrent units	5,(73,163)	646,159
Attention	AttentionLayer	5,(24,1771)	3,832,271
Transformer	Pytorch- Transformer	5,(64,1)	306,772
	TFBERT- Transformer	5,(24,24)	109,519,128

vectors into three dimensions. In the encoder input, we set to 73 cells and 163 cells in the decoder input. Finally in this model, we had 260,623 parameters for five layers and it sets to run 50 epochs.

Long Short-Term Memory (LSTM)
Firstly, in the data preparation step we applied the same with RNN. Secondly, we built LSTM with five layers as follow: inputs, encoder, decoder, dense layers, and Time Distributed layers. The encoder and decoder were LSTM layers. The vectors through LSTM were turned into three dimensions by the combination of two layers Dense and Time Distributed. The encoder had 73 cells and the decoder inputs had 163 cells. Finally, in this model, we got 836,623 parameters for five layers and it costs 50th at epoch 132 s.

Gated Recurrent Unit (GRU)
First, we applied first step with RNN and LSTM. Then, we built five layers with two inputs encoder and decoder consisting two layers GRU. After that, we combined two inputs by the layers GRU, Dense and Time Distributed. In this GRU model, we built five layers with two inputs encoder and decoder consisting two layers GRU, Dense and Time Distributed. In the encoder input, we get 73 cells and 163 cells in the decoder input. Finally, we get 646,159 parameters for five layers and it set to run 50 epochs.

Attention Mechanism
Firstly, we apply first step with RNN, LSTM, GRU. Secondly, we build five layers with two input embeddings. After that, we combine two inputs by the layers Attention, Concatenate and Time Distributed. In this model, the vectors through Attention also turn into three dimensions, so the combination of two layers Concatenate and Time Distributed. In the encoder input, we get 24 cells and 500 cells in the decoder input. Finally, in this model, we got 3,832,271 parameters for five layers and it set to run 50 epochs.

PyTorch Transformer-Model
Firstly, we applied the same first step with RNN, LSTM, and Attention. Secondly, we used PyTorch Transformer model to build five layers with inputs, output, encoder output, encoder hidden and decoder input. In the input, we got 64 cells and 1 cell in the decoder

put. Finally, in this model, we got 306,772 parameters for five layers and it set to run 50 epochs.

TFBert Model Transformer
Firstly, we applied with the same first step with PyTorch. In this TFBert model, we built five layers with inputs, TFBertModel, SlicingOpLambda, Concatenate and Dense layers. In the input, we got 24 cells and 24 cells in the decoder output. Finally, in this model, we got 109,500,696 parameters for five layers and it was set to run 50 epochs.

3.3 Model Training and Evaluation

Python programing was used to write scripts for all models. The experiments were run on Google colab research https://colab.research.google.com/. The performance of the trained models based on Accuracy as the following formula [3]. One way to measure how the algorithms predict the language is basing on the loss and accuracy performance when fitting the data. The loss is usually shown in the machine learning as basic appreciation to train the model. The models usually show best when the loss is usually approach 0 and accuracy approach 1. Accuracy is the number of correctly predicted data shown on total number of test cases. The formula of Accuracy is shown in (1),

$$Accuracy = \frac{TP + TN}{TP + TN + FP + FN} \tag{1}$$

where TP is positive samples, FP is mistake samples positive, TN is negative samples, FN is mistake samples negative. We also calculate the loss of the model base on the categorical cross entropy loss to measure the performance of the classification model whose output is a probability between 0 and 1. The standard categorical cross entropy loss is given by [4]. The formula of Loss is shown in (2),

$$L_{cce} = -\frac{1}{M}\sum\nolimits_{k=1}^{K}\sum\nolimits_{m=1}^{M} y_m^k \times \log(h_\theta(x_m, k)) \tag{2}$$

where M is the number of training examples, K is the number of classes, y_m^k is target label for training example m for class k, x is input for training example m, h_θ is model with neural network weights θ.

4 Experimental Results

While training different models, it was observed that LSTM based neural networks took the least time to execute a training epoch, RNN took the most time and GRU took slightly greater time than LSTM. An important factor affecting the performance of these models was training data set as there was a limited amount of structured data. In this paper, we compared the accuracy by (1) and loss by (2) between the model and then choose the appropriate model for translating from English to Vietnamese. In this case, we could choose Attention or PyTorch or TFBert will get the best accuracy to translate English to Vietnamese. Table 2s shows the test accuracy performance based the deep learning neural networks: RNN, LSTM, GRU, Attention, and Transformer PyTorch and Transformer TFBert. It shows that the Attention model is the best while Transformer is the second in ranks (Figs. 7–12).

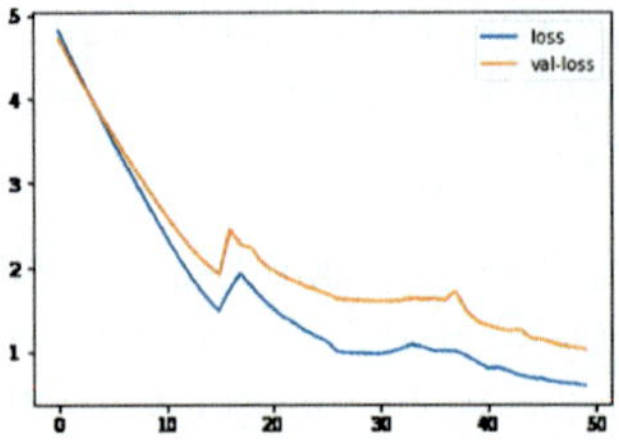

Fig. 7. RNN training and validation losses

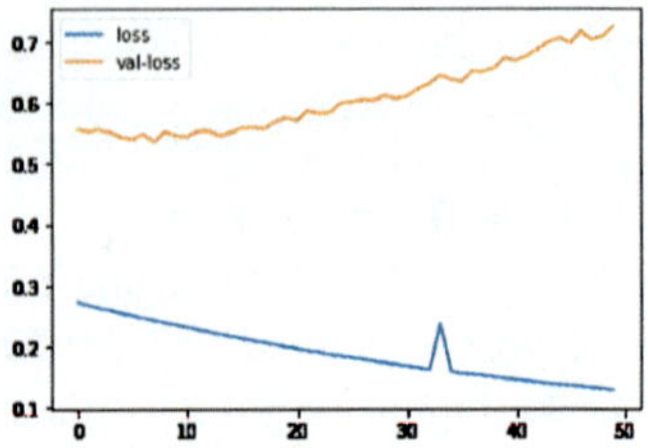

Fig. 8. LSTM training and validation losses

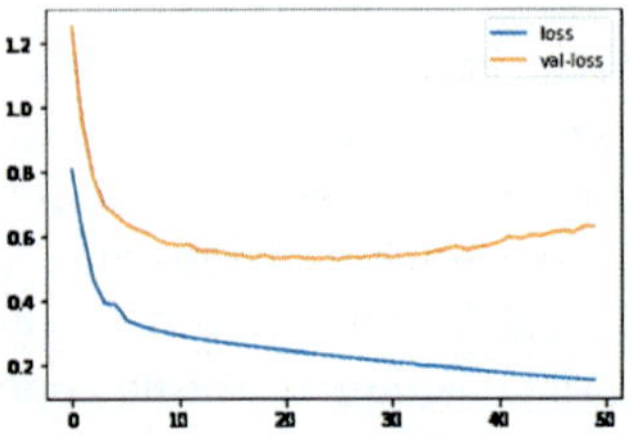

Fig. 9. GRU training and validation losses

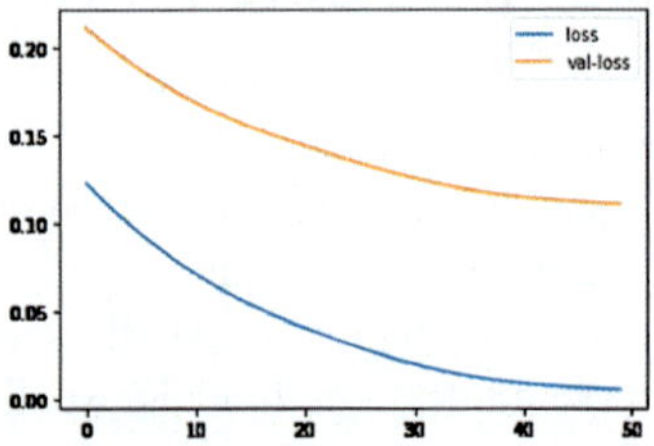

Fig. 10. Attention training and validation losses

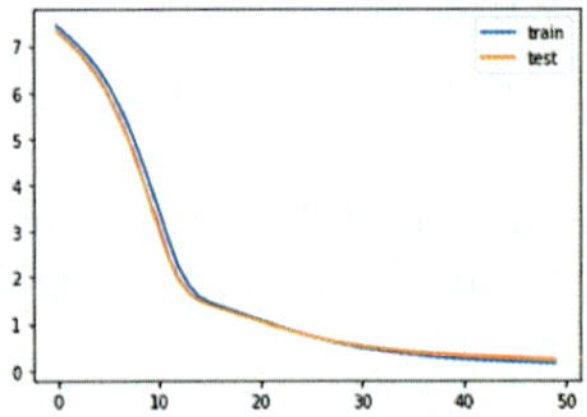

Fig. 11. PyTorch training and validation losses

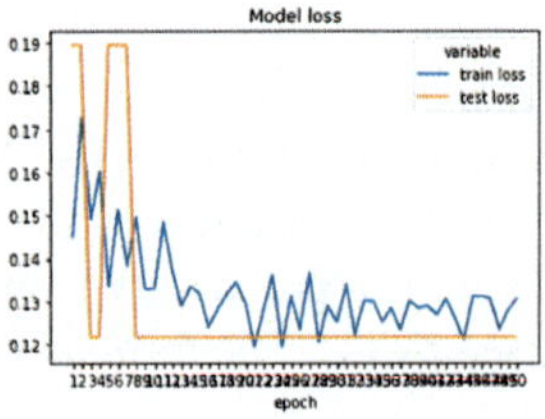

Fig. 12. TFBert training and validation losses

Table 2. Model setting.

Models	Loss	Accuracy	Time at epoch 50 (s)
RNN	0.593	0.834	54
LSTM	0.128	0.940	132
GRU	0.148	0.937	107
Attention	0.006	0.988	10
Transformer - PyTorch	0.052	0.987	11
Transformer - TFBert	0.130	0.985	71

5 Conclusion

In this paper, we showed our study on English to Vietnamese translation based Deep Learning approach. The deep learning machine translation models were trained based on Bidirectional RNN, LSTM, GRU, Attention, PyTorch Transformer, and TFBert Transformer. The performance of the models is further analyzed to reach a conclusion that TFBert Transformer, PyTorch Transformer, and Attention translators in most efficient way followed by GRU and SimpleRNN, respectively. While loss was lowest in SimpleRNN followed by GRU and it was highest in LSTM. The Attention model spent the shortest time for training and LSTM took the longest time. PyTorch Transformer and Attention had likely the same result while the first approach was very fast and cost shorted time to translate.

References

1. Tsai, C.-W., Lai, C.-F., Chao, H.-C., Vasilakos, A.V.: Big data analytics: a survey. J. Big Data, 1–3, 6–7 (2015)
2. Kambatla, K., Kollias, G., Kumar, V., Grama, A.: Trends in big data analytics, 1–3, 8–9 (2014)
3. Phan-Vu, H.-H., Tran, V.-T., Nguyen, V.-N., Dang, H.-V., Do, P.-T.: Neural machine translation between Vietnamese and English: an empiracal study. J. Comput. Sci. Cybern., 1–15 (2018)
4. Amidi, A., Amidi, S.: Recurrent neural networks cheatsheet. https://stanford.edu/~shervine/teaching/cs-230/cheatsheet-recurrent-neural-networks. Accessed 20 Dec 2020
5. Sak, H., Senior, A., Beaufays, F.: Long short-term memory recurrent neural network architectures for large scale acoustic modeling. Interspeech, 1–3 (2014)
6. Xu, S., Li, J., Liu, K., Wu, L.: A parallel GRU recurrent network model and its application to multi-channel time-varying signal classification. IEEE Access, 1–5 (2019)
7. Vaswani, A., Shazeer, N., Parmar, N., Uszkoreit, J., Jones, L., Gomez, A.N. Kaiser, L., Polosukhin, I.: Attention is all you need. https://arxiv.org/abs/1706.03762. Accessed 20 Dec 2020
8. Devlin, J., Chang, M.-W., Lee, K., Toutanova, K.: BERT: pre-training of deep bidirectional transformers for language understanding. https://arxiv.org/abs/1810.04805. Accessed 20 Dec 2020

Natural Language Processing with "More Than Words – BERT"

Saranlita Chotirat(✉) and Phayung Meesad

King Mongkut's University of Technology North Bangkok, Bangkok, Thailand
saranlita.c@email.kmutnb.ac.th, phayung.m@itd.kmutnb.ac.th

Abstract. Question-Answering (QA) has become one of the most popular natural language processing (NLP) and information retrieval applications. To be applied in QA systems, this paper presents a question classification technique based on NLP and Bidirectional Encoder Representation from Transformers (BERT). We performed experimental investigation on BERT for question classification with TREC-6 dataset and a Thai sentence dataset. We propose an improved processing technique called "More Than Words – BERT" (MTW – BERT) that is a special NLP Annotation tags for combining Part-Of-Speech tagging and Named Entities Recognition to be able for learning both pattern of grammatical tag sequence and recognized entities together as input before classifying text on BERT model. Experimental results showed that MTW – BERT outperformed existing classification methods and achieved new state-of-the-art performance on question classification for TREC-6 dataset with 99.20%. In addition, MTW-BERT also applied for question classification for Thai sentences in wh-question category. The proposed technique remarkably achieved Thai wh-classification with accuracy rate of 87.50%.

Keywords: Classification · BERT-based model · NLP Tagging · Analysis Thai Sentence

1 Introduction

The ability to ask question is important in both human and machine intelligence. The primary function of a question is to gain information. Search Engine and Question Answering (QA) systems have become important information retrieval applications. Automated question answering systems and search engines are different. Automated question answering systems receive querying inputs expressed in a human natural language that is not a keyword used in traditional search engines. In addition, they can return a paragraph or phrases containing accurate answers, rather than just list of words as corresponding answers. Therefore, it saves time for users to find out specific facts from large quantity of texts.

Knowing the possible classes of the sought-after answer narrows down the number of possible phrases or paragraphs effect to improve performance for question answering system. On the other hand, finding a suitable question type or

P. Meesad et al. (Eds.): IC2IT 2021, LNNS 251, pp. 108–116, 2021.
https://doi.org/10.1007/978-3-030-79757-7_11

identifying WH-question category (What, When, Who, Where, and How) from possible answers which specific words and/or phrases should be asked about could increase the ability to automatically generate possible questions from text. Question Classification is an important role which the aim of helping identification based on type of entity interested.

Question Classification (QC) is one of the effective ways with strong signal for QA systems and answer selection. It plays a primary role to categorize the question based on key features word. Considering key features word extraction can denote important feature in a sentence that identifies the type of question from the input by considering the set of WH-questions like what, when, who, where, and how.

Natural Language Processing (NLP) is an area of research where many researchers have presented their work and achieved better and better accuracy. Keywords or phrases that are entered to question answering systems are found in Named Entities Recognition (NER). NLP is used to identify the entities with a key name in the message, such as persons, organizations, places, dates, or other categories. The most popular of NLP used is Part-Of-Speech tagging (POS), which is the process of marking words in text based on definition and context such as NOUN, VERB, SCONJ, ADJ, etc. In an analysis Thai text, considering a sequence of words in sentences is significant because Thai sentence does not have boundary indicators to separate words and it is structural ambiguities as a part of speech depends on word order. Furthermore, previous works have suggested that applied POS and NER for helps to increase accuracy prediction because a model understand each word more.

Recently, pre-trained language models have shown to be useful in learning language representations especially BERT (Bidirectional Encoder Representation from Transformers). BERT has achieved amazing results in many natural language understandings (NLU) tasks. BERT is based on a multi-layer bidirectional Transformers [1]. More specifically, knowing possible classes of question to be asked could be one important task for automate generated questions in further so it is crucial to be able to classify question correctly from sentence. Moreover, there is not many researches to enhance BERT to improve the performance on Thai text classification.

In this work, we combine POS and NER as extra NLP annotation tags in the preprocess step to be learned both patterns of grammatical tag sequence and recognize entities together before going to classification task by BERT-based model using Transformer. The proposed method would increase the accuracy of classification, which model is more understandable for each word. The combined POS and NER annotation tags are used in the classification task by the BERT Transformer model.

We organize the remaining of the paper as follows. Section 2 presents a review of related works on Question Classification (QC), Natural Language Processing (NLP), and BERT-based model for text classification. In Sect. 3, we give details for the experimental method in the dataset, pre-processing task, and BERT classification model. In Sect. 4, we present our results on the question classification performance. Finally, Sect. 5 gives the conclusion and future work.

2 Related Work

2.1 Question Classification

Question classification (QC) is an essential part of question answering (QA) systems, which can be defined to match a question to one or several classes to resolve the answer type [2]. There are two main groups of question classification methods.

The first type is a traditional machine learning algorithms. Support vector machine (SVM) is one of the best well-known machine learning algorithms [3,4]. SVM has been introduced and used in text classification by Joachims [5]. Other researchers used Logistic Regression [6] and K-Nearest Neighbors [7]. Also, some others use deep learning that is an emerging technique for question classification. The popular deep learning question classification models have used convolution neural network (CNN), sequence-based long-term short memory (LSTM), and Bi-LSTM. Prior studies proposed a hybrid model that combined model for text classification.

To date, the best reported accuracy for learning issues involved in QC on the TREC-6 dataset is the universal sentence encoder models. This model used a Transformer that uses sentence and word level transfer to achieve the best overall performance published by the Google team [8] with an accuracy of 98.07%. The BERT-IDPT-FiT was proposed by [9], which has three steps: (1) further pre-train BERT on within-task training data (2) optional fine-tuning BERT with multitask learning and (3) fine-tune BERT for the target task. They achieved performances with an accuracy of 97.8%. [10] propose Universal Language Model Fine-tuning (ULMFiT), which pre-trains a language model (LM) and fine-tunes it on the target task using novel techniques. This method achieved CV-like transfer learning and reduced the error by 18–24% on datasets.

2.2 Bidirectional Encoder Representations from Transformers

Recently, several research efforts for investigation on pre-trained language models (Bidirectional Encoder Representations from Transformers or known as BERT) on NLP tasks. The Transformer architecture, which is based on self-attention and pointwise was proposed by [1]. BERT is designed to pre-train deep bidirectional representations from an unlabeled text by jointly conditioning on both left and right context in all layers. The first token of every sequence is always a special classification token ([CLS]). The final hidden state corresponding to this token is used as the aggregate sequence representation for classification tasks. Sentence pairs are packed together into a single sequence. They differentiate the sentences in two ways. First, they separate them with a special token ([SEP]). Second, they add a learned embedding to every token indicating whether it belongs to sentence A or sentence B. For a given token, its input representation comprises of summing the corresponding token, segment, and position embeddings.

2.3 Natural Language Processing (NLP)

Natural Language Processing (NLP) is a subfield of computer science dealing with developing methods to understand human language. Part-of-Speech tagging (POS), or entity extraction, is an essential element in the NLP task for extracting features and marks the word in a text with labels as nouns, verbs, adverbs, etc. Named Entity Recognition (NER) is an information extraction method to identify a person, organization, location, etc. Many research works have investigated preprocess tasks for improving classification models. POS and NER can reduce syntactic errors [4,6] and help a model understand each word for semantic analysis [11]. Therefore, adding a Special NLP tag into the token sequence leads to understanding each word in a sentence, integrating with the BERT model.

3 Research Methodology

3.1 Data

In this study, we used two datasets for experiments on question classification. The datasets were TREC dataset and a Thai Sentence dataset. The first dataset, TREC is a popular dataset which inferring the expected answer types of short open-domain factoid questions. TREC-6 has six labels: abbreviation, entity, description, human, location, and numeric (available at https://cogcomp.seas.upenn.edu/Data/QA/QC/). TREC question classification methods can be divided into that learn the question classification task.

Table 1. Statistics of two text classification datasets

Dataset	Classes	Average lengths	Max lengths	Train Samples	Valid Samples	Test Samples
TREC-6	6	10	37	5,000	452	500
Thai Sentence	4	15	77	563	141	176

The Second dataset is a simple sentence in Thai text collected from Thai Wikipedia. We manually categorized into 4 wh-question categories and labeled as shown in Table 2. For each type of questions, it was considered the semantic classes like when, who, where and how type questions. "When" ((เมื่อไหร่ /mˆeua r‘ai/)) category looks for consists of time and date expression. "Who" ((ใคร /krai/)/krai/) category looks for Names that are mostly Nouns of persons or organizations. "Where" ((ที่ไหน /tˆee nˇai/)) category looks for consists of location or place and prepositions words such as "in, at and near". "How much/ How many" ((เท่าไหร่/tˆao r‘ai/)) category is based upon consists of measurement meaning such as kilometer, centimeter. The annotation statistics were the following: When (220 sentences), Who (220 sentences), Where (220 sentences), and How (220 sentences). First, we randomly split the dataset to the training and testing dataset and then split training dataset again into training dataset and validation dataset with ratio 80:20.

3.2 Pre-processing

For pre-processing, we used pre-trained Bert tokenization names BERT-base-uncased for tokenization in the TREC dataset. We address a variety of Natural annotation tags added more token sequence. Stanford CoreNLP [12] was used for POS and NER. The Thai sentence dataset was used for tokenization and applied an NLP toolkit for Thai (PyThaiNLP) [13] for insert POS and NER.

Table 2. The example thai sentences dataset

Class	Simple sentences
When	สัญญาณไฟจราจรแห่งแรกของโลกติดตั้งเมื่อปีค.ศ. 1868 /săn-yaan fai jà-raa-jon hàeng râek kŏng lôhk dtìt dtâng mêua bpee kor-sŏr 1868/ (The world's first traffic light installed in A.D. 1868)
Who	พลเอกสุรยุทธ์ จุลานนท์ เป็นนายกรัฐมนตรี คนที่ 24 ของประเทศไทย /pon-àyk sù-rá-yút jù-laa-non bpen naa-yók rát-tà-mon-dtree kon têe 24 kŏng bprà-tâyt tai/ (General Surayud Chulanont is the 24th prime minister of Thailand.)
Where	การแข่งขันฟุตบอลชิงแชมป์แห่งชาติยุโรปครั้งที่ 6 จัดที่ประเทศอิตาลี /gaan kàeng kăn fút bon ching chaem hàeng châat yú-ròhp kráng têe 7 jàt kêun têe bprà-tâyt ì-dtaa-lee/ (The 6th European Football Championship was held in Italy.)
How	ท่าอากาศยานนานาชาติจูอันดาอยู่ห่างจากตัวเมืองสุาบายา 12 กิโลเมตร tâa aa-gàat-sà-yaan naa-naa châat juu an daa yòo hàang jàak dtua meuang sù-raa yaa12 gì-loh máyt/ (Juanda International Airport is far from the Surabaya city 12 kilometre.)

Based on the problem statement above, we proposed to pre-process the NLP task. We try to give more detail for a defined word with NLP tag including POS and NER as special NLP tags such as Noun Person, Noun Organization, Noun Location, Noun Date, or Num Date. Then used BERT-based classification model to the Wh-question category, and compares the accuracy classify when we applied the other model with simple sentence input and special NLP tag input.

Formally, given a sequence of tokens s = $\langle w_1, w_2, ..., w_N \rangle$, combine POS and NER as special NLP tag is to output a list of tuples $\langle ts, te, t \rangle$, each of which is a special NLP tag mentioned in s. Here, $Is \in [1, N]$ and $Ie \in [1, N]$ are the start and the end indexes of a part of speech and named entity mention; t is the entity type from a predefined category set. Figure 1 shows an example where POS and NER from the given sentence.

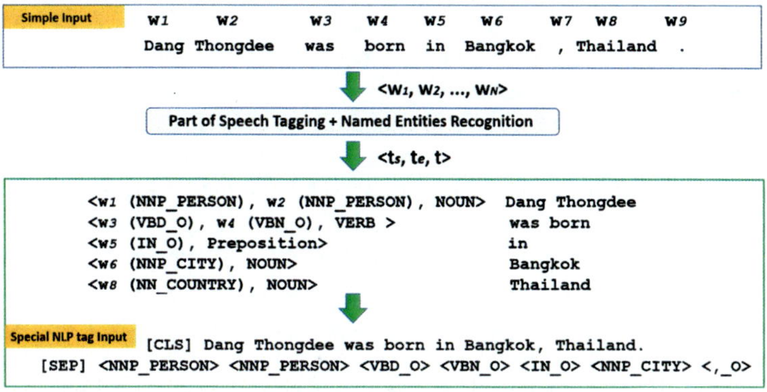

Fig. 1. An illustration of the special NLP tagging task

3.3 Classification

We compared the performance of the question classification models based on Logistic Regression (LR), Support Vector Machine (SVM), and Multinomial Naïve Bayes (MNB). We used Spacy and PyThaiNLP for text tokenization. We also used Term Frequency Inverse Document Frequency (TF-IDF) process to transform tokenized words into numeric vectors and trained with class weighted.

To fine-tune BERT question classification model, we preprocessed the data starting with adding special NLP tags and special tokens to mark beginning and ending point of sentences. Padding was also made in the case of length sentences shorter than 300 words. Next, the pre-trained BERT received the preprocessed data via the input layer. We applied BERT-base-uncased with TREC-6 dataset and BERT-base-multilingual-cased with Thai sentence dataset, which resulted in training features along with input mask, and segment id. The output of the transformer was then fed into a dense layer with ReLu activation for the intermediate layer. We used batch normalization to reduce the effect of the gradient step. Finally, we used a dense layer with softmax activations that output a probability distribution over target classes.

In the evaluation step, we set training time for a maximum of 50 epochs and early stopping with patience equal to 5. We used the Adam optimizer for deep learning with a batch size of 8 and a learning rate of 1e-5. Each training experimental set was run three times. The result of each experiment was recorded. The best performance was chosen based on the confusion matrix to calculate the weighted averages of accuracy (A), precision (P), recall (R), and F_1-Score (F_1).

4 Experimental Results

In this study, we compared the performance of the proposed method when applied to classification models with different processes. We applied POS and NER as special NLP Tags for training to traditional classification models. With the proposed preprocessing method, the best model has increased F_1 by 97.79% in case of TREC dataset and 73.06% in case of Thai sentence dataset. Table 3 shown as shows the experimental results.

We compared the performance of our method and fine-tune the model with the baselines. It found that our method by using BERT and transformer achieves state-of-the-art results on the TREC-6 dataset for a learned question classification with an accuracy of 99.20%. In addition, classifying questions from simple input has a slight decrease in accuracy of 98.80%, as shown in Table 4. Furthermore, evaluating the classifiers for both simple sentence and +Special NLP tags yielded similar performance (100%) for class type ABBR, DESC, and HUM as shown in Table 5.

The classification results for Wh-question from Thai sentence with +Special NLP tag improve accuracy increased up to 87.50%, as shown in Table 6. For "HOW MUCH" category was the most accurately predicted with the F_1score of 90.00%. Notwithstanding, "WHO" and "WHERE" categories were less than the other categories with F_1score 86.00% and 84.00%, respectively. While "WHEN" had a highest score without +Special NLP tag, as shown in Table 7.

Table 3. The result of traditional classification models

Dataset	Model	Simple Sentence				+Special NLP tag			
		A	P	R	F_1	A	P	R	F_1
TREC-6	SVM	97.00	97.09	97.00	96.96	97.80	97.84	97.80	**97.79**
	LR	96.80	96.86	96.80	96.76	97.20	97.27	97.20	97.19
	MNB	97.20	97.37	97.20	97.20	95.20	95.79	95.20	95.29
Thai Sentence	SVM	69.89	73.15	69.89	70.26	72.73	73.94	72.73	**73.06**
	LR	71.59	74.56	71.59	71.82	68.18	69.35	68.18	68.49
	MNB	71.02	73.76	71.02	71.26	68.75	71.01	68.75	69.25

Table 4. Comparison of result

Model	Accuracy
USE_T+CNN [8]	98.07
BERT-ITPT-FiT [9]	97.80
ULMFiT [10]	96.40
Simple sentence(Our model)	98.80
+Special NLP tag(Our model)	**99.20**

Table 5. The Result of TREC dataset

Class	Simple sentence			+Special tag		
	P	R	F_1	P	R	F_1
ABBR	**1.00**	**1.00**	**1.00**	**1.00**	**1.00**	**1.00**
DESC	**1.00**	**1.00**	**1.00**	**1.00**	**1.00**	**1.00**
ENTY	0.98	0.96	0.97	0.99	0.97	**0.98**
HUM	**1.00**	**1.00**	**1.00**	**1.00**	**1.00**	**1.00**
LOC	0.98	0.98	**0.98**	0.96	0.99	**0.98**
NUM	0.98	1.00	0.99	1.00	1.00	**1.00**

Table 6. The experimental results of Thai sentence dataset

Model	Accuracy
Simple sentence input	86.93
+Special NLP tag	**87.50**

Table 7. The result of Thai Sentence

Class	Simple sentence			+Special tag		
	P	R	F_1	P	R	F_1
WHEN	87.00	98.00	**92.00**	85.00	94.00	89.00
WHO	97.00	76.00	85.00	95.00	78.00	**86.00**
WHERE	75.00	84.00	79.00	78.00	91.00	**84.00**
HOW	87.00	89.00	88.00	91.00	89.00	**90.00**

5 Conclusion and Future Work

In this work, we propose a natural language processing called MTW-BERT (More Than Words BERT). The proposed method is a combination of NLP annotation tags that are Part-Of-Speech tag (POS) and Named Entities Recognition (NER). It is considered special NLP tags for deep learning to learn both patterns of grammatical tag sequence and recognized entities together. The Deep learning used is based on a BERT-based classifier and Transformer.

The experimental results showed that our solution led to a good performance in WH-question classification. The proposed model outperformed other classification methods and achieved new state-of-the-art performance on question classification datasets (TREC-6) with an accuracy of 99.20%. The proposed approach is appropriate for Thai sentence classification to several WH-question classes with an accuracy rate of 87.50%. For "WHEN" has a high score without +Special NLP tag. This because Thai sentences including a word defined to date or time, so a model could understand the word although without special NLP tags and could predict the right class. For this research, we studied only 4WH questions (When, Who, Where, and How much) because the "WHAT" class could confuse when unknown words and maybe effect to decrease the performance of prediction to other classes also. In the future work, we would involve a study on automated question generation using the MTW-BERT model to predict the possible WH-question class from Thai text.

Acknowledgment. The researcher is grateful for funded by the Office of the Permanent Secretary (OPS), Ministry of Higher Education, Science, Research and Innovation, Thailand. We thank King Mongkut's University of Technology North Bangkok, Thailand for support Information Technology.

References

1. Vaswani, A., et al.: Atteion, is all you need. In: 31st Conference on Neural Information Processing Systems (NIPS: Long Beach, CA, USA, p. 2017 (2017)
2. Xin, L., Xuan-Jing, H., Li-d, W.: Question classification using multiple classifiers. In: Proceedings of the Fifth Workshop on Asian Language Resources (ALR-05) and First Symposium on Asian Language Resources Network (ALRN) (2005)

3. Haryanto, A.W., Mawardi,E.K.: Muljono. influence of word normalization and chi-squared feature selection on support vector machine (SVM) text classification. In: 2018 International Seminar on Application for Technology of Information and Communication, pp. 229–233 (2018). https://doi.org/10.1109/ISEMANTIC.2018.8549748
4. Silva, V.A., Bittencourt, I.I., Maldonado, J.C.: Automatic question classifiers: a systematic review. IEEE Trans. Learn. Technol. **12**(4), 485–502 (2018)
5. Joachims, T.: Text categorization with support vector machines: learning with many relevant features. In: Proceedings of the 10th European Conference on Machine Learning ECML-98, Chemnitz, Germany, pp. 137–142 (1998)
6. Pranckevičius, T., Marcinkevičius,V.: Application of Logistic Regression with part-of-the-speech tagging for multi-class text classification. In: IEEE 4th Workshop on Advances in Information, Electronic and Electrical Engineering (AIEEE) (2016)
7. Li, H., Jiang,H., Wang, D., Han, B.: An improved KNN algorithm for text classification. In: 8th International Conference on Instrumentation and Measurement, Computer, Communication and Control (IMCCC), pp. 1081–1085 (2018)
8. Cer, D., et al.: Universal sentence encoder. In: Proceedings of the 2018 Conference on Empirical Methods in Natural Language Processing: System Demonstration, Brussels, Belgium, pp. 169–174 (2018). https://doi.org/10.18653/v1/D18-2029
9. Chi, S., Qiu, X., Xu, Y., Haung, X.: How to fine-tune BERT for text classification?. In: Chinese Computational Linguistics, pp. 194–206 (2019)
10. Howard, J., Ruder, S.: Universal language model fine-tuning for text classification. In: Proceedings of the 56th Annual Meeting of the Association for Computational Linguistics, pp. 328–339 (2018). https://doi.org/10.18653/v1/P18-1031
11. Pasupa, K., Ayutthaya, T.S.N.: Thai sentiment analysis with deep learning techniques: a comparative study based on word embedding, POS-Tag, and sentic features. Sustain. Cities Soc. (2019). https://doi.org/10.1016/j.scs.2019.101615
12. Manning, C., Surdeanu, M., Bauer,J., Finkel, J.,Bethard, S., McClosky, D.: The stanford CoreNLP natural language processing toolkit. In: Proceedings of 52nd Annual Meeting of the Association for Computational Linguistics: System Demonstrations, pp. 55–60 (2014). https://doi.org/10.3115/v1/P14-5010
13. Phatthiyaphaibun, W., Chaovavanich, K., Polpanumas, C., Suriyawongkul, A., Lowphansirikul, L., Chormai, P.: PyThaiNLP: Thai Natural Language Processing in Python. Zenodo (2016). http://doi.org/10.5281/zenodo.3519354

Automatically Retrieving of Software Specification Requirements Related to Each Actor

Jantima Polpinij[1(✉)] and Khanista Namee[2]

[1] Intellect Laboratory, Faculty of Informatics, Mahasarakham University, Mahasarakham 44150, Thailand
jantima.p@msu.ac.th

[2] King Mongkut's University of Technology North Bangkok, Bangkok 10800, Thailand
khanista.n@fitm.kmutnb.ac.th

Abstract. It is well-known that success or failure of any software is dependent upon requirement analysis. The most significant problem hidden in natural language software requirements may be ambiguity. This can lead to poor design and performance of the final software product and can be time-consuming at the system analysis stage. Previous studies applied natural language processing (NLP) techniques to identify these ambiguities and reduce them to improve the requirement quality. However, this problem has not yet been fully resolved. This study applies NLP to automatically extract the software system requirement specification and visualize an overview of the software that is being developed. The proposed method clusters relevant software requirement sentences of each system user by text mining technique, and then extracts 'actors' and their 'actions' from sentences using NLP techniques. The recall technique is used to evaluate the efficacy of the proposed method. Response time and relevancy of the results are significant factors for software product satisfaction.

Keywords: Requirement analysis · Ambiguity · Text mining · BM25 · Natural language processing

1 Introduction

Requirement engineering (RE) has become a popular research topic in the field of software engineering (SE), where success or failure of any software is dependent upon requirement analysis [1–4]. As a result, since 1960, requirement analysis has become an increasingly trending topic in SE.

The requirement state is a major process in the software development life cycle (SDLC) that includes the phases of planning, analysis, design, development, testing and maintenance [2]. Requirement analysis describes what the stakeholders require or expect from the system, and encompasses those tasks that determine the needs or conditions that must be met for new or altered software products. The results of requirement analysis are called software requirement specification (SRS). This describes what the software will

P. Meesad et al. (Eds.): IC2IT 2021, LNNS 251, pp. 117–126, 2021.
https://doi.org/10.1007/978-3-030-79757-7_12

do, how it will be expected to perform, and also the functionality that the product needs to fulfill all stakeholders' (business, users) needs. Therefore, the software requirements should be documented, actionable, measurable, testable, traceable, related to identified business needs or opportunities, and defined to a level of detail sufficient for system design.

In general, when analyzing small software, it is not too difficult to find and extract SRSs from software requirements gathered from many sources such as users' old system and existing related documents. Unfortunately, when working on a large software, this may not be an easy process [1, 2]. The most significant problem hidden in natural language software requirements may be ambiguity [1, 2, 4–13]. This problem has been confirmed as a key factor of software development success. Any ambiguity present in the requirement specification is considered as a defect or failure [13], leading to poor design and performance of the final software product.

Several automatic analysis methods have been proposed to improve the requirement quality. Currently, no technique is available to completely reduce ambiguity caused by lexical, syntactic and syntax ambiguities [13]. Previous studies applied natural language processing (NLP) techniques to identify these ambiguities and reduce them to improve the requirement quality. However, this problem has not yet been fully resolved.

As mentioned, the results of requirement analysis called SRS, often use case diagrams to summarize the details of a system's users (also known as '*actors*') and their interactions with the system. The interactions, often called '*actions*', represent use cases. Automatically generating '*actors*' and '*actions*' from software requirements documents is not easy to perform on a large scale. One possible solution to address this problem is to assemble relevant software requirements sentences in software requirements documents. This may help to see the overall picture of each sub-system domain, and the best way to generate a user case diagram.

As mentioned above, this study aims to automatically extract software specification requirements from software requirements documents. We apply text mining techniques to assemble relevant software requirement sentences, and then extract '*actors*' and their '*actions*' from these sentences using NLP techniques. This study does not show results in the format of diagrams. We only show the actors extracted and their actions in the system.

The structure of this paper is as follows. Section 2 describes literature review. Section 3 describes our dataset and Sect. 4 is the proposed method. Experimental results are presented in Sect. 5. Finally, conclusions are made in the last section.

2 Literature Review

2.1 Related Work

Requirement engineering is the first and crucial phase in the development of software [4]. Success or failure of any software is dependent upon requirement analysis. Manually analyzing software requirements is a time-consuming and labor-intensive task. Therefore, automated systems have been developed to gather user requirements and bridge communication gaps between users and analysts. Unfortunately, the most significant problem hidden in natural language software requirements may be ambiguity [1, 2,

4–13]. Many automatic analysis methods have been proposed to identify these ambiguities and reduce them to improve software requirement quality, but this problem has not been fully resolved. Some studies related to automatic requirement analysis using NLP, text mining, and machine learning techniques to handle language ambiguity in software requirements are illustrated as follows.

Early studies for automated analysis of software requirements were conducted by Osborne and MacNish [6] and Wilson et al. [7]. The former mentioned that ambiguity in requirement specifications causes numerous problems; for example, in defining customer/supplier contracts, ensuring the integrity of safety-critical systems, and analyzing the implications of system change requests. A direct appeal to formal specification does not solve these problems, partly because of the restrictiveness and lack of habitability of formal languages. An alternative approach, described in the paper, is to use natural language processing (NLP) techniques to aid the development of formal descriptions from requirements expressed in controlled natural language. While many problems in NLP remain unsolved, we show that suitable extensions to existing tools provide a useful platform for detecting and resolving ambiguities. Their system was demonstrated through a case study on a simple requirements specification. Wilson et al. [7] mentioned that poorly written requirements can result in software with the wrong functionality. Therefore, the Goddard Space Flight Center's (GSFC) Software Assurance Technology Center (SATC) developed an early life cycle tool for assessing requirements that are specified in natural language. The tool searches the document for terms that the SATC has identified as quality indicators, e.g. weak phrases. Reports produced by the tool are used to identify specification statements and structural areas of the requirements specification document that need to be improved.

Later, in 2005, MacDonell et al. [8] addressed the use of NLP in the requirements analysis and systems design processes. They developed a prototype toolset that can assist the systems analyst or software engineer to select and verify terms relevant to a project. Also, they described the processes employed by the system to extract and classify objects of interest from requirements documents. These processes are illustrated using a small example.

In 2015, Btoush and Hammad [11] examined the problem of extracting entity-relationship (ER) elements from natural language specifications using NLP. Their approach provided the opportunity of using natural language documents as a source of knowledge for generating an ER data model. A structural approach was used to parse specification syntactically, based on a predefined set of heuristics rules. Extracted words with their Part of Speech (POS) mapped into entities, attributes, and relationships are the basic elements of ER diagrams.

In 2016, Wang [12] proposed an approach to automatically analyze software requirement specifications (SRSs) and extract the semantic information. This approach used a machine learning and ontology based semantic role labeling (SRL) method. First, some common verbs were calculated from SRS documents in the E-commerce domain, and then semantic frames were designed for these verbs. Based on the frames, sentences from SRSs were selected and labeled manually, and the labeled sentences were used as training examples in the machine learning stage. Besides the training examples labeled with semantic roles, external ontology knowledge was used to relieve the data sparsity problem and obtain reliable results. Based on their proposed method and WordNet corpus, the senses of nouns and verbs were identified in a sequential manner through the K-nearest neighbor approach. Then, the senses of the verbs were used to identify the frame types. After training the SRL labeling classifier with the maximum entropy method, some new features were also added based on word sense, such as hypernyms and hyponyms of word senses in the ontology. Experimental results showed that their approach for automatic functional requirements analysis was effective.

In 2018, Li [14] presented an automated technique for extracting features and variability to provide reliable solutions to simplify the work of domain analysis. The main mechanism of the proposed method was NLP and machine learning techniques.

2.2 Understanding of Software Requirement Sentence

Software requirements are often written with a simple language structure and should be easy to understand. A software requirement sentence consists of two parts as '*actor*' and '*action*'. See in Fig. 1. The actor performs the action. This structure is called '*Actor-Action-Object*' [15], which is used in the Object-Oriented Software Processes through the Use Case Diagram.

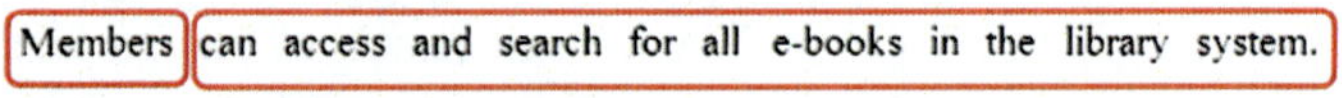

Fig. 1. An example of the position of "*actor*" and "*action*" in a requirement sentence

This structure is similar to English sentence structure that consists of two main parts: subject and predicate. The subject is the person or thing (who/what) that performs the action, while the predicate is the part of the sentence that contains a verb or a verb phrase and its complements. Therefore, if comparing between the English sentence and the software requirement sentence, the subject in the English sentence is comparable to the '*actor*' of the software requirement sentence, while the predicate in the English sentence is comparable to the '*action*' of the software requirement sentence.

3 Dataset

The dataset used for our experimental stage is a set of software requirements related to a library website written in English and contains over 50 pages. Each page contains about 40 lines. This dataset has been analyzed by system analysts, with a final SRS case diagram that details a system's users and their interactions. For a legal policy, we cannot show details of the software requirements used in this study.

4 Research Method

4.1 Searching for the System's Users (or Actors)

This section describes the process of finding and extracting all 'actors' in a software requirements collection. An 'actor' can be compared to the 'subject' in the English sentence structure. In general, the 'subject' should be a noun, pronoun, or noun group.

To find all actors, we then apply Stanford NLP Parser that works out the grammatical structure of sentences [16] to generate a parse tree of rules. Then, we extract only the sets of 'noun phrase (NP)' or 'word' that are located at the position of the sentence subject. An example is illustrated as Fig. 2.

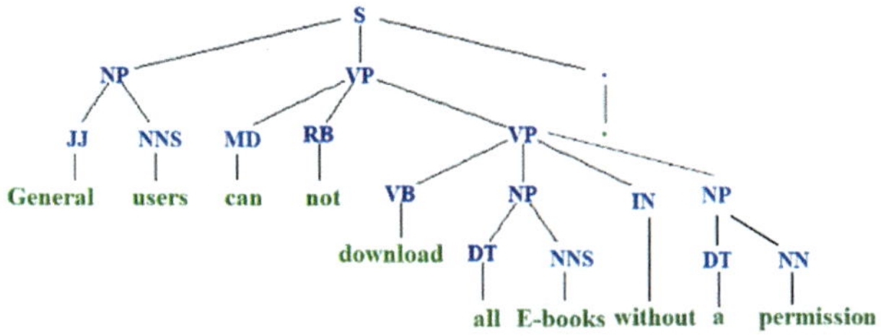

Fig. 2. An example of parsing requirement sentence to extract 'NP' considered as 'actor'

After obtaining all 'NPs' or 'words' that may be considered as actors, these NPs and word groups can make distinctions as four different actors of the system. These are administrators, librarians, members, and general users.

4.2 Using 'Actors' for Searching Their Relevant Software Requirement Sentences

After obtaining four different actors of the system, these actors are used to search their relevant software requirement sentences by keyword search technique. This is a process of preliminary grouping for relevant software requirement sentences. The overview picture of the expected results of this stage is shown as Fig. 3.

4.3 Retrieving Relevant Software Requirement Sentences of Each Actor

After some relevant requirement sentences of each actor are retrieved (mentioned in Sect. 4.2), these sentences become queries that are used to find other relevant requirement sentences of the actor. This solution helps the system analyst team to visualize a preliminary overview of the software that is being developed. Conse quently, this may increase the chances of completely analyzing and designing the software. Each process of retrieving all relevant software requirement sentences can be described as follows.

The software requirement sentences are first segmented to a meaning unit. In this study, the meaning unit of language is words. afterward, the program performs a process of stop-word removal. Finally, the software requirement sentences are represented as a vector space model, called bag of words (BOW). Later, this study applies BM25 [17] which is a ranking function for retrieving relevant software requirement sentences of each actor. BM25 is often used as a text mining technique to retrieve relevant documents based on similarity analysis.

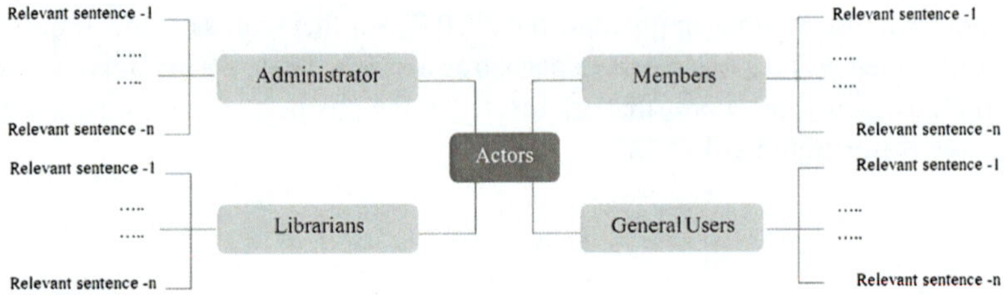

Fig. 3. The expected results of preliminary grouping of relevant software requirement sentences for each actor

In BM25 formula, it requires only *document frequency* (*df*). The *df* is number of documents where the term *t* appears. Instead of regarding the inter-relationship between the query terms within a software requirement sentence. The BM25 formula is:

$$BM25(r_1, r_2) = \sum_{i=1}^{|r_1|} idf(q_i) \times \left(\frac{tf(q_i, r_2) \times (k+1)}{tf(q_i, r_2) + k \times (1 - b + b \times \frac{|r_2|}{r_2 l_{avg}}} \right) \tag{1}$$

Let r_1 be software requirement sentence of an actor and r_2 be software requirement sentence that may be related to that r_1. Therefore, $tf(q_i, r_2)$ is the term frequency, where it is defined as the number of occurrences that the terms q-th of $r_1$appears in r_2. Indeed, $|r_2|$ is the length of $r_2$in words, while $r_2 l_{avg}$ is the average software requirement sentence length for all the software requirement sentences in the corpus. For, k and b, they are free parameters. They are used to control the weighting between $tf(q_i, r_2)$ and the normalized software requirement sentence length. Generally, the values of k and b should be in the range of $1.2 < k < 2.0$ and $0.5 < b < 0.8$ [17, 18]. This study uses the values of k and b at 2.0 and 0.8 respectively. They are the same values used in [18].

For $idf(q_i)$, it is the *inverse document frequency* of the term q-th of r_1. It can be calculated by following formula.

$$idf(q_i) = \log\left(\frac{N - df(q_i) + 0.5}{df(q_i) + 0.5} \right) \tag{2}$$

where N is the total number of sentences in the corpus, while $df(q_i)$ is the number of sentences containing the term q-th of r_1.

In general, the similarity score should be between 0 and 1. However, when using the BM25 technique to estimate the similarity score, it is possible that this technique can return a score greater than 1.0. Similarity scores should be normalized to allow a comparison of different similarity values using a single scale. Normalizing similarity scores helps to remove the mean and scale to the similarity score variance. To normalize the BM25 similarity scores in the range [0, 1], the following function also applies in this case.

$$f(x) = x/(1 + x) \tag{3}$$

where x is the similarity score generated by BM25.

4.4 Extracting of 'Actors' and Their 'Actions'

After completely retrieving all relevant software requirement sentences into their specific clusters, the software requirement sentences in each cluster are performed by Stanford NLP Parser [16] to generate their parse trees. Afterward, we extract '*verb phrases (VP)*' as actions of each actor. Some examples of the expected results are shown as Table 1.

Table 1. Some examples of the expected results.

Actors	Actions
Members	Need to register as member to get the authorization
	Need to login for access the system
	Can search e-books in the system
	...

5 The Experimental Results

This section evaluates the proposed method using recall, precision, and F-measure techniques. Recall gives a measure of how accurately our proposed method can identify the relevant data, while precision is the closeness of the measurements to each other. The F-measure is a single score that balances both the concerns of precision and recall as one number.

In this study, there are two evaluations. First, the performance of extracting all actors from the software requirement collection is evaluated, while the second evaluation measures the results of retrieving relevant software requirements of each actor.

5.1 The Evaluation of Extracting All Actors from the Software Requirement Collection

This evaluates the efficiency of extracting '*actors*' from the software requirement documents. After performing this process, the NPs or word groups extracted are distinct, and all actors of the system can be obtained. They include *administrators*, *librarians*, *members*, and *general users*. Results are shown in Table 2.

Table 2. The results of extracting actors.

Results	Recall score
Extracted actors	1.00

In Table 3., the results are good in terms of recall. However, some errors can be found. For example, consider the following phrase, 'the processes can be initiated'. The word ‘processes’ is not an actual actor but it is extracted because it is an NP that is located at the position of the subject of the sentence. Therefore, some constraints should be provided to handle this issue and improve the quality of extracting actors from the software requirement documents.

5.2 The Results of Retrieving Relevant Software Requirements of Each Actor

After testing via the recall, precision, and F-measure, the results of retrieving the relevant software requirements relating to each actor are presented in Table 3.

Table 3. The results of retrieving relevant software requirements of each actor

Actors	Recall	Precision	F-measure
Administrators	1.00	0.77	0.87
Librarians	1.00	0.81	0.89
Members	1.00	0.71	0.83
General users	1.00	0.74	0.85

Consider Table 3. Although the results are satisfactory, they still have some errors. Consider the following software requirement sentence.

“The members can search e-books from the system, and they can also download these e-books.”

In the example sentence above, ‘*the members*’ is an ‘*actor*’ in the system, while the pronoun ‘*they*’ indicates ‘*the members*’. Unfortunately, our method cannot recognize the meaning of the word ‘*they*’ and to which ‘*actor*’ the word ‘*they*’ refers. Consequently, the proposed method of retrieving relevant software requirements of each actor can lead to failure analysis.

The failure rate recognized in this study requires improvement. Two solutions are suggested for further analysis. A large number of language patterns is also necessary. However, developing language patterns with only the training set is not sufficient, and a variety of sentence patterns is required.

6 Conclusion

Requirement analysis is a major process in the software development life cycle that describes the functionality the product needs to fulfill all stakeholders’ (business, users) needs. Requirement analysis encompasses tasks that determine the needs or conditions needed to meet new or altered software products. On a small software, it is not difficult to find and extract SRSs from software requirements. By contrast, for large software, this is never an easy process. The most significant problem hidden in natural language software requirements may be ambiguity. This problem has been confirmed as a key factor of software development success because ambiguity can lead to poor design and performance of the final software product. Previous studies applied NLP techniques to identify these ambiguities and reduce them to improve the quality of software requirements. This issue is the challenge in this study. We present a method of automatically retrieving software requirements of each actor. Doing this may be an easier way to perform requirement analysis tasks. The main mechanisms of the proposed method are NLP and text mining techniques. Then, we apply NLP techniques to extract all actors of the system, and text mining techniques to find the relevant software requirements of each actor. Recall, precision, and F-measure techniques are used to evaluate the efficacy of the proposed method. Response time and relevancy of the results are significant factors regarding satisfaction.

Acknowledgement. This research was funded by King Mongkut's University of Technology North Bangkok. Contract no KMUTNB-64-DRIVE-34. We would like to deliver our greatest appreciation for their support.

References

1. Kamsties, E.: Understanding Ambiguity in Requirements Engineering. Engineering and Managing Software Requirements, pp. 245–266 (2005)
2. Kerry, E., Delgado, S.: Applying software engineering practices to produce reliable, high-quality and accurate automated test systems. IEEE Autotestcon (2009)
3. Badariah, S., Shamsul, S., Ghani, A.A.A.: A new maturity model for requirements engineering process: an overview. J. Software Eng. Appl. **5**, 340–350 (2012)
4. Mishra, A., Awal, A., Elijah, J., Abdulg, A., Gana, Rabiu, I.: Automation of requirement analysis in software engineering. Int. J. Recent Innov. Trends Comput. Commun. **5**(5), 1173–1188 (2017)
5. Liu, G., Huang, S., Piao, X.: Study on requirement testing method based on alpha-beta cut-off procedure. In: International Conference on Internet Computing in Science and Engineering, vol. 1, pp. 396-402 (2008)
6. Osborne, M., Macnish, C.K.: Processing natural language software requirement specifications. In: Processing Natural Language Software Requirement Specifications (1996)
7. Wilson, W.M., Rosenberg, L.H., Hyatt, L.E.: Automated analysis of requirement specifications. In: Proceedings of the 19th International Conference on Software Engineering, pp. 161–171 (1997)
8. MacDonell, S.G., Min, K., Connor, A.: Autonomous requirements specification processing using natural language processing. In: 14th International Conference on Intelligent and Adaptive Systems and Software Engineering (2005)
9. Verma, K., Kass, A.: Requirements analysis tool: a tool for automatically analyzing software requirements documents. In: Sheth, A., et al. (eds.) ISWC 2008. LNCS, vol. 5318, pp. 751–763. Springer, Heidelberg (2008). https://doi.org/10.1007/978-3-540-88564-1_48
10. Vinay, S., Shridhar, A., Orashanth, D.: An Approach towards Automation of Requirements Analysis (2009). https://www.researchgate.net/publication/44259706
11. Btoush, E., Hammad, M.: Generating ER diagrams from requirement specifications based on natural language processing. Int. J. Database Theory Appl. **8**(2), 61–70 (2015)
12. Wang, Y.: Automatic semantic analysis of software requirements through machine learning and ontology approach. J. Shanghai Jiaotong Univ. (Sci.) **21**(6), 692–701 (2016). https://doi.org/10.1007/s12204-016-1783-3

13. Vimalraj, J.T., Seema, B.: Identification of ambiguity in requirement specification using multilingual word sense. Int. J. Adv. Res. Comput. Commun. Eng. **5**(6), 386–388 (2016)
14. Li, Y.: Feature and variability extraction from natural language software requirements specifications. In: Proceedings of the 22nd International Systems and Software Product Line Conference, vol. 2, pp. 72–78 (2018)
15. Mishra, A., Sureka, A.: Automatic detection of semantic inconsistency between BPMN process model and SBVR rule model. Computer Science (2015)
16. Stanford Parser. https://nlp.stanford.edu/software/lex-parser.shtml. Accessed 01 Nov 2020
17. Lee, C.Y., Hu, D.D., Feng, Z.Y., Yang, C.Z.: Mining temporal information to improve duplication detection on bug reports. In: IIAI 4th International Congress on Advanced Applied Informatics, pp. 551–555 (2015)
18. Yang, C.Z., Du, H.H., Wu, S.S., Chen, X.: Duplication detection for software bug reports based on bm25 term weighting. In: Conference on Technologies and Applications of Artificial Intelligence, pp. 33–38 (2012)

Reservoir Inflow Time Series Forecasting Using Regression Model with Climate Indices

Jakkarin Weekaew[1], Pakorn Ditthakit[2], and Nichnan Kittiphattanabawon[1](✉)

[1] School of Informatics, Walailak University, Nakhon Si Thammarat 80160, Thailand
{wjakkari,knichcha}@mail.wu.ac.th

[2] School of Engineering and Technology, Walailak University, Nakhon Si Thammarat 80160, Thailand

Abstract. The problem of reservoir inflow forecasting plays a critical role in reservoir management. However, reservoir inflow forecasting must be necessarily accurate and timely. This paper presents practical machine learning (ML) technique and the optimal lead time for reservoir inflow forecasting. The two well-knows ML: Support Vector Regression (SVR) and Random Forests (RF), were used to predict water inflow volume into the reservoir. Both methods will improve the efficiency of the reservoir inflow more accurately by using reservoir information data (amounts of rainfall and inflow) and climate indices (Sea Surface Temperatures (SSTs) and ocean indices). All historical monthly time series reservoir information for 18 years (between 1998 and 2015) is collected from the Hui Nam Sai Reservoir Project, located in Nakhon Si Thammarat Province, Thailand. Our experimental results showed Random Forest with climate indices gave the best performance method. We also found that the top three optimal lead time for reservoir inflow forecasting were the 10-month (t+10), 2-month (t+2) and 6-month (t+6) ahead with giving the least of RMSE of 4.41, 4.58 and 4.62, respectively. Similarly, when using ASE as a criterion, the top three optimal lead time for reservoir inflow forecasting were the 3-month (t+3), 10-month (t+10), and 5-month (t+5) with giving the least of ASE of 3.11, 3.14, and 3.16 respectively.

Keywords: Support vector regression · Random forests · Reservoir inflow forecasting · Time series · Climate indices

1 Introduction

Over the past century, water management has been playing a vital role in irrigation in many countries. Effective water management in reservoirs will reduce the severity of natural disasters such as floods and droughts. Adhering to water balance management principles is to create a balance between water demand in various activities and the water budget. The estimation of reservoir inflow in different objectives is directly impacted to the reservoir operation from these requirements. However, one of the major problems is the inflow forecasting is still inaccurate, which directly affects the accuracy in predicting the amount of water budget. If this problem is resolved, it drives to decrease flooding

P. Meesad et al. (Eds.): IC2IT 2021, LNNS 251, pp. 127–136, 2021.
https://doi.org/10.1007/978-3-030-79757-7_13

during the rainy season. Besides, the occurrence of water shortages during the dry season will be reduced.

The reservoir inflow forecasting techniques can be classified into 4 types: 1) Physical model is the use of real physical processes to simulate various steps. 2) Time series model is a statistical method to predict the amount of water inflowing the reservoir. For example, the Autoregressive Moving Average (ARMA) and Autoregressive Integrated Moving Average (ARIMA) was used to predict the monthly amount of water flowing into the reservoir [1]. 3) Hydrological Model is mathematical model to explain the relationship between rainfall and runoff. For examples, the use of SWAT model to forecast the runoff under changes in hydrological characteristics of Nam Oon reservoir [2]. However, there are still problems in analyzing large amounts of data which directly affect the accuracy of forecasting, and 4) Machine Learning-based model is computer algorithms to improve itself through experience or the past data sets automatically. For examples, forecasting the amount of water inflowing the reservoir by using the Multi-layer perceptron (MLP) model for the reservoir operation. This technique has been applied in several earlier research papers such as [3–5]. The climate phenomenon indices were introduced directly in machine learning models for inflow forecasting. They made this potentially valuable data available for decision making in reservoir operations. Similarly, [6] developed statistical model for dam inflow forecasting. This research used monthly dam inflow and fourteen climatic indices, such as Southern Oscillation Index (SOI) El Niño southern oscillation (ENSO), etc. These methods, multiple linear regression (MLR), support vector machines (SVM) and artificial neural networks (ANN), were used to forecast the monthly reservoir inflow in South Korea. The monthly climate indices and reservoir inflow data sets from 1998 to 2012 were used for training stage and those data sets between 2013 and 2016 were applied for testing stage. The ensemble methods, bayesian model averaging (BMA), simple model averaging (SMA), and naive forecast (NF), were compared among their performance. The results show BMA was more accurate and useful than SMA and NF. [7] investigated the performance of the six algorithms to forecast the amount of inflow of the Soyang River Dam. They employed time series data of the weather and the dam inflow data from 1980 to 2019. The findings indicated that the MLP model generated the best results in forecasting dam inflow with giving the root mean squared errors (RMSE) of 77.218 m^3/s, mean absolute error (MAE) of 29.034 m^3/s, correlation coefficient (r) of 0.924, and determination coefficient (r^2) of 0.817. In addition, they studied the performance of two combined machine learning models. It was found that GB_MLP model can predicts the dam inflow more accurate than the RF_MLP. In addition, [8] compared the efficiency of 3 artificial intelligence (AI) techniques, including SVR, the Back Propagation Neural Network (BPNN) and the Long Short-Term Memory (LSTM) neural network model. Four data sets about reservoir operations, i.e., the amount of water flowing into the reservoir, amount of water demand, the reservoir upper and lower edge of rule curve, were utilized. The data sets were collected every 2–4 h. The model's performance was investigated using Root Mean Square Error (RMSE). Forecasting results at monthly level with descending error order are as follows: LSTM = 161.712 MCM, BPNN = 1,122.827 MCM, and SVR = 1,932.814 MCM. Their processing time took 00:01:56, 00:05:47, and 74:05:56, respectively. However, LSTM techniques use very high memory resources, therefore not suitable for general situation.

Apart from the abovementioned methods, random forests (RF) is also another approach that has been commonly employed to predict the amount of inflow. [9] stated that this approach has appeared as key competitors to boosting and support vector machines. It is superior to the aforementioned techniques as it easy and fast to apply, generate highly accurate predictions and can cope with overfitting problem even with a very large number of input variables. Many researchers have applied random forest in the literatures. For instance, [5] compared the ability of RF, artificial neural network (ANN), and SVR for forecasting 1 month-ahead reservoir inflows in the USA and China. The result indicated that the random forest approach was better than the other two techniques. Similar result has been found by [10], who examined an efficient model for predicting lake water level variations in China. According to the root mean square error (RMSE) and determination coefficient (r^2), the random forest model gave the best performance for daily forecasting by comparison with artificial neural networks, support vector regression, and a linear model. With an average RMSE of 0.25 m for five stations within the lake, the highest accuracy was attained if the model inputs applying discharge series at 4-day-ahead and the previous week's average water level. From the previous works, SVR is a technique that is popularly used in many fields such as forecasting electricity demand [11]. In science [12] and in the management of water resources [4, 5, 13]. It also uses a little memory resources when compared to some artificial neural network techniques [14, 15]. RF is a supervised machine learning method and influential white-box classification and regression model. Almost the result of several works indicated this method better than simple methods such as ANN SVR, etc. RF is not only great accurate method for classification or regression but also, it can solve the overfitting problem for a very large number of input parameters [5, 16].

Although there are an extensive studies and applications of using machine learning models to predict reservoir inflow with global climate change indices in the developing country. Therefore, this current study contributes using the SVR and the RF techniques by first to evaluate the performance of these two methods to forecast the amount of inflow of the reservoir in advance on a monthly basis. Second, to compare the performance of machine learning methods are SVR and RF with the 12 month-lag of climate phenomenon indices and reservoir inflow. The remainder of the paper is organized as follows. Section 2 provides the experimental method and explains the data used in the analysis. Section 3 reports the empirical findings and discussion. Section 4 presents the conclusion.

2 Machine Learning Techniques

2.1 Support Vector Regression (SVR)

Support Vector Regression is an application based on the Support Vector Machine (SVM) by Vapnik [17]. The goal of SVR is to solve the problem of regression equations in a feature space, which uses the function f (x_i), describes the nonlinear association between feature x_i and objective value that y_i can indication as in Eq. (1).

$$f(x_i) = w * \varphi(x_i) + b \tag{1}$$

Where $\varphi(x_i)$ is the variation function and w is the coefficient vector and w and b present the weight and bias, which are constant values in the case of classification. Data type (classification case). The C variable is a variable used to measure the significance between complexity and losses. The w and b values can be initiate from Eq. (2).

$$fR(w) = \frac{1}{2}\|w\|^2 + C\sum_{i=0}^{n} L_{\in}(y_i f(x_i)) \tag{2}$$

A main problem is discovery the suitable dividing line in the case that the data is not non-linear, using soft-margin technique, unable to discover the appropriate dividing line between the data. A general solution to this problem is the use of kernel functions, which are existing in many studies [8, 18]. The regression function was shown in Eq. (3) where $k(x_i, x)$ is the kernel function.

$$f(x) = \sum_{i=1}^{l}\left(\varphi_i - \varphi_i^*\right)k\left(x_{i,}x\right) + b \tag{3}$$

In this research, I use ε - insensitive, which is popular today and use to test with the kernel type are Gaussian RBF and Linear kernel as shown in the following equation:

Linear kernel: $k(x, x_i) = x, x_i$.

Gaussian RBF: $k(x, x_i) = exp - \frac{|x - x_i|}{2\emptyset^2}$.

2.2 Random Forests (RF)

The random forests (RF) method as introduced by Breiman [16] is a nonparametric, white-box classification and regression model. Classification and regression trees (CARTs) are techniques to partition the parameter space based on a set of rules embedded in a decision tree (see Fig. 1 below), where each node separations according to a result rule as Fig. 1. The mechanism in a regular tree-growing development is to recursively carry out a binary split of a data set ("Parent Node") into two sub data sets ("Child Nodes") based on any user-defined dividing rule. The objective of all methods is to prevent overfitting in analysis process [5].

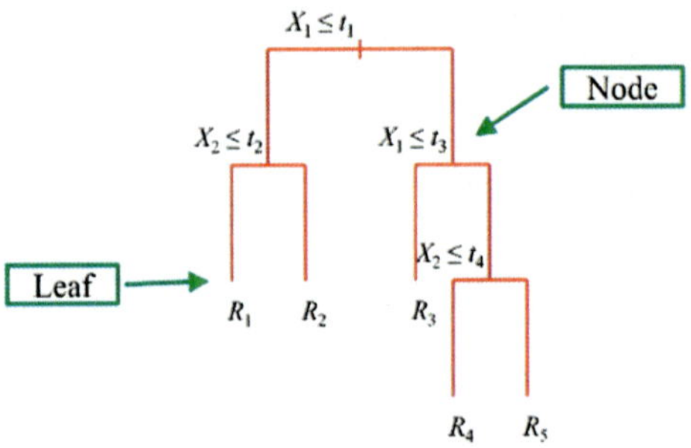

Fig. 1. Decision tree mechanism [16].

The Gini Index and Information Gain Ratio are the most well-known feature ranking methods. This research, Random forest method uses the Information Gain Ratio as a variable selective measure.

3 Method

3.1 Research Framework

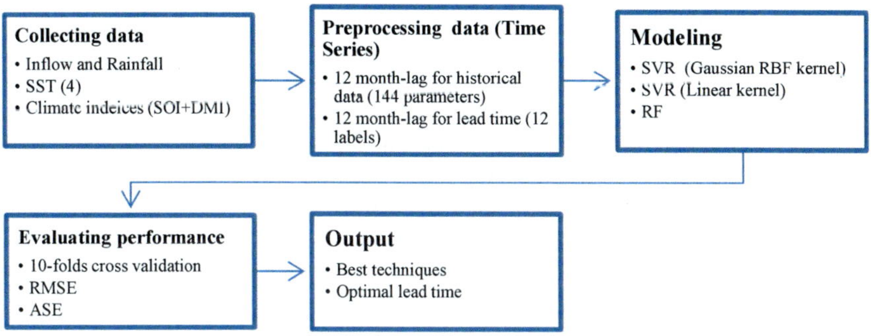

Fig. 2. Research framework.

The proposed framework has five main steps: 1) Collecting data, 2) Preprocessing Data (Time series), 3) Modeling, 4) Evaluating performance, and 5) Output. In the first stage, the collected variables consist of the amount of inflow, amount of rainfall, 4 SST parameters, and Climate indices (SOI and DMI). The second stage is included of two substages; the 12 month-lag for historical data preparing and the 12 month-lag for lead time (12 labels) preparing. In order to compare the performance of the model in the next stage, the model of this present study was created by three machine learning techniques which are 1) SVR with Gaussian RBF; 2) SVR with Linear kernel and 3) RF. In the fourth stage, we evaluate the performance of modeling by using 10-folds cross validation, RMSE and ASE. Finally, the best model and the optimal lead time were selected as shown in Fig. 2.

3.2 Study Area

Huai Nam Sai Reservoir was created in the fiscal year 1992 by the Royal Irrigation Department (RID). It is located at the coordinate 47NNJ 891-722, Section 4924I, Chauat District, Nakhon Si Thammarat Province. The reservoir is an embankment dam with a width of 8.00 m, the length 946 m, the height 40 m, and its capacity of 80 MCM. This medium-sized reservoir is under the upper Pak Phanang Irrigation and Maintenance Project, Irrigation Office 15, Royal Irrigation Department, Ministry of Agriculture and Cooperatives. It delivers irrigating water to three irrigation areas: 1) Water supply system in Khuan Khanun settlement, 2) Water supply system to serve people in Khuan Meechai, the royal initiative project, and 3) Khlong Mai Dee Weir irrigation system including irrigation area of 33,913.83 acre This reservoir directly benefits 20,948.62 acre in various villages, including six sub-districts in Cha-uat District, Nakhon Si Thammarat province. For example, agriculture helps increase crop productivity, such as rice and off-season kitchen plants. It is also a breeding ground for fish and helps prevent flooding in the Pak Phanang lowland area. The monthly rainfall and the reservoir inflow information between 1998 and 2015 (18 years), giving 216 data set, was collected from the Royal Irrigation Department (RID). Those statistical values present in Table 1.

Table 1. The summary of statistical value of monthly rainfall and reservoir inflow

Number of data set	Monthly rainfall (MCM)			The amount of reservoir water inflow (MCM)		
	Min	Max	Avg.	Min	Max	Avg.
216	0	1,017	5.17	0	58.34	8.5

3.3 Data Used

Three data sources from Climate Prediction Center (CPC), the National Weather Service (NWS) [19] used in the research are 1) Sea surface temperature (SST) that consist of 4 variables, and 2) Ocean indices that consists of Dipole Mode Index (DMI) and Southern Oscillation Index (SOI). Those two Ocean indices relate to El Nino or La Nina events in the Pacific Ocean, and 3) the monthly reservoir hydrological data consist of rainfall (R) and reservoir inflow (Inf). Before the analysis step, the 12 lag-month time series before and ahead was prepared, that is, the historical time series data are t−1 to t−12 and t+1 to t+12 for predicting monthly reservoir inflow. After the data preparation step, there were 144 parameters between 1998 and 2015, as shown in Table 2.

Table 2. The information of input variables

Data used	Input parameters	Data sources
Monthly reservoir inflow	Rainfall reservoir inflow	Upper Pak Phanang Operation and Maintenance Project, Irrigation Office 15, Royal Irrigation Department, Thailand
Sea Surface Temperature (SST)	NINO1+2, NINO3, NINO3.4, and NINO4	US National Oceanic and Atmospheric Administration (NOAA)
Ocean indices	Dipole Mode Index (DMI)	
	Southern Oscillation Index (SOI)	Japan Agency for Marine-Earth Science and Technology (JAMSTEC)

3.4 Experimental Setup

The six experiments were set up for our study, as presented details in Table 3. Two machine learning techniques, i.e., SVR (Gaussian RBF kernel, and Linear kernel), RF, were applied. The influence of climate indices, as mentioned in Sect. 3.2, was investigated.

Table 3. The experimental setup

Methods	Algorithms	Climate Indices	Symbol
Support Vector regression	Gaussian RBF kernel	Without climate indices	SVR-RBF
Support vector regression	Gaussian RBF kernel	With climate indices	SVR-RBF-CI
Support vector regression	Linear kernel	Without climate indices	SVR-Linear
Support vector regression	Linear kernel	With climate indices	SVR-Linear-CI
Random Forest	Gain Ratio Information weighted	Without climate indices	RF-GR
Random Forest	Gain Ratio Information weighted	With climate indices	RF-GR-CI

3.5 Performance Evaluation

The model's performance evaluation used two statistical methods: Root Mean Square Error (RMSE), and Absolute Error (ASE). The model is very effective when the RMSE and ASE values approach 0. We used 10-fold cross-validation for evaluating the models.

4 Results and Discussion

This section shows the results of comparing the performance of three machine learning techniques with investigating the influence of climate index parameters and determining the suitable lead time for reservoir inflow forecasting. Tables 4 and 5 present RMSE and ASE values in unit of million cubic meter (MCM) obtained from six experiments. Huay Nam Sai Reservoir has the capacity at normal level of 80 MCM. The results show that RF-GR-CI was the best method for reservoir inflow forecasting due to giving the least values of RMSE of 4.41 MCM and ASE of 3.14 MCM at 10-month ahead. The investigating results are as follows. There was no significant effect of having climate index parameters on the performance for the cases of SVR-RBF versus SVR-RBF-CI and SVR-Linear versus SVR-Linear-CI. Those cases gave the almost similar value of RMSE and ASE. However, climate index parameters influenced the cases of RF-GR versus RF-GR-CI. As we can see, RMSE and ASE's lower values were obtained for all forecasting lead time. Considering the suitable lead time for reservoir inflow forecasting, the different model's performance index (RMSE or ASE) gave different results. Using RMSE as a criterion, the most efficient for reservoir inflow forecasting were 10-month ahead (t+10), followed by 2-month (t+2) and 6-month (t+6) ahead. Those gave the least RMSE values of 4.41 MCM, 4.58 MCM, and 4.62 MCM, respectively as shown in Table 4. In addition, using ASE as a criterion, the most efficient for reservoir inflow forecasting were the 3-month (t+3), 10-month (t+10) and 5-month (t+5) ahead. Those gave the least ASE values of 3.11 MCM, 3.14 MCM and 3.16 MCM, respectively as shown in Table 5. When simultaneously considering two model's performance indices, we found our developed model for reservoir inflow gave the best efficiency for forecasting 10-month ahead (t+10). This long lead time was very helpful for the policymaker to plan for reservoir operation. The development in forecasts in this research after used SSTs and climate indices supports these previous study [20]. However, future studies should clarify the individual contributions of SSTs and the climate indices.

Table 4. RMSE (MCM) values obtained from six experiments

Methods	Lead time (months)											
	T+1	T+2	T+3	T+4	T+5	T+6	T+7	T+8	T+9	T+10	T+11	T+12
SVR-RBF	6.09	6.10	6.12	5.90	6.13	6.01	5.97	6.05	5.92	5.75	6.03	5.88
SVR-RBF-CI	6.09	6.10	6.12	5.90	6.13	6.01	5.97	6.05	5.92	5.75	6.03	6.12
SVR-Linear	5.45	5.51	5.48	5.22	5.36	5.16	5.27	5.35	5.34	5.07	5.31	5.34
SVR-Linear-CI	5.45	5.51	5.49	5.36	5.17	5.27	5.27	5.35	5.35	5.07	5.31	5.34
RF-GR	5.43	5.37	5.34	5.28	5.40	5.05	5.25	5.25	5.28	4.83	4.95	5.08
RF-GR-CI	4.69	**4.58**	4.77	4.70	4.70	**4.62**	4.72	4.77	4.78	**4.41**	4.71	4.76

Table 5. ASE (MCM) values obtained from six experiments

Methods	Lead time (months)											
	T+1	T+2	T+3	T+4	T+5	T+6	T+7	T+8	T+9	T+10	T+11	T+12
SVR-RBF	4.15	4.12	4.09	4.09	4.10	4.09	4.14	4.09	4.08	4.07	4.15	3.99
SVR-RBF-CI	4.15	4.12	4.09	4.09	4.10	4.09	4.14	4.09	4.08	4.07	4.15	4.16
SVR-Linear	3.53	3.67	3.58	3.57	3.54	3.44	3.58	3.56	3.63	3.58	3.62	3.55
SVR-Linear-CI	3.53	3.67	3.59	3.54	3.45	3.57	3.57	3.56	3.63	3.58	3.62	3.55
RF-GR	3.65	3.62	3.50	3.69	3.65	3.51	3.68	3.55	3.59	3.52	3.53	3.57
RF-GR-CI	3.16	3.19	**3.11**	3.29	**3.16**	3.18	3.25	3.23	3.21	**3.14**	3.22	3.22

5 Conclusions

This study investigated the performance of three techniques, namely SVR (Gaussian RBF kernel) SVR (Linear kernel) and RF with/without climatic indices are compared in forecasting for monthly reservoir inflow. The dataset from two parts are the reservoir information data from Huai Nam Sai Reservoir, that located in southern of Thailand, and the climate indices from US National Oceanic and Atmospheric Administration (NOAA) and Irrigation and Maintenance Project Irrigation Office 15, Irrigation Department, Thailand.

In summary, the key findings of this study can be summarized as follows:

1. The RF algorithm was a suitable alternative to the SVR (Gaussian RBF kernel) and SVR (Linear kernel) algorithm for Huai Nam Sai Reservoir inflow forecasting based on available information and data presented. This reservoir is located in the southern Thailand.
2. The climatic indices can improve the forecasting performance of RF for every advance time between 1-month and 12-month. Based on the error metrics, the RMSE and ASE were decreased significantly by about 4.88–14.60% and 9.54–13.62%, respectively. However, the climate indices not suitable for SVR both Gaussian RBF and Linear kernel.
3. The optimal lead time is 10-month as RMSE with giving the least of RMSE was 4.41 and with giving the second least of ASE was 3.11.

Acknowledgements. The research funding from Ministry of Higher Education, Science, Research and Innovation is gratefully acknowledged.

References

1. Valipour, M., Banihabib, M.E., Behbahani, S.M.R.: Parameters estimate of autoregressive moving average and autoregressive integrated moving average models and compare their ability for inflow forecasting. J. Math. Stat. **8**(3), 330–338 (2012)
2. Tongsiri, J., Kangrang, A.: Prediction of future inflow under hydrological variation characteristics and improvement of Nam Oon reservoir rule curve using genetic algorithms technique. Mahasarakham Univ. J. Sci. Technol. **37**(6), 775–788 (2018)
3. Chiamsathit, C., Adeloye, A.J., Bankaru-Swamy, S.: Inflow forecasting using artificial neural networks for reservoir operation. Proc. Int. Assoc. Hydrol. Sci. **373**, 209–214 (2016)
4. Lin, G.F., Chen, G.R., Huang, P.Y.: Effective typhoon characteristics and their effects on hourly reservoir inflow forecasting. Adv. Water Resour. **33**(8), 887–898 (2010)
5. Yang, T., Asanjan, A.A., Welles, E., Gao, X., Sorooshian, S., Liu, X.: Developing reservoir monthly inflow forecasts using artificial intelligence and climate phenomenon information. Water Resour. Res. **53**(4), 2786–2812 (2017)
6. Lee, D., Kim, H., Jung, I., Yoon, J.: Monthly reservoir inflow forecasting for dry period using teleconnection indices: a statistical ensemble approach. Appl. Sci. **10**(10), 3470 (2020)
7. Hong, J., et al.: Development and evaluation of the combined machine learning models for the prediction of dam inflow. Water **12**(10), 2927 (2020)

8. Zhang, D., et al.: Modeling and simulating of reservoir operation using the artificial neural network, support vector regression, deep learning algorithm. J. Hydrol. **565**, 720–736 (2018)
9. Biau, G.: Analysis of a Random Forests Model. J. Mach. Learn. Res. **13**, 1063–1095 (2012)
10. Li, B., Yang, G., Wan, R., Dai, X., Zhang, Y.: Comparison of Random Forests and other statistical methods for the prediction of lake water level: a case study of the Poyang Lake in China. Hydrol. Res. **47**, 69–83 (2016)
11. Sarhani, M., El Afia, A.: Electric load forecasting using hybrid machine learning approach incorporating feature selection. In: Proceedings of the International Conference on Big Data Cloud and Applications, BDCA 2015, Tetuan, Morocco (2015)
12. Ivanciuc, O.: Applications of support vector machines in chemistry, vol. 23, pp. 291–400 (2007)
13. Domingos, S., João, F.L., Paulo, S.G.: An intelligent hybridization of ARIMA with machine learning models for time series forecasting. Knowl. Based Syst. **175**, 72–86 (2019)
14. Ren, J., Ren, B., Zhang, Q., Zheng, X.: A novel hybrid extreme learning machine approach improved by k nearest neighbor method and fireworks algorithm for flood forecasting in medium and small watershed of loess region. Water **11**(9), 1848 (2019). https://doi.org/10.3390/w11091848
15. Yang, S., Yang, D., Chen, J., Zhao, B.: Real-time reservoir operation using recurrent neural networks and inflow forecast from a distributed hydrological model. J. Hydrol. **579**, 124229 (2019)
16. Tyralis, H., Papacharalampous, G., Langousis, A.: A brief review of random forests for water scientists and practitioners and their recent history in water resources. Water **11**(5), 910 (2019)
17. Drucker, H., Surges, C.J.C., Kaufman, L., Smola, A., Vapnik, V.: Support vector regression machines. In: Advances in Neural Information Processing Systems, vol. 1, pp. 155–161 (1997)
18. Mariette, A., Khanna, R.: Efficient Learning Machines Theories, Concepts, and Applications for Engineers and System Designers. Apress (2015)
19. Climate Prediction Center (CPC), the National Weather Service (NWS). "Monthly Atmospheric and SST Indices." https://www.cpc.ncep.noaa.gov/data/indices/. Accessed 28 Dec 2020.
20. Amnatsan, S., Yoshikawa, S., Kanae, S.: Improved forecasting of extreme monthly reservoir inflow using an analogue-based forecasting method: a case study of the Sirikit Dam in Thailand. Water **10**(11), 1614 (2018)

ADGAS: An Advanced Data Generation for Anomalous Signals

Thasorn Chalongvorachai and Kuntpong Woraratpanya(✉)

Faculty of Information Technology, King Mongkut's Institute of Technology Ladkrabang, No. 1, Chalong Krung 1, Ladkrabang, Bangkok 10520, Thailand
62606065@kmitl.ac.th, kuntpong@it.kmitl.ac.th

Abstract. Anomaly detection using deep learning approaches is still challenging, especially in the case of data limitations. A small number of samples for training deep learning models typically result in poor performance of detection and classification. Previously, data augmentation was one of the methods used to solve this problem. The data augmentation with rotation, permutation, time warping, and their combination can increase the performance of anomaly classification. However, this method is still limited and does not guarantee that the generated output data have adequate varieties and keep original characteristics of data. Our recent work, data augmentation and generation for anomalous time series signals (DAGAT) was proposed to expand the space of possible augmented data by implementing vanilla augmentation on various domains in conjunction with variational autoencoder (VAE). Nonetheless, the DAGAT still has barriers, which are an uncontrollable number of target results, a missed opportunity of integrating multiple augmentation characteristics in latent space, and a possibility of including any bad data for training in VAE. To overcome these limitations, this paper proposed an advanced data generation for anomalous signals (ADGAS). By focusing on the quality of generated data, one more quality classifier (QC) was added as a prepossessing step of VAE. In this way, the experimental results showed that convolutional neural networks (CNNs), used as a performance tester, trained with the generated datasets of ADGAS achieved better accuracy in classifying anomalous events when compared to models trained with a combination of rotation, permutation, and time warping data augmentation methods.

Keywords: Data generation · Anomaly detection · Deep learning · Variational autoencoder · Convolutional neural networks

1 Introduction

In real-world, anomaly detection and classification on signals are difficult, since anomalous datasets are rare to discover, difficult to record, and inaccessible publicly, for instance, patient medical records, machinery failures, etc. [1–3]. A lack of anomalous datasets becomes a great barrier for anomalous research

P. Meesad et al. (Eds.): IC2IT 2021, LNNS 251, pp. 137–147, 2021.
https://doi.org/10.1007/978-3-030-79757-7_14

since there is inadequate information to learn and notice anomalous incidents, especially in a deep learning field. A too-small number of training sets lead to low generalization, which results in overfitting models [4]. Indeed, the lack of anomalous datasets is not a new problem. In the past, the approach to solve this issue was the use of outlier detection techniques [5]. These techniques learn through normal data only and expect that any samples, which are outside a learned area, are identified as anomalies. However, this approach is unable in classifying anomalous classes, since it recognizes only the normal area of data distribution and the learned model does not know what kind of outliers are.

In order to solve this limitation, most papers [6–9] applied data augmentation to increase the number of training samples. The data augmentation technique usually distorts, modifies, and reforms original data slightly to generate new sample sets. This technique can help increase the generalization and reduce the overfitting problem of learned models in deep learning [10].

Um et al. [11] proposed a method of generating various augmented samples with rotation, permutation, time-warping, etc., and applied on wearable sensor data to train a deep learning model. This method can improve Parkinson's disease classification performance due to increasing the number of training sets.

Even though data augmentation can be a solution for an increase in training sets, it could not provide a variety of data. Moreover, the existing methods did not have any tools to ensure that the quality of augmented data is good enough to be training sets. Recently, data augmentation and generation for anomalous time series signals (DAGAT) [12] was proposed to solve the previously mentioned problem.

The concept of DAGAT framework consisted of three functionalities: (i) generation of augmented data on different domains to reach a variety of samples, (ii) usage of VAE for integrating augmented data on different domains to transform them to latent space, and (iii) selection of multiple augmentation characteristics in the latent space. With this concept, we can select in-between types of augmented data. Nevertheless, DAGAT still has some limitations, for instance, it cannot integrate multiple augmentation characteristics per sample in the latent space, some bad samples are possibly trained in VAE, and the number of samples is uncontrollable.

Therefore, an advanced data generation for anomalous signals (ADGAS) has been proposed. ADGAS is an improved framework of DAGAT as shown in Fig. 1. In order to ensure the performance of augmented data, the implementation of ADGAS framework has been tested to generate sensor data on the testbed dataset. The performance of ADGAS outperformed the baseline method [11].

The remainder of this paper is structured as follows: Sect. 2 explains typical data augmentation and a DAGAT framework. Section 3 describes the procedure of ADGAS in details. Section 4 reports the classification performance of ADGAS. Section 5 summarize an overall performance of ADGAS and discusses further possible improvement in future work.

2 Related Works

2.1 Typical Data Augmentation

In 2017, Um et al. [11] proposed data augmentation techniques on wearable sensor data for Parkinson's disease classification and evaluated the performance using CNNs. Methods to perturb multivariate sensor data consist of jittering, scaling, magnitude warping, cropping, time-warping, permutation, and rotation techniques. Furthermore, these techniques can be combined to create various augmented data.

The experiments were made with a dataset of Parkinson's disease patients that were recorded via wearable sensors. They contain classes of different symptoms. Results showed that implementing rotation (Rot), rotation and time-warping ($Rot + TimeW$), and rotation with time-warping and permutation ($Rot + TimeW + Perm$) methods can achieve accuracy up to 82.62% to 86.88%.

Though these methods can be useful for training deep learning models, there are numerous spaces of possible augmented data that have not been discovered yet. Additionally, using only a similar augmentation approach for an entire dataset does not generalize enough. In other words, the dataset for training should contain various methods in data modification. Moreover, the finished augmented dataset possibly contains some bad data that sample shape does not comply with the original data or might not or lowly have characteristics of data.

2.2 DAGAT Framework

In 2020, we proposed a framework of data augmentation and generation for anomalous time-series signals (DAGAT) [12]. The concept of this framework consists of three functionalities. (i) Vanilla augmentation [11] is applied to anomalous time-series signals on various domains, i.e., upsampling, fast Fourier transform, and time-series decomposition, to generate a variety of anomalous samples, and then all of them are transformed back to the time domain. (ii) VAE is used to transform the augmented samples from (i) to latent space; this makes it easy to visualize characteristics of the augmented samples, and (iii) the last functionality consists of three procedures: data picker (DP) selects more samples from the latent space, and then signal fragment assembler (AFS) combines two characteristics of selected samples to create a new sample. Lastly, the quality classifier (QC) compares a generated sample to the original one and evaluates the score for decision-making. This makes a DAGAT framework efficient in producing the good quality of generated datasets for deep learning purposes. Note that domain transformation can be adjusted freely, for instance, many time-to-frequency transforms can be implemented in this framework.

3 Proposed Method

This paper proposed a framework called an advanced data generation for anomalous signals (ADGAS). This framework was redesigned to eliminate DAGAT

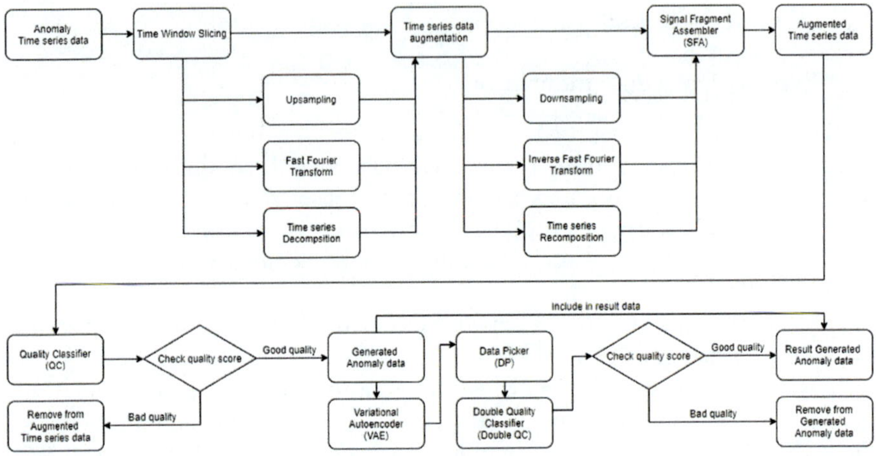

Fig. 1. ADGAS framework.

problems, i.e., the uncontrollable number of results, miss opportunity of including multiple augmentation characteristics in latent space, and some bad samples probably included for training in VAE. The framework of ADGAS is illustrated in Fig. 1 and its processes are explained in Subsect. 3.1–3.9.

3.1 Time Window Slicing

ADGAS framework initiates with time window slicing to increase the number of anomalous samples by moving the time window over an interesting portion of signals with determined window size and step. This can expand the unit of a signal sample before feeding it to further steps and helps create possible data.

3.2 Time Series Data Augmentation

By using techniques in Subsect. 2.1, ADGAS can expand the number of samples with higher varieties before proceeding. In this framework, jittering, scaling, magnitude warping, time-warping, and permutation techniques were used.

3.3 Upsampling and Downsampling

As mentioned in Subsect. 2.2, DAGAT not only used typical data augmentation but also applied these techniques on different domains. This makes it able to generate synthetic samples, while the typical augmentation cannot do. Upsampling and downsampling is one of the techniques to transform signals to a different domain. The purpose of using this technique is to transform time-domain signals to the upsampling-domain signals to increase their sampling rate [13]. This can provide a possible space available for augmentation with methods described in Subsect. 2.1. After completing the augmentation process, the augmented signals are downsampled back to the time domain.

3.4 Fast Fourier Transform and Its Inverse

Fast Fourier transform (FFT) [14] is used in DAGAT and ADGAS frameworks. One-dimensional FFT and its inverse (IFFT) are defined in (1) and (2), respectively.

$$Y(k) = \sum_{j=1}^{n} x(j) W_n^{(j-1)(k-1)} \tag{1}$$

$$x(j) = \frac{1}{n} \sum_{k=1}^{n} Y(k) W_n^{-(j-1)(k-1)} \tag{2}$$

where $W_n = e^{(-2\pi i/n)}$, n is the length of signal, $Y(k)$ is frequency-domain signal and $x(j)$ is time-domain signal. Here, FFT is used as a tool to transform a signal to a frequency domain, and then augment that signal, after that IFFT is taken to transform it back to the time domain.

3.5 Time Series Decomposition and Its Recomposition

Time series decomposition used in this framework decomposes a signal into three different components, consisting of trend (T_t), seasonal (S_t), and residual (R_t). These components can be represented as additive and multiplicative models as expressed in (3) and (4), respectively.

$$y_{t_a} = T_t + S_t + R_t \tag{3}$$

$$y_{t_m} = T_t \times S_t \times R_t \tag{4}$$

where y_{t_a} and y_{t_m} are the time series of additive and multiplicative models, respectively [15]. Once the augmentation process is complete, each component is recomposed back to a time-domain sample.

3.6 Signal Fragment Assembler

After retrieving all augmented signals from the previous process, signal fragment assembler (SFA) divides the signals into fragments i_s and j_s before arbitrary concatenating each fragment to create a new sample (n_s). This approach can combine two characteristics of data augmentation techniques to increase the varieties of the generated samples. SFA can be defined by (5).

$$n_s = i_s || j_s \tag{5}$$

where $||$ is the concatenation operator. When compared to the DAGAT framework, SFA is placed in front of the VAE to assemble two characteristics of data into one before proceeding to the quality classifier. The advantage of this placement is to provide data with higher varieties for feeding into VAE and allows it to eliminate poor quality data.

3.7 Quality Classifier

Quality classifier (QC) is important in this framework to ensure that the generated samples have a good quality. The quality of generated samples is evaluated by histogram overlapping. In this step, the generated sample and the original one are transformed to histogram charts. The similarity score is calculated by the difference of overlapping area between two histogram charts. QC computes the percentage of intersect area for scoring. The range of acceptable scores can be adjusted freely depends on applications. The QC selects merely generated anomaly samples within the determined threshold.

Notice that when we compare ADGAS with DAGAT, the QC is placed in front of VAE as a selector in selecting only the good quality of the augmented samples. The modification of data flow in the ADGAS framework can ensure that training sets, which are feed into VAE, are good enough for dataset generation.

One more QC, called double QC, is placed behind of data picker, described in Subsect. 3.9. The role of double QC is to solve the problems of the uncontrollable number of outputs and output quality. The double QC evaluates the quality of generated samples, selected by a data picker. If the quality is not acceptable, the data picker picks another new sample from latent space randomly. This process continues in this fashion until it retrieves the determined number of samples that is set at the beginning Fig. 2.

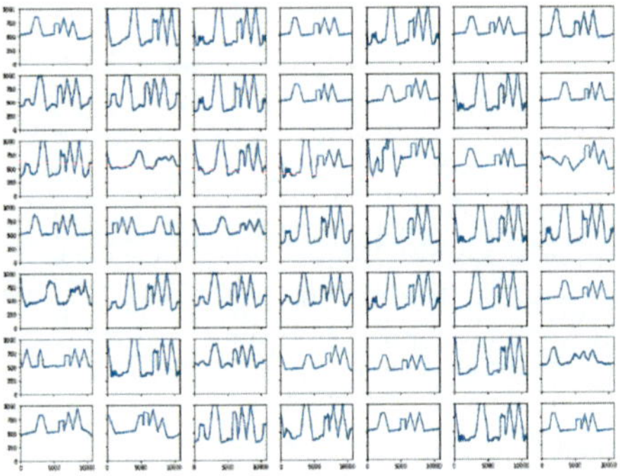

Fig. 2. An example of generating anomalous signals from ADGAS framework.

3.8 Variational Autoencoder (VAE)

This paper uses VAE to work as a model that can create latent space from learned samples [16]. This can be useful for data generation. The objective function of VAE can be expressed as (6).

$$logP(X) - D_{KL}[Q(z|X)||P(z|X)] = E[logP(X|z)] - D_{KL}[Q(z|X)||P(z)] \quad (6)$$

where X is the signal sample, z is the latent space, $P(X)$ is the probability distribution of signal samples, $P(z)$ is the probability distribution of latent space, $Q(z|X)$ is the projection of signal onto latent space, $P(X|z)$ is the projection of signal given by latent space, and D_{KL} is the KL divergence function.

If the total number of augmented data from the previous step does not reach the required number, VAE is trained using good quality data. With this approach, we have a latent space for augmented samples and can generate unlimited data. We also can gather an undiscovered sample from the learned latent space. In other words, we can retrieve a sample that is in-between various signal characteristics.

3.9 Data Picker

Data picker (DP) is a process of arbitrarily selecting samples on the latent space. The key factor of arbitrary selection is specifying a location and a circle area. Responding to this, DP randomly sets a center of a circle area on the latent space as (x_s, y_s). Further, with (a, b) parameters that are random numbers between 0 and 1, DP can randomly define an angle and radius of the circle area as stated in (7) and (8), respectively. This yields final coordinate output of (x_l, y_l) that is newly generated samples as defined in (9) and (10).

$$\theta = 2\pi a \tag{7}$$

$$r_l = r_d\sqrt{b} \tag{8}$$

$$x_l = r_l \times cos(\theta) + x_s \tag{9}$$

$$y_l = r_l \times sin(\theta) + y_o \tag{10}$$

4 Experiments and Results

4.1 Dataset Explanation

To evaluate our generated datasets for model training, we used a secure water treatment (SWaT) dataset that gathers data from a testbed of a small-scale real-world water treatment system [17]. This dataset contains the records of sensors and actuators of cyber and physical attacks that can interrupt or damage the system. Here, we brief one of the attack scenarios as follows. In this testbed, there are two important sensors used for attack simulation, i.e., a sensor for detecting the overflow of the raw water tank and another one for monitoring the underflow of the reverse osmosis (RO) water tank. The attack scenario can happen, when the attacker attempted to keep the motorized valve of the raw water tank on continuously, yielding the overflow of water. On the other hand, when the attacker attempted to keep the pump of the reverse osmosis (RO) water tank on continuously, yielding the underflow of water. In our experiments, two separate attack scenarios were selected for binary anomaly classification. The former is the attack scenario on the level sensors of the raw water tank (RAW) and the later is the attack scenario of the reverse osmosis (RO).

4.2 Experimental Setup

In this experiment, we used a one-dimensional convolutional neural network (1D-CNN) [18], as a performance tester, to evaluate the generated datasets. The model of 1D-CNN consists of 7 layers of convolution with 16, 32, 64, 64, 64, 64, and 64 features, respectively. 3×1 kernel, 2 strides, and rectified linear unit (ReLU) activation function were used in every layer. Also, dropout was applied at the rate of 0.1 [19] and batch normalization [20] was applied for every convolution layer to reduce overfitting, increase generalization, standardize inputs, and reduce generalization error. All flatten features were connected to a fully connected layer with 5,312 nodes and ReLU was set as the activation function. The final layer has 1 node with a sigmoid function for binary classification.

For baseline settings, each record contains 10,800 samples (3 h) that were sliced from a dataset with a 2-second time window before applying rotation (Rot), rotation and permutation ($Rot + Perm$), and rotation with permutation and time warping ($Rot + Perm + TimeW$) techniques. Note that this experiment also used the arbitrary values of amplitude, phase, and frequency in the data augmentation process, similar to [11].

For ADGAS settings, data generation initiates with a 2-second time window slicing, the same length as the previous settings. Jittering, scaling, magnitude warping, time warping, and permutation samples with arbitrary values of amplitude, phase, and frequency. The Data picker randomly chooses samples within 5 units of the radius, and then calculates the similarity score. The acceptable score should be in a range of 80%–95%. Note that cropping and rotation techniques did not include in this experiment, since the cropping makes the length of output data different from others, and the rotation results in a poor score at the QC state.

Here, 5-fold cross-validation was used in all experiments; that is, 5,000 synthetic anomaly records were equally divided into five groups. 50% of generated anomaly records and 50% of real normal records were used for training. In a testing phase, 50% of the real anomaly records and 50% of the real normal records were used by slicing data from an original dataset. Results were reported on average accuracy and F1-score for evaluation analysis.

4.3 Experimental Results

The results of the anomaly event classification are shown in Table 1. Accuracy and F1-score presented that the 1D-CNN model trained with datasets generated by ADGAS outperformed the Rot (R), $Rot + Perm$ (RP), and $Rot + Perm + TimeW$ (RPT) augmentation techniques in most cases.

Although accuracy and F1-score of ADGAS and baseline methods might vary depends on the generated datasets, the trend of scoring reveals that the varieties of augmentation have an impact on the performance of classification models as shown in Table 1. Based on observation of accuracy and F1-score of ADGAS and baselines, we can see that the more techniques we combine, the better performance we get. This happened in most cases. In addition, the better generated datasets are from an improvement of the framework. The use of two-stage QCs is an important factor that can help improve the better generated datasets. The first QC is placed in front of VAE to select a good quality of samples for its model. Another QC is placed at behind of data picker to perform as the double QC to ensure that only the good quality of generated samples was selected. As a result, the accuracy and F1-score of ADGAS were improved significantly. Furthermore, a byproduct of usage of the two-stage QCs is that we can set the specific number of generated samples.

Note that the DAGAT framework did not include as our baseline for performance comparison, because it could not reach the required number of generated samples. This phenomenon happened as the DAGAT framework classifies records after all generated processes are already complete, resulting in an uncontrollable desire number of records.

Table 1. A comparison of anomaly classification evaluated in terms of accuracy and F1-score of 1D-CNN model trained with eight different datasets generated by four different techniques: ADGAS, *Rot* (R), *Rot* + *Perm* (RP), and *Rot* + *Perm* + *TimeW* (RPT)

Dataset[a]	ADGAS[b]		R[b]		RP[b]		RPT[b]	
	Acc.	F1-score	Acc.	F1-score	Acc.	F1-score	Acc.	F1-score
RAW	**84.53%**	**0.8035**	75.39%	0.6665	77.34%	0.6797	77.22%	0.7746
RO	**75.94%**	0.7280	68.83%	0.7697	72.14%	**0.7798**	75.14%	0.7491

[a]Normal and abnormal test sets, in the testing phase, are from the original datasets, RAW and RO.

[b]Each technique generates two synthetic datasets, thus resulting in eight datasets in total for training.

5 Conclusion

This paper has proposed an improved version of DAGAT, called ADGAS, that can solve the existed problems: an uncontrollable number of outputs, not including multiple augmentation characteristics in latent space, and the bad data trained by VAE. Three modified sections of the DAGAT framework have been implemented as follows: (i) Inserting the signal fragment assembler (SFA) as a pre-processing step of VAE, (ii) adding the double QC as a post-processing of DP to ensure in quality of samples for training VAE, and (iii) placing DP into a loop, while picking data from the latent space, and selecting only samples in the latent space. Experimental results showed that 1D-CNN trained with generated datasets from ADGAS provided better performance in classification, when compared with the models trained with augmented datasets from baseline methods. This achievement comes from the following main contributions:

- Placing SFA in front of VAE can increase the variety of data, leading to generalization in a trained model.

- Creating a loop of DP can define the specific number of outputs.
- Adding double QC as a post-processing phase of DP helps improve the quality of selected samples in latent space.

In future work, various transformations will be implemented in ADGAS to create a variety of patterns in augmented states. Furthermore, the quality classifier with other evaluation methods will be applied for ADGAS to improve the efficiency of the scoring process, yielding the better quality of generated datasets.

References

1. Ahmad, S., Lavin, A., Purdy, S., Agha, Z.: Unsupervised real-time anomaly detection for streaming data. Neurocomputing **262**, 134–147 (2017)
2. Shaikhina, T., Khovanova, N.A.: Handling limited datasets with neural networks in medical applications: a small-data approach. Artifi. Intell. Med. **75**, 51–63 (2017)
3. Khot, L., et al.: Evaluation of technique to overcome small dataset problems during neural-network based contamination classification of packaged beef using integrated olfactory sensor system. LWT - Food Sci. Technol. **45**, 233–240 (2012)
4. Ying, X.: An overview of overfitting and its solutions. J. Phys. Conf. Ser. **1168**(2), 022022 (2019)
5. Malik, K., Sadawarti, H., Kalra, G.: Comparative analysis of outlier detection techniques. Int. J. Comput. Appl. **97**, 12–21 (2014)
6. Mikołajczyk, A., Grochowski, M.: Data augmentation for improving deep learning in image classification problem. In: 2018 International Interdisciplinary PhD Workshop (IIPhDW), pp. 117–122 (2018)
7. Wang, F., Zhong, S., Peng, J., Jiang, J., Liu, Y.: Data augmentation for EEG-based emotion recognition with deep convolutional neural networks. In: MMM (2018)
8. Huang, L., Pan, W., Zhang, Y., Qian, L., Gao, N., Wu, Y.: Data augmentation for deep learning-based radio modulation classification. IEEE Access **8**, 1498–1506 (2020)
9. Kyu, P.M., Woraratpanya, K.: Car damage detection and classification. In: Proceedings of the 11th International Conference on Advances in Information Technology, IAIT2020 (2020)
10. Shorten, C., Khoshgoftaar, T.: A survey on image data augmentation for deep learning. J. Big Data **6**(1), 1–48 (2019)
11. Um, T.T., et al.: Data augmentation of wearable sensor data for Parkinson's disease monitoring using convolutional neural networks. In: Proceedings of the 19th ACM International Conference on Multimodal Interaction (2017)
12. Thasorn Chalongvorachai, K.W.: Dagat: data augmentation and generation for anomalous time series signals. In: 9th Joint Symposium on Computational Intelligence (JSCI9). Bangkok, Thailand (2020)
13. Karioja, K., Laurila, J.: Signal upsampling in machine diagnostics - an experimental approach (2016)
14. Oberst, U.: The fast fourier transform. SIAM J. Control Optim. **46**, 496–540 (2007)
15. Dagum, E.B.: Time series modelling and decomposition. Statistica **70**(4), 433–457 (2010)
16. Kingma, D.P., Welling, M.: Auto-encoding variational bayes. CoRR abs/1312.6114 (2014)

17. Goh, J., Adepu, S., Junejo, K.N., Mathur, A.: A dataset to support research in the design of secure water treatment systems. In: CRITIS (2016)
18. Kiranyaz, S., Avci, O., Abdeljaber, O., Ince, T., Gabbouj, M., Inman, D.: 1D convolutional neural networks and applications: a survey. ArXiv abs/1905.03554 (2019)
19. Park, S., Kwak, N.: Analysis on the dropout effect in convolutional neural networks. In: ACCV (2016)
20. Ioffe, S., Szegedy, C.: Batch normalization: Accelerating deep network training by reducing internal covariate shift. ArXiv abs/1502.03167 (2015)

Object Detection with Convolution Neural Networks and Custom-SmallerVGGNet

Quang Nhat Tran(✉) and Siranee Nuchitprasitchai

Faculty of Information Technology and Digital Innovation, King Mongkut's University of Technology North Bangkok, Bangkok, Thailand
s6207011866512@email.kmutnb.ac.th, siraneen.n@itd.kmutnb.ac.th

Abstract. Computer vision has reached significant levels of Object detection. Whether performed via computer vision or deep learning, there are three main results that we to wish obtain, including a list of bounding boxes (or x and y coordinators) for each object in the image, a class label linked with each bounding label, and the probability or confidence score associated with each bounding box and class label. Object detection is becoming even more popular with the breakthrough of deep learning, and especially Convolutional Neural Networks (CNN) which are comprised of two elements, namely convolution layers and pooling layers. With these successes in deep learning, we have more choice on selecting a method and technology to experiment and improve accuracy. There are several datasets such as MNIST, Animal, CIFAR-10, SMILE and ImageNet; these datasets would be a based foundation for researcher to do the experiment. However, there is an approach which can achieve to get a dataset quickly. This paper will introduce two methodologies for object detection and determine the future research directions of researching with object detection in deep dive. First, the traditional object detection pipeline, which will focus on these techniques such as sliding windows, image pyramid and classification. Second, the based network of an object detection framework, this will used certain classification network as a based network in a deep learning object detection such as SmallerVGGNet, Faster R-CNN, SSD or YOLO. Lastly, we experimented using custom SmallerVGGNet and received a positive outcome in so far as greater than 95% of accuracy.

Keywords: Object detection · CNN · Faster R-CNN · SSD · YOLO · Custom SmallerVGGNet

1 Introduction

There are extant differences between object detection and image classification. For example, when executing image classification, it transfers the input image to a neural network and the obtained result is a single class label, or with a probability as well. An example is shown in Fig. 1.

For object detection, whether performed using deep learning or another computer vision technique, the result will seek to exactly localize each object in the image. Take Fig. 1; it can be seen that the process of object detection when comparing the traditional

P. Meesad et al. (Eds.): IC2IT 2021, LNNS 251, pp. 148–158, 2021.
https://doi.org/10.1007/978-3-030-79757-7_15

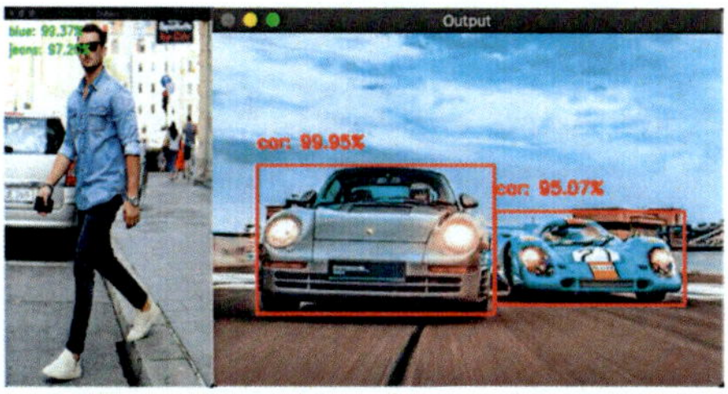

Fig. 1. Example of image classification (left) and object detection (right).

way and modern way using deep learning will produce different output. The history of object detection was begun in 2001 by Viola and Jones [1], known hence as Viola Jones Detectors. Soon after in 2005, HOG detection (Histogram of Object Gradients) was introduced by Dalal and Triggs [2], followed by extensions to HOG detection by Felzenszwalb 2008 [3]. These three works were milestone in traditional object detection, as distinct from "neural network" based approaches. All of these techniques are programming-based build on handcrafted features. The next milestone in object detection is commonly known as the 'deep learning era'. In general, this subdivides into two stages, namely "two-stage detection" and "one-stage detection". There is a trade-off of speed over accuracy between the two. Two-stage detection is more concerned with percentage of accuracy which oscillates between 80 and 90%, and which contains several technologies such as R-CNN (Region-based Convolutional Neural Networks [4]), Fast R-CNN [5] and Faster RCNN [6], however the accuracy of experiments involving such technologies experiment has yet to reach 95%. This paper will therefore investigate means to make two-stage detection achieve such levels of accuracy. One-stage detection focusses on the speed of processing while maintaining accuracy at a certain level. It includes several techniques, including YOLO [7] (You Only Look Once), SSD (Single Shot object Detection) [8] and Retina-Net [9]. However, accuracy levels are still around 85–90% based on the research paper [10]. In this research paper, the major is to increase the accuracy based on the existing framework, approximately reaching to 97%. Furthermore, the work is not related to any kind of existing dataset [11], the model can detect in several set of items base on the flexible source of data which can retrieve via Bing search API. It also leads to a real-time object detection [12] in future work.

2 Related Works

2.1 Compare the Process Between Traditional and Modern CNN Methodologies

The traditional methodology mainly focusses on manual feature extraction, before using machine learning to classify the data points and then provide final output. The modern methodology is mainly focused on deep learning neural networks; each network layer collects a set of data points and uses the output as input for the next layer and it will learn and improve its accuracy by themselves without human interaction.

In conclusion, the modern methodology improves performance over the traditional approach.

2.2 Method 1: The Traditional Object Detection Pipeline

The first method is not a pure end-to-end deep learning object detector. First of all, we instead utilize a technique from the above-mentioned Viola Jones Detectors research, namely "sliding window" (Fig. 2).

Fixed size sliding windows [13], which slide from left-to-right and top-to-bottom to localize objects at different locations, in order to detect the edge of the image for features extraction.

The basic example is explained by the coding and figure below; this technique needs a lot of time to find and label an object.

Mathematics behind is to find the max sum in an array with defined step or stride:

Array: arr[] = {1,2,3,4,5} with size: n = 5, then Step / stride k = 2

$f_{obj} = \max\sum k\ n(\text{arr,n,k})$, Then max sum will calculate based on:

```
for (int i = 0; i < n - k + 1; i++) {int current_sum = 0;
for (int j = 0; j < k; j++) current_sum current_sum + arr[i + j];
    max_sum = max(current_sum, max_sum);}
```

Code:

```
Line 1: image = cv2.imread(args["image"])
Line 2: (winW, winH) = (128, 128)
Line 3: for resized in pyramid(image, scale=1.5):for (x, y, window) in sliding_window(resized, stepSize=32, windowSize=(winW, winH)):
    if window.shape[0] != winH or window.shape[1] != winW:
        continue; clone = resized.copy()
        cv2.rectangle(clone, (x, y), (x + winW, y + winH), (0, 255, 0), 2)
```

Result:

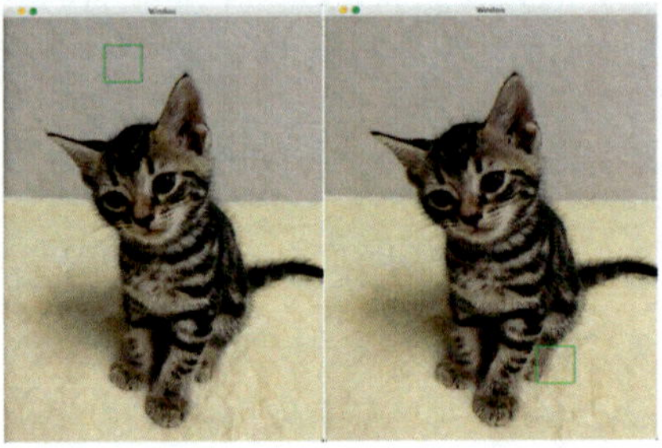

Fig. 2. Sliding window for image classification

Line 1: Given an input image, OpenCV [12] (a library of programming functions targeted at real-time computer vision) is used to read an image.
Line 2: Set the loop and, use the OpenCV library to go through the image from top left to bottom down, and determine the pixel position, which is interested shows a cat's face.

Secondly, The image pyramid (Fig. 3) computed is similar to that featured in HOG detection, which detect objects at varying scales when mixing with the sliding window technique; thus we can scale up or down the image giving more opportunity to detect the object.

The objective of this technique is to reduce noise, as well as to enable image analysis, and enhancement. This can be applied in many real-life applications such as detecting blurred objects; e.g. a car registration plate in heavy snow or rain fall.

The two most popular inside techniques are firstly Gaussian Pyramids [13] where images are weighted down using a Gaussian average such that each pixel containing a local average corresponds to a neighborhood pixel on a lower level of the pyramid. Secondly there are Laplacian Pyramids [13] which implement a multi-scale, multi-orientation band-pass filter bank used for example, in image compression and texture synthesis. The mathematics can be explained as follows:

The input image with the size of $g_0 = (2^n+1) * (2^n+1)$

The process will copy the image to another scale called $g_1 = (2^{n-1}+1) * (2^{n-1}+1)$.

Until $g_i = (2^{n-i}+1) * (2^{n-i}+1)$

This example below is intended to demonstrate how Gaussian Pyramids work in Python and the result in image classification output.

Code:

```
for (i, resized) in enumerate(pyramid_gaussian(image, downscale=2)):
    if resized.shape[0] < 30 or resized.shape[1] < 30:break
    cv2.imshow("Pyramid Layer {}".format(i + 1), resized)
    cv2.waitKey(0)
```

Result:

Fig. 3. Pyramid image classification result

Thirdly, there is simple, traditional classification via Convolutional Neural Networks (CNNs), at each stop of the sliding window + image pyramid We extract the Region of Interest (ROI), feed it into a CNN, and obtain the output classification for the ROI. If the classification probability of label L is higher than some threshold T, we mark the bounding box of the ROI as the label (L). Repeating this process for every stop of the sliding window and image pyramid, we obtain the output object detectors. Finally, we apply non-maxima suppression (NMS) [13] to the bounding boxes yielding our final output detections.

The purpose of the NMS is to remove or filter out the object proposal that greater than the defined threshold. And then give us a final result with correct object detection.

In order to do that we need calculate the Intersection over Union (IoU) [13] of all proposals, and remove each which has an IoU greater than the threshold N.

NMS simply filter and remove the unnecessary bounding box as demonstrated in the code below (Fig. 4).

Code:

```
for(startX, startY, endX, endY) in boundingBoxes:
        cv2.rectangle(orig, (startX, startY), (endX, endY), (0, 0, 255), 2)
        pick = non_max_suppression_slow(boundingBoxes, 0.3)
```

Result:

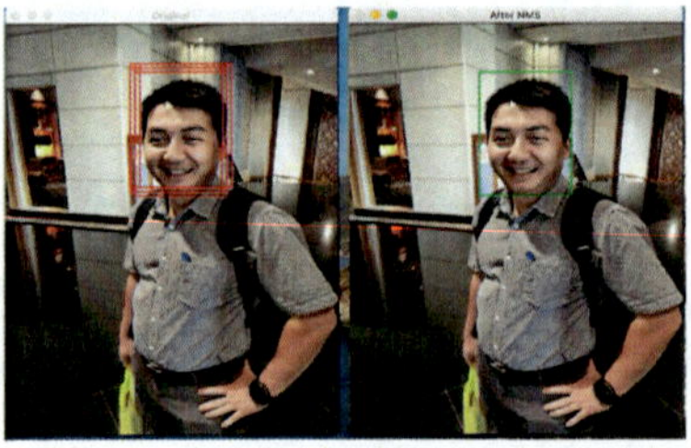

Fig. 4. The result of hand-craft technique using openCV to detect face, using my own picture.

However, this method does not require full deep learning from end to end. For a small data set, this technique will save a lot of time. However, it is not automatic and requires human interaction. We will discuss more on Deep Learning with CNN [14] in the next methodology and the related works.

2.3 Method 2: Based Network of an Object Detection Framework

The second deep-learning method for object detection allows the classification network to be treated as a base network, e.g., in a deep learning object detection framework such as SmallerVGGNet, Faster R-CNN, SSD, or YOLO. The benefit here is that we can create a complete end-to-end deep learning-based object detector.

The Neural Network Layers of a Deep Learning Object Detector for Example of VGG16 [14]

There are many components, sub-components, or even sub-sub-components of deep learning in such object detectors; there are 2 definitions we need to clarify in object detection:

1. The object detection framework such as Fast R-CNN, SSD, YOLO
2. The based network which is fit into object detection framework at a Fig. 7

Based networks are commonly CNN architectures, which include: SmallerVGGNet (This will be demonstrated in later section below), ResNet, MobileNet, DenseNet, ImageNet [13]. CNN is a popular base neural network for Object Detection. The way it works is quite similar to the above-mentioned sliding window. But instead of sum max and expand the window size, the convolution network has a filter to extract features of interest; the filter moves from left to right until it reaches the last, it then continues to shift one row down and left to right until the last row is reached. The following Fig. 5 which illustrates this concept.

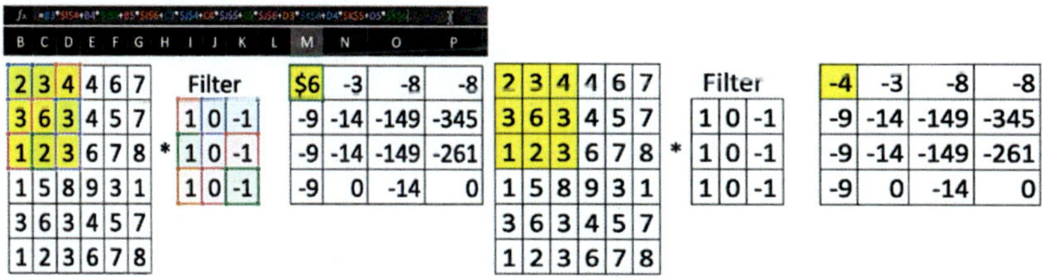

Fig. 5. Left- the input image will be multiple with the filter, after multiple each of cell the result will be on the right.

In conclusion, we can formulate the CNN input and output as follows:

Input has a size n * n multiplied by the filter size m * m; this will give us the output size (n − m + 1) * (n − m + 1). Similar to this, these other CNN architectures will do the same way on filtering the image matrix, just the different on how they connect each layer together and make a reasonable number of scaling and feature extraction.

$$O[n, m] = (I * f)[n, m] = \sum i \sum j\, f[i, j]I[n - j, m - k]$$

O: output matrix
I: input matrix with size n * n
ff: the filter matrix with size m * m
i, j: represent the index of the matrixes, respectively.

3 Proposed Framework

Personal Stuff Detection Model (Fig. 6) with Keras, CNN with SmallerVGGNet
Model Overview:

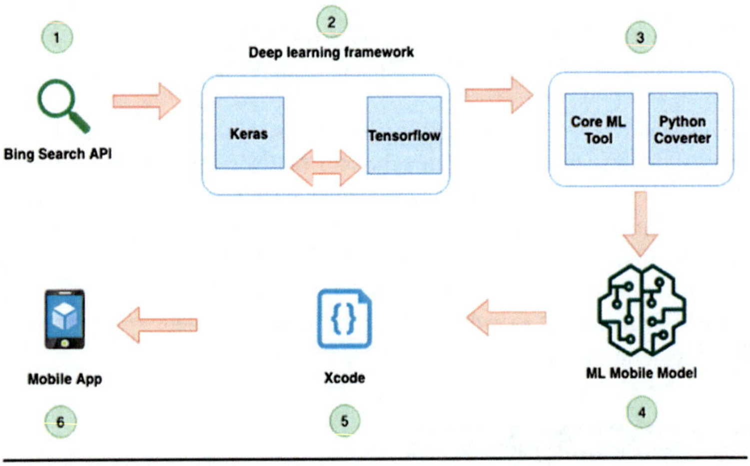

Fig. 6. Experiment Process from scratch to a simple mobile application for experiment

Step 1: using the Bing search library to quickly download the data set, in this experiment we choose to use around 1000 images.

URL ="https://api.cognitive.microsoft.com/bing/v7.0/images/search"
search = requests.get(URL, headers=headers, params=params)

Step 2: design a deep learning framework to train the set of data which was collected in step 1. This paper will compare the base SmallerVGGNet with the one customized from SmallerVGGNet in order to achieve more accuracy.
Step 3: after the training is completed from step 2, the model is generated as expected. However, it cannot be used with another coding language and so must be converted into the Core ML model, which can be used more with mobile applications.
Step 4: After successful conversion, we have a Core ML model
Step 5: Use XCode to make a native mobile application with Swift
Step 6: Run the application on mobile and make a final test with mobile

The scope in this propose framework is to build a pre-trained model which can be able to detect the certain labeled objects. With the improvement in mobile application, it can be able to make deep learning framework to be close to a real time application instead of manual script detection. In order to bring the academic work with more accuracy percentage to an applicable application in several domains.

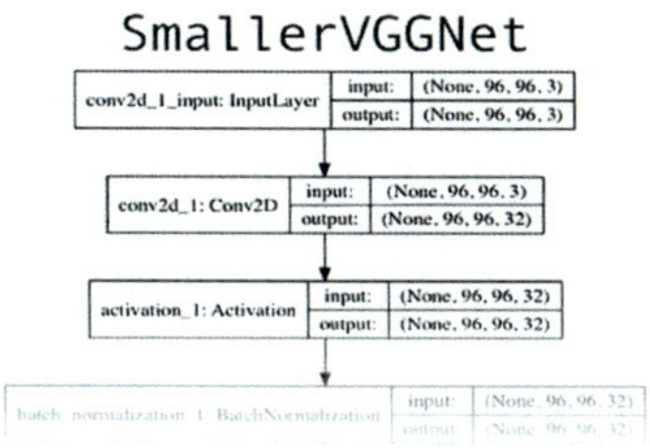

Fig. 7. SmallerVGGNet with image size 96 × 96 × 3

Experiment 1: Deep Learning Architecture with a SmallerVGGNet
Architecture:

SmallerVGGNet in Keras Model with a Result in Fig. 8 and Fig. 9

```
#CON=>RELU=>POOL
model.add(Conv2D(32,(3,3),padding="same",input_shape=inputShape))
model.add(Activation("relu"))
model.add(BatchNormalization(axis=chanDim))
model.add(MaxPooling2D(pool_size=(3,3)))
model.add(Dropout(0.25))
#(CON=>RELU)*2=>POOL
model.add(Conv2D(64,(3,3),padding="same"))
model.add(Activation("relu"))
model.add(BatchNormalization(axis=chanDim))
model.add(Conv2D(64,(3,3),padding="same"))
model.add(Activation("relu"))
model.add(BatchNormalization(axis=chanDim))
model.add(MaxPooling2D(pool_size=(2,2)))
model.add(Dropout(0.25))
#(CON=>RELU)*2=>POOL
model.add(Conv2D(128,(3,3),padding="same"))
model.add(Activation("relu"))
model.add(BatchNormalization(axis=chanDim))
model.add(Conv2D(128,(3,3),padding="same"))
model.add(Activation("relu"))
model.add(BatchNormalization(axis=chanDim))
model.add(MaxPooling2D(pool_size=(2,2)))
model.add(Dropout(0.25))
#FC=>RELUlayers
model.add(Flatten())
model.add(Dense(1024))
model.add(Activation("relu"))
model.add(BatchNormalization())
model.add(Dropout(0.5))
#softmaxclassifier
model.add(Dense(classes))
model.add(Activation("softmax")
```

From the chart above, it can be seen that the experiment reduces the loss in certain points and improve the accuracy to 97%, and with the record of 26 min (average 15s to complete one epoch). A further study will focus on further experiments on custom VGGNet; we can mix and create a new Net architecture to prove the more accuracy and this model can be utilize by another application.

SmallerVGGNet

Epoches	Ratio	Time	loss(%)	accuracy(%)	val_loss(%)	val_accuracy(%)
90	28/28	15s 547ms/step	0.1876	0.9404	0.6243	0.8442
91	28/28	15s 550ms/step	0.1753	0.9426	0.6908	0.8225
92	28/28	15s 529ms/step	0.1566	0.9483	0.8355	0.7835
93	28/28	15s 551ms/step	0.1743	0.9406	0.683	0.8442
94	28/28	15s 548ms/step	0.1705	0.9539	0.9271	0.8095
95	28/28	15s 546ms/step	0.1758	0.9433	1.2772	0.7143
96	28/28	15s 551ms/step	0.1463	0.9483	1.8427	0.7056
97	28/28	15s 548ms/step	0.2106	0.9415	0.8135	0.8052
98	28/28	15s 547ms/step	0.2109	0.9224	0.6511	0.7965
99	28/28	16s 569ms/step	0.1835	0.9404	1.2166	0.7316
100	28/28	16s 562ms/step	0.2161	0.9308	0.6953	0.8312

Fig. 8. Result of 100 epochs

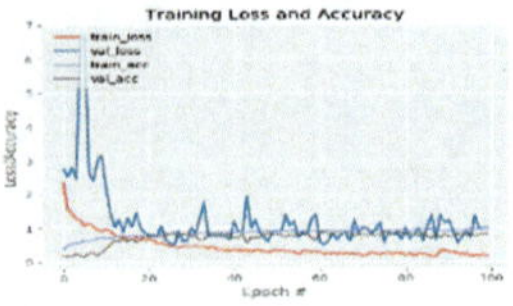

Fig. 9. Result of training

Experiment 2: Deep Learning Architecture with a Custom SmallerVGGNet Architecture, the Result Illustrated in Fig. 10 and Fig. 11

Go through the architecture of SmallerVGGNet, this experiment will add up some layer and technique in order to make the accuracy overpass 95% to achieve 97%.

The improve including:

Step 1: add up two convolution layers with 256 filter.

```
model.add(Conv2D(256, (3, 3), padding="same"))
model.add(Activation("relu"))
model.add(BatchNormalization(axis=chanDim))
model.add(Conv2D(256, (3, 3), padding="same"))
model.add(Activation("relu"))
model.add(BatchNormalization(axis=chanDim))
model.add(MaxPooling2D(pool_size=(2, 2)))
model.add(Dropout(0.25))
```

Step 2: compile the model and make an optimization

```
model.compile(loss='binary_crossentropy', optimizer ='adam', metrics = ['accuracy'])
```

CustomSmallerVGGNet

Epoches#	Ratio	Time	loss(%)	accuracy(%)	val_loss(%)	val_accuracy(%)
90	28/28	28s 1s/step	0.2075	0.9336	0.5868	0.8398
91	28/28	29s 1s/step	0.2226	0.9291	0.5794	0.8312
92	28/28	32s 1s/step	0.2429	0.9291	0.4927	0.8398
93	28/28	29s 1s/step	0.1833	0.9386	0.5864	0.8312
94	28/28	32s 1s/step	0.1199	0.9651	0.6956	0.8009
95	28/28	30s 1s/step	0.1538	0.9399	0.6815	0.8268
96	28/28	27s 977ms/step	0.1383	0.9471	0.7213	0.8355
97	28/28	27s 971ms/step	0.1707	0.9438	0.9608	0.7576
98	28/28	26s 943ms/step	0.2011	0.9393	0.8826	0.7835
99	28/28	27s 956ms/step	0.156	0.9494	0.7211	0.8009
100	28/28	29s 1s/step	0.099	0.971	0.8608	0.7879

Fig. 10. Custom model in 100 epochs

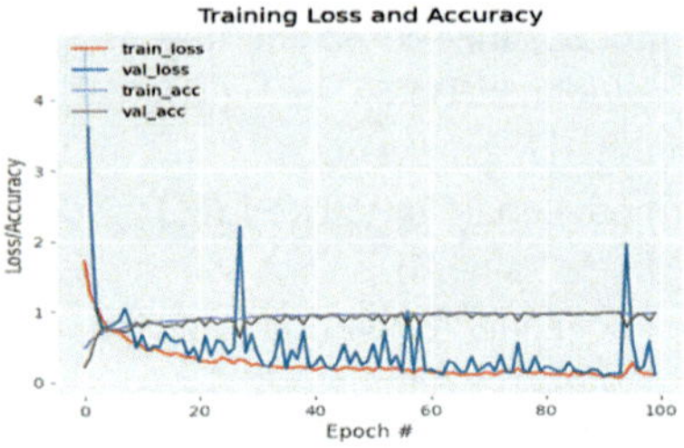

Fig. 11. Result of training

It can be seen that with same number of random seed ratio, the custom model takes a little bit more time to complete in each epoch, however it provides more accuracy in the certain pattern. In case of adding more layer, we found that the time consuming will be double in each step and not provide any increments in accuracy percentage.

For future production, with this pattern in GPU processing, the speed will not be the impact in comparison with the standard one. At this point, the pattern is optimized and cannot add more convolution layer or filter, otherwise it makes the training processing is more time consuming and no more additional point for accuracy improvement.

Practical Result with custom SmallerVGGNet can be shown in the Fig. 12.

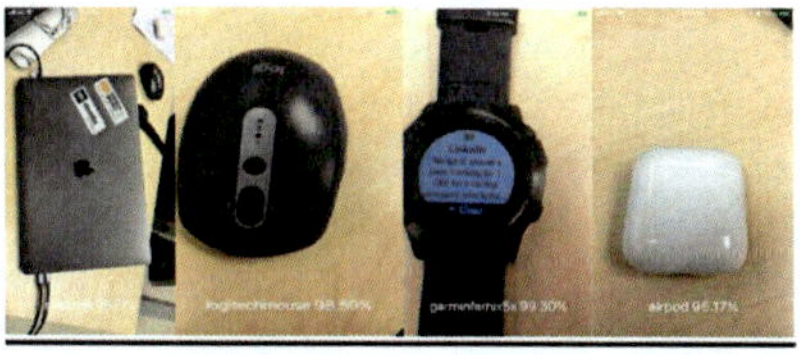

Fig. 12. Model works quite accuracy more than 95% with some objects reach to 99%

4 Conclusion

Image Classification and Object detection with a traditional methodology is a good base of knowledge for practitioners, however with the huge amounts of data which generate by the internet and social networks, it may prove difficult to run the program and modify the function every time we have to detect some certain things. It can be seen that the accuracy may improve with different techniques, and we may incur loss of speed (20s for one epoch), whilst gaining more accuracy. Our future work will focus on a single stage of detection and merge these two stages together to improve the speed and maintains the same accuracy at around 97%. This paper is focusing on the accuracy on top of SmallerVGGNet model, it is technically more flexible for researcher to find a custom dataset which are not belong to the source based such as Kaggle like another research paper. The model is adding up with 6 convolution and 2 pooling layers. The dataset is custom design and achieve from the Bing search cognitive API, which contain more 1100 pictures in all format and size. Then the process is auto resizing to 96 × 96 pixels with no manually retouching. Object detection with deep learning and CNN may be a new solution and a guideline for object detection researchers. There are a lot of work in the future that related to object detection, and we need to improve more on the performance and light-weight net layer with more accuracy.

References

1. Viola, P., Johnes, M.J., Snow, D.: Detecting pedestrians using patterns of motion and appearance. In: 9th IEEE International Conference on Computer Vision (ICCV 2003), Nice, France, 14–17 October 2003
2. Dalal, N., Triggs, B.: Histograms of oriented gradients for human detection. In: IEEE Computer Society Conference on Computer Vision and Pattern Recognition (CVPR 2005) (2005)

3. Girshick, R.B., Felzenszwalb, P.F., McAllester, D.: Object detection with grammar models. In: NIPS 2011: Proceedings of the 24th International Conference on Neural Information Processing Systems, December 2011, pp. 442–450 (2011)
4. Chen, C., Liu, M.-Y., Tuzel, O., Xiao, J.: R-CNN for small object detection. In: Lai, S.-H., Lepetit, V., Nishino, Ko., Sato, Y. (eds.) ACCV 2016. LNCS, vol. 10115, pp. 214–230. Springer, Cham (2017). https://doi.org/10.1007/978-3-319-54193-8_14
5. Girshick, R.: Fast R-CNN. In: IEEE International Conference on Computer Vision (ICCV) (2015)
6. Ren, S., He, K., Girshick, R., Sun, J.: Faster R-CNN: towards real-time object detection with region proposal networks. IEEE Trans. Pattern Anal. Mach. Intell. **39**, 1137–1149 (2017)
7. Du, J.: Understanding of object detection based on CNN family and YOLO. J. Phys. Conf. Ser. **1004**, 012029 (2018). 2nd International Conference on Machine Vision and Information Technology (CMVIT 2018) 23–25 February 2018, Hong Kong
8. Zhang, S., Wen, L., Bian, X., Lei, Z., Li, S.Z.: Single-shot refinement neural network for object detection. In: Proceedings of the IEEE Conference on Computer Vision and Pattern Recognition (CVPR), pp. 4203–4212 (2018)
9. Zhang, H., Chang, H., Ma, B., Shan, S., Chen, X.: Cascade RetinaNet: maintaining consistency for single-stage object detection. In: BMVC (2019)
10. Liu, W., et al.: SSD: single shot MultiBox detector. In: Leibe, B., Matas, J., Sebe, N., Welling, M. (eds.) ECCV 2016. LNCS, vol. 9905, pp. 21–37. Springer, Cham (2016). https://doi.org/10.1007/978-3-319-46448-0_2
11. Yu, F., Seff, A., Zhang, Y., Song, S., Funkhouser, T., Xiao, J.: LSUN: construction of a large-scale image dataset using deep learning with humans in the loop. arXiv:abs/1506.03365 (2015)
12. Chandan, G., Jain, A., Jain, H., Mohana: Real time object detection and tracking using deep learning and OpenCV. In: 2018 International Conference on Inventive Research in Computing Applications (ICIRCA), Coimbatore, pp. 1305–1308 (2018). https://doi.org/10.1109/ICIRCA.2018.8597266
13. Zou, Z., Shi, Z., Guo, Y., Ye, J.: Object detection in 20 years: a survey. arXiv:1905.05055 (2019)
14. Qassim, H., Verma, A., Feinzimer, D.: Compressed residual-VGG16 CNN model for big data places image recognition. In: 2018 IEEE 8th Annual Computing and Communication Workshop and Conference (CCWC), Las Vegas, NV, pp. 169–175 (2018). https://doi.org/10.1109/CCWC.2018.8301729

HFMD Skin Rash Detection Using Convolutional Neural Networks

Naqibullah Vakili(✉), Nipat Phattarakijtham, Jonathan H. Chan, and Worarat Krathu

School of Information Technology, King Mongkut's University of Technology Thonburi, Bangkok 10140, Thailand
{naqibullah.vakili,nipat.phattarakijtham}@mail.kmutt.ac.th, {jonathan,worarat.kra}@sit.kmutt.ac.th

Abstract. Skin rash problems are common and often temporary if they can be treated in time. Hand-Foot-Mouth disease (HFMD) is an extremely contagious viral infection common in infants, and can quickly develop into a severe problem. The infection can spread quickly through close contact with an infected person. HFMD usually involves the hands, feet and mouth, but sometimes can be detected elsewhere. To assist in diagnosing HFMD, we built a dataset of various skin conditions from readily available web sources. We propose the use of a deep convolutional neural networks (CNN) to distinguish HFMD rash from other skin conditions and normal skin. This study presents a promising application of CNN with a precision rate of 0.92%, 0.96%, and 0.88 for HFMD, non-HFMD, and normal skin, respectively. Our proposed model can facilitate a proper and early detection of HFMD rash to assist in containing HFMD outbreaks regionally.

Keywords: Hand, Foot and Mouth Disease · HFMD · Multiclass Classification · Deep Learning · Convolutional Neural Network

1 Introduction

The Hand-Foot-Mouth Disease (HFMD) is among the oldest recorded viruses that infect humans. It was first reported in New Zealand in 1957 and has since become considerably frequent worldwide. Mainly caused by Enterovirus 71 EV71 infection, HFMD is a common contagious disease that immediately develops into a severe problem, mainly affecting infants, preschool children, and sometimes adults [1]. Children under the age of 5 years are at greater risk of disease infection because they are less likely to have the necessary antibodies to protect them against viruses causing HFMD infection [2]. Recently, the outbreak of HFMD has increased and over a million cases have been reported, posing a severe threat to public health, mainly in the Asia Pacific region, including Malaysia, Thailand, Singapore, Japan, Australia, and across China [1, 3, 4]. HFMD is transmitted through close personal contact, virus particles being present in mucus and saliva in the air from coughing and in an infected person's feces. In schools, an individual class can be closed for many days for clean-up. This may lead to an economic burden, especially in low-income communities.

P. Meesad et al. (Eds.): IC2IT 2021, LNNS 251, pp. 159–168, 2021.
https://doi.org/10.1007/978-3-030-79757-7_16

Three inactivated monovalent EV-A71 vaccines have been licensed in China. The vaccine effectiveness for full vaccination was estimated to be 91.1% in non-severe cases and 73.3% in severe HFMD disease cases. However, HFMD cases have increased in most countries in the Asia Pacific region [5]. This has prompted concerns that, without intervention, the disease's public health impact and spread will continue to intensify [6]. With the illness continuing to spread more in Asian countries, efficient and timely diagnosis of HFMD will be vital in mitigating the infection rate.

The most defining symptoms of HFMD is the skin rash that it causes, as the other signs closely mimic other illnesses. The distinctive pattern of inflammation and rash progression across the body, usually the hands, feet, and mouth, are critical signs that healthcare providers use to visually diagnose the disease. Without immediate medical attention, the disease may be fatal for infants in the short term. Recognizing HFMD disease at the initial stage of the illness will increase the chance of getting treatment and mitigate fatal consequences [11].

Several works have used data mining and machine learning in automatic HFMD disease diagnostics, identifying the HFMD virus from clinical data [2, 4]. Aryan et al. produced the only published paper on classifying HFMD from digital images [7]. They classified HFMD rash from normal skin without any inflammation using image processing and image recognition techniques.

This study aims to leverage deep convolutional neural networks (CNNs) to classify and distinguish HFMD from other skin conditions and normal skin, assisting the medical practitioner and patient alike in detecting and possibly diagnosing the infection. Since visual data is considered the primary source for diagnostics of any skin rash, a system capable of HFMD virus characterization can be implemented in many fields. It could be deployed as an application to be used in schools, airport security, and telemedicine to prevent transmission of the disease, or potentially as a phone application in developing nations, where healthcare assistants are rare, to aid the recognition of HFMD.

The rest of the paper is organized as follows. In the first section, we discuss related work. Our dataset, CNN, and experimental setup are presented in Sect. 3, followed by results and an evaluation of misclassifications. Sections 4 and 5 present conclusions and propose future work.

2 Related Work

Several researchers have used various skin rash detection methods to identify skin diseases. Here we briefly review some recent techniques that have used transfer-learning for the same purpose. Convolutional Neural Networks are computer models inspired by the biological neurons present in our brain. These models have been proven to be very useful, accurate, and reliable in many tasks. They have already obtained near-human performance in several challenging natural image stratification and computer vision tasks. Furthermore, CNNs have recently gained enormous attention and have been increasingly implemented in medical study and radiology [8–10].

Toward automated skin disease detection, Nasr-Esfahani et al. [8] employed a CNN model to identify melanoma with a predictive accuracy of 81%. Recently, to diagnose the most common skin conditions seen in health care, a CNN-based model has been developed by Google, with results "on par with U.S board-certified dermatologist" [10, 11]. CNNs have been applied to other diseases as well, including diabetic retinopathy, lymph node metastasis, and breast cancer [12–14].

A CNN needs a large dataset for model training. However, it is often a challenge to acquire adequate labeled medical images due to the expertise required and high labor intensity for dataset curation. If only a small image dataset is possible, superior strategies, e.g., data augmentation or transfer learning, are regularly applied to improve classification performance. Transfer learning uses a pre-trained classifier to work on a different but related problem [11, 12].

Compared to creating a CNN from scratch, transfer learning needs significantly less training data and reduced runtime. Recently, comparative research on a small dataset has shown that transfer learning is more helpful than other techniques investigated [9, 11]. In another study, Kermany et al. used transfer learning to implement a CNN classifier with performance comparable to that of human experts in distinguishing age-related macular degeneration and diabetic macular edema [10]. Following their steps, Haenssle et al. [15] the winner of the ISBI 2016 challenge, developed an Inception v4 model and tested it against 58 dermatologists after assessing the binary classification performance on two test datasets of only 100 and 300 images. Although the specificity and sensitivity of the proposed model were higher than the dermatologists' mean performance, the model's performance on the publicly available International Symposium on Biomedical Imaging 2016 Challenge test dataset was below that of the challenge's two winning entries. Moreover, several CNN-based methods were developed in response to the need to classify skin images for timely treatment and diagnosis of diseases with a shortage of dermatologists.

Thus, with a lack of tools for visual identification of HFMD and an increasing number cases of HFMD in the Asia Pacific region countries associated with contact-related transmissions and spread through populations, models that can accurately recognize HFMD rash are urgently required. Improved HFMD rash identification ability would help primary care providers efficiently address various new challenges posed by the reemergence of HFMD into communities in the Asia Pacific region [16].

Moreover, as mentioned earlier in the first section, most of the research in the field of HFMD used data mining and machine learning to detect and classify the disease from clinical data. Liu et al. [4] used a Random Forest model on 16 variables with structured and unstructured clinical data to predict HFMD.

The literature review found that such an expert method is needed to distinguish HFMD rash from other skin conditions and normal skin.

3 Material and Architectures

3.1 Data Collection

As there are no public resources available containing extensive archives of HFMD images, we created the dataset for the proposed methodology from public sources. We used the Google Images service to parse images from the web, and the following queries were performed: "HFMD rash", "HFMD inflammation", normal skin and other diseases. The dataset contains rash images from 13 Non-HFMD rash condition states, including

Measles, Scabies, Eczema, Psoriasis, Erythema, Impetigo, Utricaria, Atopic Dermatitis, Lupus, Ringworm, Bowen's Disease, Keratosis, and Dermatofibroma, which are grouped as non-HFMD skin rashes, in addition to HFMD skin rashes. Images of normal skin were also added to the dataset. Table 1 presents the distribution of the dataset. In total, there are 1,479 images of Non-HFMD rashes, 366 HFMD images, and 347 normal images present in the dataset. Our dataset's images were classified into three classes: HFMD, Non-HFMD, and Normal skin.

Table 1. Distribution of image samples.

Disease	Number of images	Disease	Number of images
Measles	112	Scabies	211
Eczema	238	Psoriasis	115
Erythema	103	Impetigo	86
Utricaria	77	Atopic Dermatitis	109
Lupus	68	Ringworm	125
Bowen's Disease	109	Keratosis	96
Dermatofibroma	65	HFMD	366
Normal	347	**Total**	**2,192**

For the non-HFMD section, it is necessary to have images similar in appearance to HFMD as well as images that significantly differ in order to provide overall diversity to the dataset [11].

Due to the shortage of data, it is hard to obtain a high-quality model by training it from scratch. To overcome this challenge, we used the transfer learning method. To mitigate the shortage of proper HFMD images in the dataset for training and testing, we applied image augmentation techniques such as minor adjustments, horizontal flips, and random rotations to create HFMD images' duplicates. The evaluation showed that our final classifier, adapted from a pre-trained model, presented better performance without augmentation.

3.2 Preprocessing

Since we had no control over the images and camera resectioning in acquisition, preprocessing was performed. As the images were collected from different sources, the dimensions of the images varied. To use the ResNet classifier, the images were resized into $224 \times 224 \times 3$ because of ResNet requirements. The images were normalized by subtracting the mean RGB value of the ImageNet dataset, as suggested in [17], and the pre-trained CNN was optimized on the images.

3.3 General Architecture of Convolutional Neural Networks

Initially, the Convolutional Neural Network is a set of stacked layers, including linear and nonlinear processes. These layers learn cooperatively. Typically, a CNN model consists of the convolutional-layer, Max-pooling layer, followed by a standard multilayer neural network, namely a fully-connected layer and a loss layer at the backend.

ResNet-50 and some other deep ResNet classifiers were first proposed in 2015 by Microsoft Asia and demonstrated outstanding performances in many fields. ResNet models won ImageNet ILSVRC (Large Scale Visual Recognition Challenge) 2015 competitions in image detection, localization, and classification. Also, ResNet give state-of-the-art performance on ImageNet localization, detection, and other tasks [9]. The ResNet model consists of 5 steps, where each step contains Convolutional (Conv) layer, Rectified-Linear-Unit (ReLU), and Normalization (Batch Norm) layers. Step 1 consists of the convolution layer and max pooling. A convolutional layer applies a filter called a kernel, which passes over an image to build a feature map. Initially, kernel size is 7 × 7 with 64 output channels and a stride of 2. Max pooling reduces image dimensionality by retaining the maximum value in each patch of a feature map, which the kernel carries over. Steps 2–5 are residual blocks. Residual blocks look like main road networks without gates in their skip connections to permit data flow from the first layers to the final layers. Following step 5, there is an average-pooling and a fully-connected layer—average pooling reduces image dimensionality by taking each matrix square's average value. The fully connected layer takes the result of the convolution/pooling and outputs the image classification's final probabilities [9, 11, 18].

3.4 Classification

The latest version of Fastai and PyTorch were used to perform deep learning classification. Originally, the input images were preprocessed, then fed to a ResNet-50 model. Finally, classification was performed using the CNN classifier. For each training step, the dataset was divided randomly into training, validation and test sets with a 60:20:20% split ratio, respectively, where 60% of the images were used for training the model, 20% were used to validate the model and 20% were used to test the model. A stratified 5-fold cross-validation was applied to train, evaluate and test the model. Hyperparameters are represented in Table 2. These parameters were chosen empirically according to a series of experiments conducted on the whole dataset, which present the best results of classification.

Table 2. Hyperparameter choices.

Parameter	Optimization algorithm	Learning rate	Momentum	Batch size	Activation function	Iterations
Value	SGD	0.001	0.9	8	linear	15

As presented in Table 2, the stochastic gradient descent algorithm was applied in our classifier to learn the best set of weights and neural network biases that decrease the loss function. While learning, the stochastic gradient descent algorithm operates by randomly picking a small number of training inputs—the batch size was set to 8. The learning rate, which describes the rate at which a function passes over the search space, was set to 0.001. A small learning rate leads to more accurate results, but it needs more training time. The momentum is an additional factor to define how fast the stochastic gradient descent algorithm converges on the optimum point. It was set to 0.9. Based on the training experience and balancing the performance with required processing time, the number of iterations, or epochs, was set to 15.

3.5 Metrics

To evaluate the model, accuracy, precision, recall, and F1-score were used to define the performance of the proposed model. These are described through the following Eqs. (1), (2), (3), and (4).

Accuracy is the proportion of predictions our system got correct [19].

$$Accuracy = \frac{(TP + TN)}{(TP + TN + FP + FN)} \tag{1}$$

Precision is the proportion of positive classifications that were correct [19].

$$Precision = \frac{TP}{TP + FP} \tag{2}$$

Recall is the proportion of actual positives that were recognize correctly [19].

$$Recall = \frac{TP}{TP + FN} \tag{3}$$

F1-score is a consolidation of the precision and the recall metrics to summarize the model's predictive performance [19].

$$F1 - Score = \frac{2 \times Precision \times Recall}{Precision + Recall} \tag{4}$$

where: TP = True positive; FP = False positive; TN = True negative; FN = False negative classifications, respectively.

3.6 Results

To evaluate the proposed model's effectiveness, we used a combination of accuracy, precision, recall, and F1-score. Results across all our experimental configurations are presented in Table 3. Our deep learning model identifies three image classes (Healthy skin, HFMD, and Non-HFMD). As presented in Table 3, our model was able to achieve good results when implemented to classify the diseases of HFMD from other skin conditions and healthy skin.

Table 3. Multiclass testing performance metrics.

Classes	Precision	Recall	F1-score
HFMD	0.92	0.89	0.91
Non-HFMD	0.96	0.96	0.96
Normal	0.88	0.92	0.90

Our model was able to achieve higher accuracy on binary classes (HFMD and Non-HFMD), with an accuracy of 0.956 on the validation dataset. Testing performance measurement Sensitivity, Specificity, and Accuracy for binary class are presented in Table 4. More details about the results of binary classification were reported in our earlier work [20].

Table 4. Binary class testing performance metrics.

Images classified	Accuracy	Sensitivity	Specificity
	0.951	0.813	0.971

3.7 Evaluation of Images that Were Classified Incorrectly

To verify our results, we manually analyzed some misclassified images. Figure 1a presents an example of a false positive prediction, whereas Fig. 1b an example of a false negative sample. The left photo shown in Fig. 1a is a mouth measles rash, misclassified by the model as a HFMD rash. This example presents the difficulty of distinguishing the HFMD rash from other skin diseases. Figure 1b on the right presents sample images of HFMD that were misclassified as normal skin. As most of the false positive and false negative identifications were made on images comparable in appearance to the HFMD rash such as scabies and measles, training images of both the HFMD infection and these similar subgroups are needed to reduce this problem.

To further illustrate the difficulty of classification, we present a couple of sample images that were correctly classified by the Convolutional Neural Networks (see Fig. 1c, Fig. 1d). In contrast to Fig. 1a and 1b, a correctly identified HFMD image of the mouth is shown in Fig. 1c and a non-HFMD case of the hand is shown in Fig. 1d. The HFMD cases primarily involve the hand (Fig. 1b and 1d), the foot, or the mouth (Fig. 1a and 1c). However, the rashes can appear in other parts of the body as well, such as on the leg, as shown in Fig. 1e and Fig. 1f. As presented in these images though, these HFMD cases of the leg were misclassified as scabies and measles, respectively. In order to rectify such problems, more skin rash data will need to be collected.

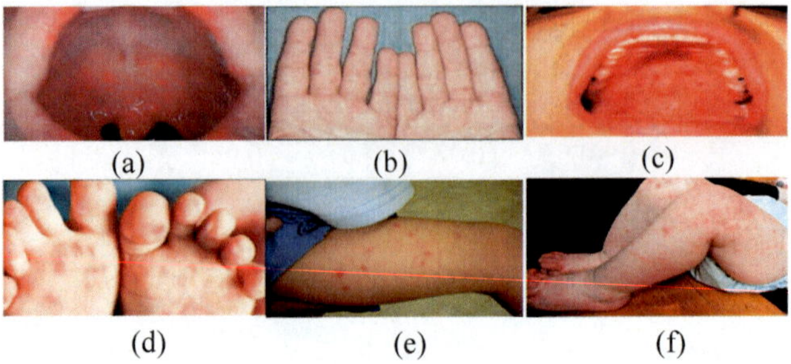

Fig. 1. Sample images of false-positive prediction (a), false negative prediction (b), correct HFMD classification (c), correct Non-HFMD classification (d), false-negative prediction (e), false-negative prediction (f).

4 Conclusions

This research proves that it is possible to solve the challenge of HFMD rash image detection using deep learning. With the spread of HFMD continuing to increase in Asia Pacific countries, an adequately trained model capable of detecting HFMD rashes is vital in combatting the outbreak. Our research shows that deep learning solutions perform on par and even better on the task than classifiers to identify the color, texture, and shape features of HFMD [7], even provided a minimal number of labeled images. Moreover, an advantage of our proposed method is that it does not need manual pre-processing of images. According to the result, the ResNet50 model proved to be accurate and efficient in detecting HFMD disease from other skin conditions and normal skin. Our proposed model, achieved an average accuracy of 0.951 and an average sensitivity of 0.813 on test datasets for binary classes. Also, our model achieved higher accuracy for classifying three classes (HFMD, Non-HFMD, and Normal skin) with the precision of 0.92, 0.96, 0.88, respectively, on the test dataset.

We can improve performance with the addition of more images. Moreover, our classifier only focuses on the rash's appearance and structure and does not consider the distribution of the rash or part of the body.

5 Future Work

Further research should include information about different simultaneous symptoms such fever, poor appetite, feeling unwell and painful sores, which were not included in this model due to the lack of these data. When available in the future, such information can be easily combined into our model to further improve diagnosis efficiency. We have various ages, genders, and body parts across images but do not have a wide variety of skin colors. Another possible future development of this research is collecting more ethnically diverse images for model retraining. By combining more images from various racial groups, our prototype model can be further improved. More importantly, it can be more readily implemented in telemedicine to assist health professionals and primary care providers in diagnosis.

As mentioned earlier, this study's ultimate goal is to deploy our model as a phone application. Nowadays, 45.4% of the world population owns a smartphone. For the growing number of newly qualified doctors who use smartphones in the field, a phone-based application could serve as a tool in the diagnosis of the HFMD virus.

Another possible direction of additional research is applying a multi-task system to conduct a hierarchical diagnosis of HFMD disease and provide visual information to describe the diagnosis, address the limitations, and improve the explainability of an HFMD disease diagnostic system.

Further research can also train a model for a particular body part to identify HFMD and Non-HFMD skin conditions specifically in the hand, foot, or mouth. It may be beneficial to have multiples such models to run together, which will produce more reliable results.

Acknowledgments. The authors would like to thank Nicola Lawford for proofreading and providing technical feedback on this work.

References

1. Aswathyraj, S., Arunkumar, G., Alidjinou, E.K., Hober, D.: Hand foot and mouth disease (HFMD): emerging epidemiology and the need for a vaccine strategy. Med. Microbiol. Immunol. **205**(5), 397–407 (2016)
2. Wang, J.F., et al.: Hand, foot and mouth disease: spatiotemporal transmission and climate. Int. J. Health Geogr. **5**, 10–25 (2011). https://doi.org/10.1186/1476-072X-10-25. PMID: 21466689; PMCID: PMC3079592
3. Zahao, J., Wang, Y.: The transmission of Hand, Foot, and Mouth Disease in East and Southeast Asia. International Journal of Health Geographic. BioRxiv, p. 612580 (2019)
4. Liu, G., et al.: Developing a machine learning system for identification of severe hand, foot, and mouth disease from electronic medical record data. Sci. Rep. **7**, 16341 (2017). https://doi.org/10.1038/s41598-017-16521-z
5. World health organization. https://www.who.int/biologicals/vaccines/enterovirus_71/en/. Accessed 24 Jan 2021
6. WHO. https://iris.wpro.who.int/bitstream/handle/10665.1/5521/9789290615255_eng.pdf. Accessed 24 Jan 2021
7. Aryan, M.F., Krathu, W., Arpnikanondt, C., Tassaneetrithep, B.: Image recognition for detecting hand foot and mouth disease. In: Proceedings of the 11th International Conference on Advances in Information Technology (IAIT2020), Bangkok, Thailand (2020)
8. Nasr-Esfahani, E., et al.: Melanoma detection by analysis of clinical images using convolutional neural network. In: 38th Annual International Conference of the IEEE Engineering in Medicine and Biology Society (EMBC), pp. 1373–1376. IEEE, Orlando, FL (2016)
9. Yadav, S.S., Jadhav, S.M.: Deep convolutional neural network based medical image classification for disease diagnosis. J. Big Data **6**(1), 1–18 (2019). https://doi.org/10.1186/s40537-019-0276-2
10. Kermany, S.D., et al.: Identifying medical diagnoses and treatable diseases by image-based deep learning. Cell **172**(5), 1122–1131 (2018)
11. Glock, K., Napier, C., Louie, A., Gary, T., Gigante, J., Wang, Q.: Measles Rash Image Detection Using Deep Convolutional Neural Network. ArXiv: 2005.09112 [eess.IV] (2020)

12. Mobiny, A., Singh, A., Nguyen, H.: Risk-aware machine learning classifier for skin lesion diagnosis. J. Clin. Med. **8**(8), 1241 (2019). https://doi.org/10.3390/jcm8081241. PMID: 31426482; PMCID: PMC6723257
13. Burlina, P.M., Joshi, N.J., Ng, E., Billings, S.D., Rebman, A.W., Aucott, J.N.: Automated detection of erythema migrans and other confounding skin lesions via deep learning. Comput. Biol. Med. **105**, 151–156 (2018)
14. Mishara, S., Chaudhary, S., Imaizumi, H., Yamasaki, T.: Assessing Robustness of Deep learning Methods in Dermatological Workflow. ArXiv: 2001.05878v2 (2020)
15. Haenssle, H.A., et al.: Man against machine: diagnostic performance of a deep learning convolutional neural network for dermoscopic melanoma recognition in comparison to 58 dermatologists. Ann. Oncol. **29**(8), 1836–1842 (2018)
16. Esteva, A., et al.: Dermatologist-level classification of skin cancer with deep neural networks. Nature **542**, 115–118 (2017)
17. Krizhevsky, A., Sutskever, I., Hinton, G.E.: Imagenet classification with deep convolutional neural networks. In: Proceedings of the 25th International Conference on Neural Information Processing system- Volume 1 (NIPS 2012), pp. 1097–1105. Curran Associates Inc., Red Hook, NY, USA (2012)
18. Sharma, S.: DermaDetect: A computer vision and deep learning approach for an accurate diagnosis of skin conditions and rashes (2019). https://doi.org/10.13140/RG.2.2.11636.91522
19. Hasan, M., Barman, S.D., Islam, S., Reza, A.W.: Skin cancer detection using convolutional neural network. In: Proceedings of the 2019 5th International Conference on Computing and Artificial Intelligence, pp. 254–258. ACM (2019)
20. Vakili, N., Phattarakijtham, N., Chan, J.H., Krathu, W.: Hand foot and mouth rash detection using deep convolution neural network. In: 9th Joint Symposium on Computational Intelligence (JSCI9), Bangkok, Thailand (2020)

Comparative Study Considering Garbage Classification Using In-Depth Learning Techniques

Nittaya Muangnak[1(✉)], Chaiwat Sirawattananon[1], Meet Ganpatlal Oza[2], and Rathanit Sukthanapirat[1]

[1] Faculty of Science and Engineering, Kasetsart University, Sakon Nakhon 47000, Thailand
nittaya.mu@ku.th

[2] Department of Computer and Communication Engineering, Manipal University Jaipur, Jaipur, Rajasthan 303007, India

Abstract. As population grows, there is an exponential increase in garbage generated. This garbage contains a high percentage of plastic that is recyclable. Therefore, it is necessary to segregate different types of garbage. To minimize the impact of incorrect garbage separation, we propose using an automated system to correct the waste classification. We studied in-depth learning techniques by using comparison-based experiments. A ResNet-152 and ResNet-50 were investigated to obtain the best-fit classification model for garbage classification. The model was trained on two datasets, TrashNet and local datasets which contains 5,326 images of four garbage categories (metal, pet, plastic, and trash). The accuracies for the pre-training hyperparameters with classification time on average were 98.99% with 0.27 s and 98.81% with 0.72 s for ResNet-152 and ResNet-50, respectively. The experiment testing results on both ResNet-152 and ResNet-50 yielded outperformed accuracy respectively by 95.72% and 94.15%. Using computer vision and IoT-based system, we implemented the selected model having more steady learning rate, ResNet-50, on the IoT devices to simulate the automated garbage segregation machine. The Raspberry Pi camera connected on a microcontroller captures one by one image of the garbage when the motion sensor triggers it. The captured image is then sent to the classification model by returning a predicted garbage class displayed on the output screen. Our future works based on the output generating a garbage category. The final classification model would be integrated with a fully automated sorting machine by automatically moving to its respective bin using a motorized sliding tray.

Keywords: Garbage classification · Waste sorting system · Image-based classification · Machine learning · Deep learning

1 Introduction

Waste management is one of the main issues worldwide. In Thailand, such problems have a broader environmental impact. According to a survey of the amount of waste in

P. Meesad et al. (Eds.): IC2IT 2021, LNNS 251, pp. 169–180, 2021.
https://doi.org/10.1007/978-3-030-79757-7_17

2016 [1], Thailand has a total amount of solid waste approximately 28.71 million tons per year. It was recyclable about 44% of the total amount of solid waste. The segregation of recyclable garbage is vital to minimize the impact of improper waste disposal (e.g., plastic, glass, metal and general) and collaborate the long-term waste management.

Consequently, waste separation should be promoted, especially recyclable waste which is the majority type of garbage. There existing works introducing the automated garbage classification using artificial intelligence and computer vision. The application of supervised-based learning approaches to constructing the classification model based on computer vision and deep learning are investigated.

Existing novels in image-based garbage classification techniques using supervised-based learning method were described in the following articles. Gundupalli et al. 2017 [2] reviewed on automatic sorting and recycling process of source-separated municipal solid waste (MSW) techniques. Spectral images were applied by classification algorithm to differentiate types of source-separated solid waste into metal, plastic, paper, glass, and wood. The waste management system has been polished by Abdallah et al. [3] to show the comparative studies of various AI applications in solid waste management (SWM). The literature mainly focused on efficiency, software platform, pros/cons of AI models considered ANN, SVM, LR, DT, GA, and hybrid. The limitation could be listed as follows: lack of data, the numerous AI models and their rapid evolution distraction, a key issue in the black-box nature of AI models, and most of the studies were direct applied AI models to solve particular waste problems. Hao-nan Guo et al. [4] studied machine learning-based classification techniques for municipal solid waste management. The comparative methods were summarized grouped by applications where the highest accuracy was 99.95% (CNN + SVM). Organic solid and recyclable waste management based on these following machine learning methods yielded a high classification accuracy. However, there would be limitations of the image dataset's quality and features, insufficient data, and ill-defined parameter tuning of the models. Moreover, Flores and Tan [5] studied and represented the object detection algorithm based on digital image processing to review the strength and weaknesses of existing supervised machine learning algorithms.

The novel model so-called RecycleNet was introduced by Bircanoglu et al. [6]. The neural-based network connection layers implemented using Keras Tensorflow were skipped inside the dense blocks through the comparison-based methods: the deep residual network (ResNet-50), MobileNets, Inception-v4, DenseNets, and Xception. The number of parameters in the 121 layered networks was reduced to improve the prediction time faster. The downsized sampling 512 × 384 pixels of 2,527 TrashNet images were pre-trained using the convolutional model with the unbalanced splitting dataset. The model provided six type-prediction of recyclable waste: paper, glass, plastic, metal, cardboard, and trash. The model was deployed to implement the smart bin application enabling the autonomous detecting and sorting recyclable waste. Even though the waste sorting model so-called RecyClick was investigated by Acharekar [7], the idea of recycling at a click was involved in providing an application program interface of segregation application. The application allowed users to upload waste images to classify the images into three classes: e-waste, metal, and plastic, through the sorting model. The model was pre-trained using augmented pre-processed image dataset (data augmentation: 268 × 188 shifted different orientations). The proposal approaches were object detection based on machine learning techniques with the comparison-based method FAST R-CNN, SPP-NET, and R-CNN. The highest performance model taking 10 h on training and 2 s on predicting was FAST R-CNN. The intelligent waste sorting system was also mentioned in Adedeji [8] to categorize solid waste into glass, metal, paper, and plastic. Deep learning convolutional neural network was adopted to pre-trained the classification model using image dataset 224 × 224 pixels. ResNet-50, CNN, SVM, ML, and SGDM models were determined with selected high accuracy 87% setting 12 epochs of parameter fine-tuning. Aral [9] has improved the higher accuracy up to 95% from applying different sorting garbage models against Adam and Adadelta optimizer. This study also worked on TrashNet image dataset (1,984 images in total). Cross-validation on pre-trained with 70:17:13 of training, testing, and validation dataset were investigated DenseNet121, DenseNet169, InceptionResnetV2, MobileNet, and Xception models. The most outperform model was DenseNet121 with 95% accuracy and InceptionResNetV2 with 94% accuracy by Adam and Adadelta optimizer. However, as mentioned earlier, some of the approaches partially provided garbage classification by costly conditions on automatic sorting approaches. We then introduce an affordable approach to obtain an outperformed automatic garbage classification reminding.

Therefore, garbage classification approaches are proposed by examining in-depth learning algorithms in comparison. A ResNet-152 and ResNet-50 were investigated to obtain the best-fit classification model for garbage classification. To cover both universal and local consumers, the model was trained on two datasets, TrashNet and self-collected image datasets which contains four garbage categories: metal, pet, plastic, and trash. A parameter fine-tunes and processing time on model training and testing for each algorithm could also be considered. An outperform model could be selected to implement with the robotic automation simulation for further hardware system design in discarding solid waste precisely to the right trashcan. We also expect to perform the integration of the classification model with a fully automated sorting machine by automatically moving to its respective bin using a motorized sliding tray. The garbage disposed by the students and campus staff could also be used to turn collected points use in the university redemption.

2 Experimental Setup and Evaluation Schemes

This section describes the image data collections as well as the evaluation schemes used in our experiments including the transfer learning parameters are described below.

2.1 Garbage Image Data Collections

We have used deep learning approaches for our experiments as transfer learning improves voluntary classification task. The data is obtained from two sources which are (1) TrashNet dataset [10] containing 2,527 images of different garbage categories and (2) local dataset for 2,799 images. TrashNet dataset was taken using Apple iPhone 7 Plus, 5S, and SE by placing the object on a whiteboard poster with sunlight or room lighting. The garbage images from 6 types (cardboard, glass, metal, paper, plastic, and trash) were recategorized and annotated into four classes, which were PET (clear plastic bottle), plastic (general plastic bottle), metal, and trash. Another dataset, the garbage objects were obtained manually using image sensing module with the same conditions as TrashNet using Apple iPhone11 and web camera, Logitech C922 Pro Stream Webcam 1080p. The total number of 5,326 images grouped by garbage types is represented in Table 1. The examples of garbage images are illustrated in Fig. 1.

Table 1. Classification of the images classified by garbage type.

Collections	Number of Images				Total
	PET Plastic	Plastic	Metal	Trash	
TrashNet	133	183	297	1,915	2,527
Local dataset	425	1,425	583	366	2,799
Total	558	1,608	929	2,231	5,326

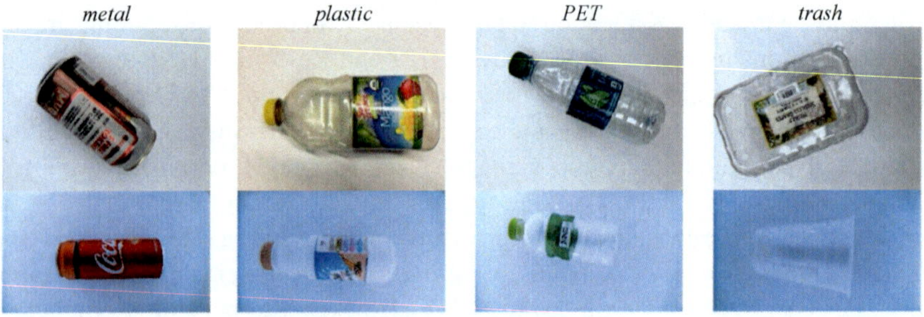

Fig. 1. The examples of the garbage image collections, TrashNet (up) and local dataset (bottom).

The data collections overall 5,326 images were taken into data augmentation techniques to manifest significant features and to increase the scale of the data. The images from both datasets have been downsized sampling to 224 × 224 pixels. We then applied six augmentation techniques, including horizontal flip, Gaussian blur, median blur, rotate 90-degree and 270-degree angle, and converting images into grayscale and HIS color spaces. Figure 2. illustrates the examples images after augmentation. Total 31,956 augmented images were divided into training and testing for 70 and 30 percent, respectively.

2.2 The In-Depth Transfer Learning Parameters

A machine learning technique based on the multiple layers of Artificial Neural Network (ANN) so-called deep learning was applied to obtain the garbage classification model. It can be used to process large and complex data sets efficiently. The model itself consists of the basic structure of the nodes in which input variables are fed into the network by assigning weight factors through the activation function. Bias values and factors

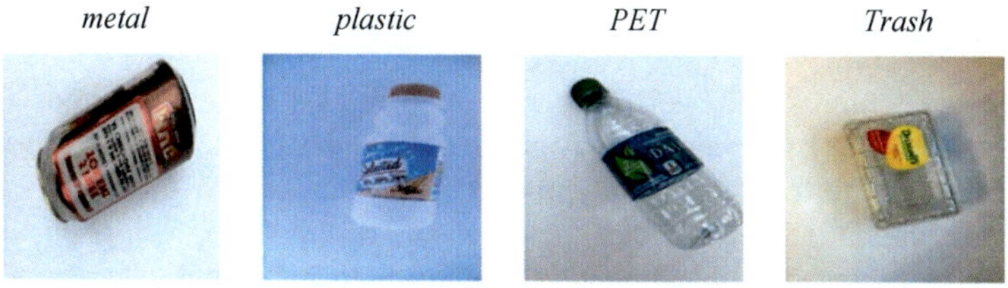

Fig. 2. The examples of six augmented images.

were combined to optimize all the weighted input variables providing the best-fit output variables.

The activation function is represented, as in (1)

$$y = g\left(w_0 + \sum\nolimits_{i=1}^{p} w_i x_i\right) \tag{1}$$

and Fig. 3 demonstrates the simplistic artificial neuron concept.

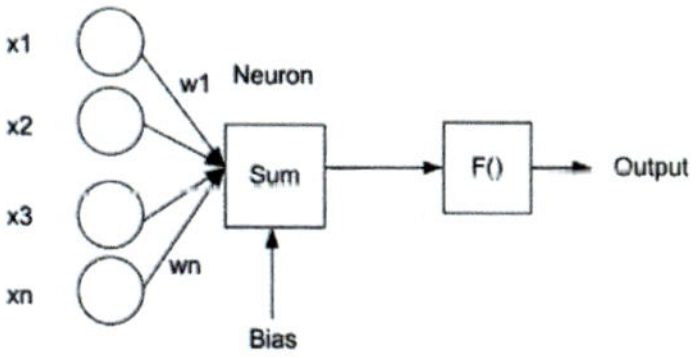

Fig. 3. A simplistic artificial neuron.

Both datasets would be used to train the creation of a numerical garbage classification model using the ANN algorithm. This research involved the built-in functions, Keras, training with the Google TensorFlow development kit. Model performance was performed by randomly dividing training and testing datasets based on stratified random technique several times until the best fit was achieved.

The parameter setting of the learning model could be described as follows. The classifier model was built based on pre-trained deep residual networks (ResNet) 50 layers [11] and 152 layers [12]. The first input layer was applied convolutional zero-padding on pre-trained the fully connected layer. The training image dataset is represented by the 3-dimension array sized 224 × 224 × 3 pixels, which is the number of neurons used to configure all parameters in the input layers. The pre-trained load weights on ImageNet were chosen as the model activation function and randomly closing the neurons. The Softmax function was required to define transfer learning parameters on the classifier activation function. The parameter setting for the hidden layer has been fully connected and fine-tuned the number of nodes to 100,352. Finally, the output layer parameter setting has only four labelled labels to classify into the four garbage classes. The learning errors and performance plots during the training process are shown below in Fig. 4.

2.3 Performance Metrics

We applied and compared the performance of the two classification methods, ResNet-50 and ResNet-152. Two evaluations were used for this study as follows. First is the evaluation scheme for the classification model. The latter was evaluated on testing the model using unseen dataset to yields the efficiency of the garbage classification. The accuracy in percentage was determined to manifest both evaluation schemes.

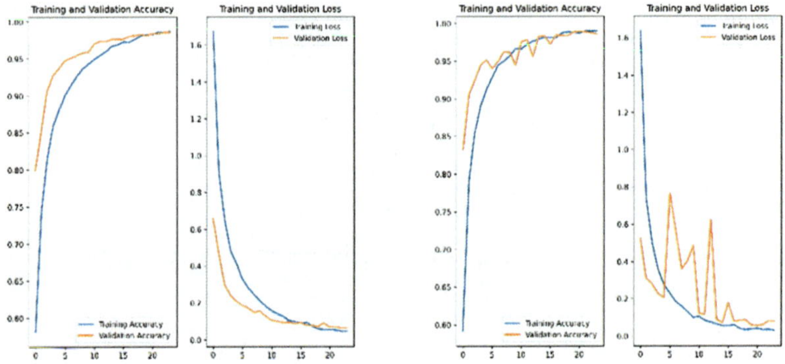

Fig. 4. The learning rate illustrating by training accuracy and loss with 24 epochs, ResNet-50 (left) and ResNet-152 (right)

3 Garbage Classification Using In-Depth Learning Techniques

In this section, this article describes the proposed methods which consists of the classification model construction and the model validation as shown below in details.

3.1 The Classification Model Based on Deep Learning Algorithm

We have applied ResNet-50 as our model as transfer learning proves valuable. The dataset is obtained from two sources, TrashNet and the local dataset for training the model. The building block of the learning model, the deep residual networks (ResNet) is illustrated in Fig. 5 (left).

The Deep Residual Network (ResNet) introduces a skip connection to fit the input among hidden layers without modification of the input. The Skip connection enables a deeper network. The transfer learning helps us to modify the model according to our data and yield better results. Figure 5 (right) illustrates the standard architecture of ResNet-50 showing the skip-connection.

In this study, the parameters referred to in the garbage classification framework represented the dimensions of the network. Once the weighted input variables have been activated to minimize bias to obtain classified types of garbage. The dimensional reduction of the hidden layers of the convolutional neural network was performed to provide the best-fit parameters. A dropout layer followed by a dense (fully connected)

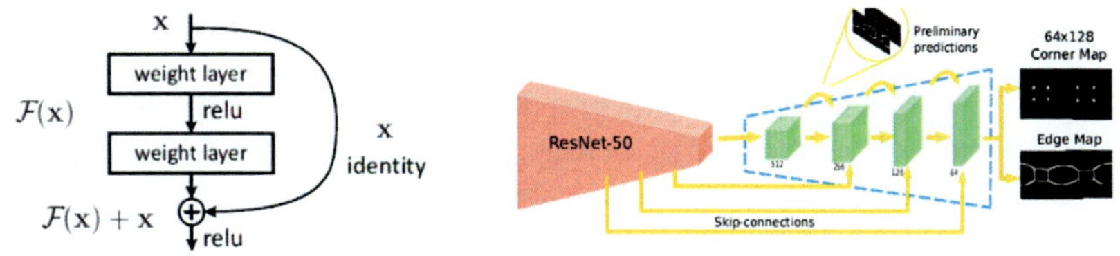

Fig. 5. Building block of deep residual network: ResNet (left) and the standard architecture of ResNet-50 (right). [PanoRoom: From the Sphere to the 3D Layout - Scientific Figure on ResearchGate. Available from: https://www.researchgate.net/figure/PanoRoom-architecture-from-ResNet-50-we-build-a-single-decoder-that-jointly-predicts_fig1_327303108 [accessed 26 Dec 2020]].

layer generated a confidence score for each class. Finally, the final model using an Adam optimizer was compiled and fine-tuning with a low learning rate. The network tuned these parameters to produce comparable results with state-of-the-art approaches. The classification model picked a parameter-based action which involves predicting the unseen dataset of garbage images at each time step.

While the methods of the convolutional learning show promising results, the image post-processing in term of the prediction confidence value was not reaching the pilot training cut-off (this would be discussed in the next sub-section). The model was trained for 24 epochs and received an accuracy of around 98.81% with a training loss of 0.0346.

The ResNet-152 consisting of 152 layers was also applied to obtain the classification model using the same datasets. It also introduces skip connection to fit the input parameter from the hidden layers without any modification of the input. The standard architecture of ResNet-152 [12] is illustrated in Fig. 6.

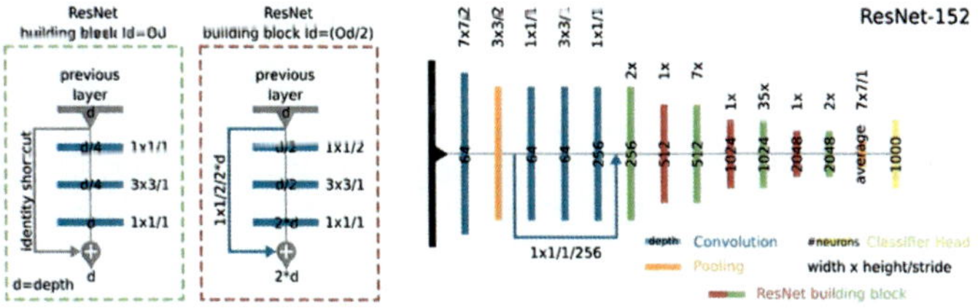

Fig. 6. The standard architecture of ResNet-152. [Available from: https://www.researchgate.net/figure/Conceptual-overview-of-the-ResNet-building-block-and-the-ResNet-152-architecture-43_fig7_341576780].

4 Experimental Results and Discussions

This section provides the experimental results including the discussions of the study as follows. The experimental results were obtained by the comparison of the two classification methods, ResNet-50 and ResNet-152. First result is the efficiency of the classification model. The evaluation on testing the model using unseen dataset to yields the efficiency of the garbage classification is shown below.

4.1 Numerical Results

We compared our proposed method with [6] and [8]. To evaluate the efficiency of classification model, the proposed approaches were applied on 5,326 images from the TrashNet and the local image collections. The numerical results shown in Table 2. is the accuracy and loss on training, validating, and testing.

Table 2. The accuracy and loss on training, validating, and testing.

Algorithm	Accuracy (%)			Loss (%)
	Training	*Validating*	*Testing*	
ResNet-50	98.81	98.50	94.15	0.0346
ResNet-152	98.99	98.56	95.72	0.0298

The ResNet-50 and ResNet-152 model were pre-trained for 24 epochs. The accuracies were reported completely high approximately 99% with a minimum training loss of 0.0298.

Meanwhile, the performance measurement was performed with the selected model having more steady learning rate which is ResNet50 by testing on 2 experiment tests. While the unseen dataset for 3,218 images was devoted to test on the model efficiency (the first testing module), 136 items of garbage objects were feed into the IoT-based sorting machine prototype (the second testing module) for model testing performance. The ResNet-50 classification accuracies and processing times on testing are recorded completely high as 98.53% with 0.27 and 89.77% with 0.72, respectively. The ResNet-152 classification accuracies and processing times on testing are recorded completely high as 99.65% with 1.20 and 91.79% with 2.53, respectively. One more reason why we pick the ResNet-50 model is it consumed less time than the latter one on the second testing module. The model could be unable to correctly predict a few cases. The experiment results are shown in Table 3 and Table 4.

Table 3. The numerical results on the unseen dataset.

Model	Real class	Prediction confidence Value				Predicted rate (%)	Time (s)
		Class1 (can)	*Class 2 (plastic)*	*Class 3 (PET)*	*Class 4 (Trash)*		
ResNet-50	Can	0.9871	1.24e-06	0.0042	0.0086	98.72	0.2680
	Plastic	6.55e-06	0.9862	0.0105	0.0033	98.55	0.2674
	PET	0.0033	0.0025	0.9877	0.0065	98.84	0.2675
	Trash	0.0132	0.0009	0.0068	0.9790	98.03	0.2670

(continued)

Table 3. *(continued)*

Model	Real class	Prediction confidence Value				Predicted rate (%)	Time (s)
		Class1 (can)	*Class 2 (plastic)*	*Class 3 (PET)*	*Class 4 (Trash)*		
ResNet-152	Can	0.9992	0.0000	0.0004	0.0003	100.00	1.1989
	Plastic	0.0007	0.0002	0.9955	0.0036	99.71	1.2047
	PET	0.0000	0.9986	0.0013	0.0001	100.00	1.1960
	Trash	0.0084	0.0003	0.0038	0.9874	98.90	1.2147

Table 4. The numerical results on the garbage object items inserting into the waste sorting machine.

Model	Real class	Prediction confidence value				Predicted rate (%)	Time (s)
		Class1 (can)	*Class 2 (plastic)*	*Class 3 (PET)*	*Class 4 (Trash)*		
ResNet-50	Can	0.9425	2.63e-07	0.0006	0.0570	93.75	0.9425
	Plastic	2.05e-10	0.9176	0.0031	0.0793	91.18	2.05e-10
	PET	8.63e-06	2.05e-10	0.9985	0.0015	100.00	8.63e-06
	Trash	0.0647	0.0189	0.1714	0.7450	74.14	1.9229
ResNet-152	Can	0.9806	0.0002	0.0009	0.0183	100.00	2.5292
	Plastic	0.0000	0.9492	0.0496	0.0011	94.28	2.5123
	PET	0.0000	0.0000	1.0000	0.0000	100.00	2.5531
	Trash	0.0381	0.0645	0.1768	0.7205	72.88	2.5122

4.2 Discussions

This section provides the discussion on the study findings. Getting started by the image dataset, the correlation between the image features of the input and test images affects the accuracy. The study provided two experiments from different testing dataset acquisition, so the input images were diversity. The parameters referred to the garbage classification framework were prominent in the performance of the best-fit dimensions. The classification model based on the neural network model has enabled the parameter-based optimization of the image-based garbage classification. Figure 7. shows the outstanding results even in partial object was taken.

While the comparative method proposed by Adedeji and Wang [8] was shown to be 87% accurate with approximately 0.48 of validation loss by 12 epochs, our proposed method was performed better than the results in [8] with the same iterations at 97% accuracy with 0.11 of validation loss. The better optimizer functions used in this study preferred Adam optimizer, while the lower accuracy of the method proposed in [8] was accounted for by the use of the Stochastic Gradient Descent (SGD) optimizing function.

Fig. 7. The outstanding results in partial object taken provides precise prediction.

However, the methods of convolutional learning show slightly promising results in [8], they require careful handling of the action and state space, which can restrict the models' expressive capacity and learning capability. Keep in mind that standalone CNN based models aim for their expressive ability and in-depth learning modelling capability as same as our based approach. The post-processing in term of the prediction confidence value was not reaching the pilot training cut-off as shown the lowest accuracy result of trash category.

However, the garbage classification found some limitations. For example, the image collection should be considered to provide both with and without bottle lid (see Fig. 8). It is necessary to retrain the transferred learning using the inclusion of addition image collection to provide sufficient feature dimension.

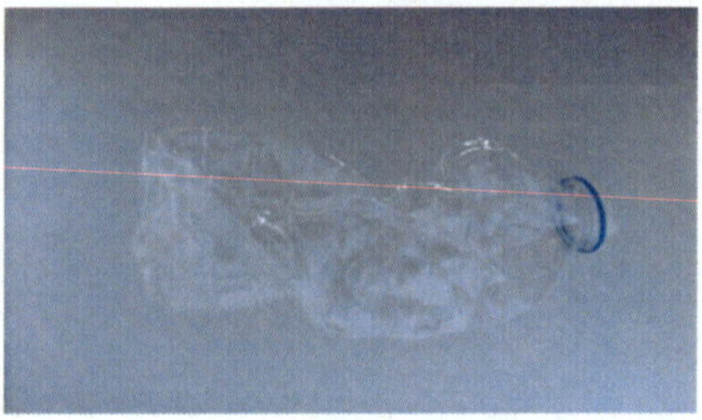

Fig. 8. The unexceptional results predicting PET (without lid) to trash.

Since the proposed approach would be expected to perform by designing the IoT-based waste sorting machine used in the university not in the production environment, the image data collection was obtained under the same conditions as the designed sorting machine by using high-quality cameras, piece by piece of waste on a white background.

5 Conclusions

The summarization of the research findings is provided in this section. The additional perspectives relating to this article is also discussed as follows.

cross-sections of aluminum products so that scrap was minimized and overall extrusion rate achieved. The third stage was to select the best billet for each product, though not presented in the paper. This global approach, selecting a set of billets for a given set of products, was further considered in Masri and Warburton [3, 4]. The problem was formulated as an integer linear programming model. Due to a large number of possible billet sizes and a large number of products, only simple extrusion operation was considered, multiple pushes within the same billet were not included. For a pair of billet size and customer order, resulting scrap was calculated a priori. Hajeeh [5] formulated a mathematical model to include scraps from log cutting. Even though the goal was to identify cutting patterns within aluminum logs so that overall scrap was minimized, possible billet sizes were pre-defined. Again, wastes from each billet assignment were calculated before optimization.

The second approach was focused around minimizing scraps between consecutive billets. When billets are extruded successively, material interface around both billets must be removed from the aluminum profile, leading to the third type of scrap mentioned in Fig. 1. A Finite Element simulation tool was developed by Reggiani, Segatori, Donati and Tomesani [6] to predict charge weld location between two consecutive billets that must be scrapped. Later, Oza and Gotowala [7] studied the effect of ram speed, billet and die temperatures on traverse weld length using HyperXtrude, a commercial CAD software for extrusion. An extensive empirical study was conducted by Ferras, Almeida, Silva, Correia and Silva [8] to identify which process parameters contributing the most to material scrap.

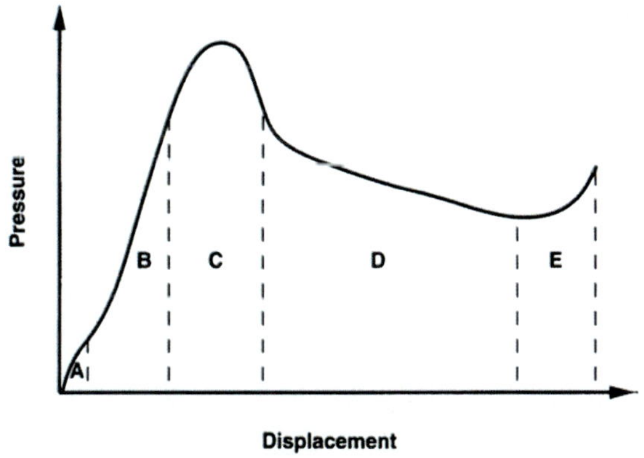

Fig. 2. Extrusion pressure profile

Sheppard [9] had shown that extrusion pressure within one stroke is not constant, as shown in Fig. 2. This pressure profile leads to non-uniform metal flow velocity into and out of the extrusion die. Thus, aluminum profile density varies along its length. Slight change in operating conditions, such as changing billet temperatures, will lead to different profile density. While most research assumed average profile density to be constant, most likely it is not true in practice. Human operators do adjust their process conditions for various reasons. And, as will be shown in Sect. 4, slight change in profile density can lead to significant differences in scrap.

3 Research Methodology

3.1 Problem Formulation

In creating an extrusion plan for a customer order, a few process parameters are known or given. They are denoted in Fig. 1 and summarized as follow.

- Aluminum log: weight *G*, in kg and its density *b*, in kg/cm.
- Customer order: profile length *L* and order quantity *n*.
- Extrusion press constraints: B_{max} and B_{min} respectively, minimum amount of backend remained after extrusion, *e*, all in kg.
- Extrusion die: number of openings in the die, *a*, amount of material remained in die pot *P*, in kg, and profile density *w*, in kg/m
- Runout table: cutting saw position L_s from the die, maximum and minimum extrusion length that can be supported by runout table R_{max} and R_{min} respectively, gripper sizes *h* and *t* to straighten profiles.

The extrusion plan developed should describe how aluminum logs are cut into billets, how billets are extruded to certain length(s) and cut into customer order length and eventually meet customer order quantity. In this research there are two possible setups considered. Figure 3 demonstrates the case of two extrusions per billet, whereas consecutive billets are not shown. In the case of multiple extrusions per billet, process planner must determine billet sizes to be used as the first and consecutive billets B_1 and B_2 respectively. For each billet, how many extrusions C_1 and C_2 are performed and in each extrusion how many pieces of product, q_1 and q_2 with length *L* are made within each profile. The extrusion lengths L_{E1} and L_{E2} from each billet should be constant for simplicity in profile cutting. All *C*'s are at least 1.

Schematics of multiple billets per extrusion are shown in Fig. 4. Here at most three different billet sizes, B_1, B_2 and B_3, can be used to form a set of billets that will result in single extrusion. The second billet B_2 can be used D_2 times within each billet set while the first billet B_1 and the last billet B_3 are used only once. D_2 can be zero or positive integers. Each billet results in q_1, q_2, and q_3 pieces of product within each profile respectively. As it can be observed in Fig. 3, the first billet in the next set, B_1', can be smaller than B_1 by the amount of material remained in the die pot. The extrusion length L_E can be computed for each billet set.

To improve material utilization, yield is an obvious objective. However, it can be simplified in two steps. First, yield is the ratio of product output made to aluminum log input. Say, if an aluminum log can be used to make 100 pieces of product length *L*, and customer order quantity is 260 pieces, three aluminum logs will be used. Even though the third aluminum log may be used partially to meet customer order, in this research, each aluminum log will be used entirely. This will not change the optimal solution, should partial log cutting is allowed. Secondly, since the entire aluminum logs are always used, the yield will be proportional to how many pieces of products are made with the same amount of aluminum log input. With proper constraints to use up aluminum logs, the objective function is to maximize the total number of products made.

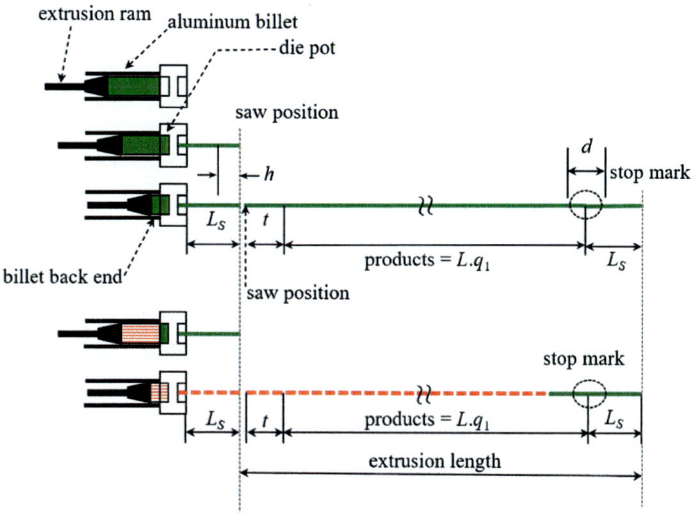

Fig. 3. Aluminum extrusion process: 2 extrusions per billet ($C_1 = 2$)

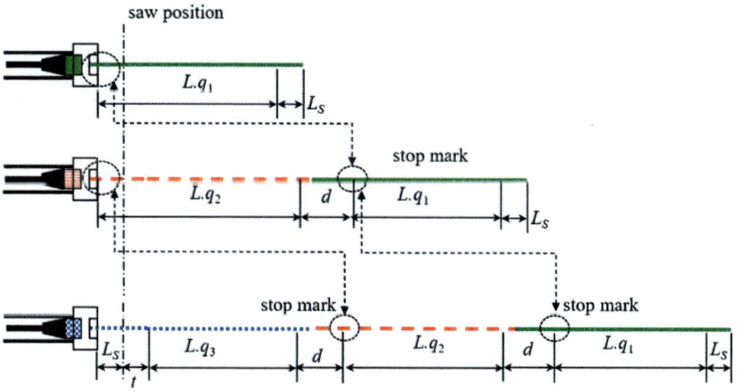

Fig. 4. Multiple billets per extrusion (3 billets per extrusion).

Assumptions for the problem formulation are summarized as follows.

- Process parameters as specified in the problem formulation section are known or pre-determined.
- Saw used for profile cutting on the runout table is at fixed position L_s from the die.
- Material losses due to cutting saw is negligible.
- Desired product length, L and minimum required quantity n are known input variables for billet selection.

- Average aluminum profile density, w, is pre-estimated. Usually, a value close to the upper range is used to ensure that enough material remained in the back end after each extrusion.
- Process parameters such as extrusion speed, temperature, etc. are pre-determined. No significant adjustment is made during extrusion.
- For multiple pushes per billet case, extrusion length is uniform for one billet size. At most two billet sizes are used to meet a customer order.
- For multiple billets per extrusion case, extrusion length should be uniform for each set of billets. At most three billet sizes are used to form a billet set.

3.2 Optimization Model for Multiple Extrusions per Billet

The optimization model for multiple extrusions per billet case can be formulated as follows. The objective function (1) is to maximize the total number of products made from the entire logs. The first parenthesized term is the number made from the first log while the second term is for the number of products made from all consecutive logs. Equations (2) and (6) are mass balance equations for the first and second billets which can be derived from Fig. 3. Equations (3) and (7) are to ensure that billet sizes selected are within limits for the extrusion press. Equations (4)–(5) and (8)–(9) are to impose constraint on extrusion lengths for the first and consecutive billets respectively. The number of consecutive billets required, n_{b2}, to meet customer order quantity can be calculated from (10). The total number of consecutive billets, g_1 and g_2, that can be packed into the first and next aluminum logs are calculated in (11), (12). Equation (13) is used to determine how many aluminum logs, M_2, aside from the first log, are used to meet customer order quantity. All decision variables are denoted in (14)–(16).

$$maximize \quad (aC_1q_1 + aC_2q_2g_1) + aC_2q_2g_2M_2 \tag{1}$$

Subject to

$$B_1 = e + P + aw\{C_1(L_s + Lq_1 + t) + L_s + h\} \tag{2}$$

$$B_{min} \leq B_1 \leq B_{max} \tag{3}$$

$$L_{E1} = L_s + t + Lq_1 + h \tag{4}$$

$$R_{min} \leq L_{E1} \leq R_{max} \tag{5}$$

$$B_2 = e + aw\{C_2(t + L_s + Lq_2) + h\} \tag{6}$$

$$B_{min} \leq B_2 \leq B_{max} \tag{7}$$

$$L_{E2} = L_s + t + Lq_2 + h \tag{8}$$

$$R_{min} \leq L_{E2} \leq R_{max} \tag{9}$$

$$n_{b2} = \left\lceil \frac{n - aC_1q_1}{aC_2q_2} \right\rceil \tag{10}$$

$$g_1 = \left\lfloor \frac{G - B_1/b}{B_2/b} \right\rfloor \tag{11}$$

$$g_2 = \left\lfloor \frac{G}{B_2/b} \right\rfloor \tag{12}$$

$$M_2 = \left\lceil \frac{n_{b2} - g_1}{g_2} \right\rceil \tag{13}$$

$$C_1, C_2 \geq 1, integer \tag{14}$$

$$q_1, q_2, \; n_{b2}, g_1, g_2, M_2, \geq 0, integer \tag{15}$$

$$B_1, B_2, L_{E1}, L_{E2} \geq 0 \tag{16}$$

3.3 Optimization Model for Multiple Billets per Extrusion

Similarly, for multiple billets per extrusion setup, the objective function (17) is to maximize the total number of products made. Here $a(q_1 + D_2q_2 + q_3)$ represents the total number of products made per billet set. Equations (18)–(20) are mass balance equations for the first billet set which can be derived from Fig. 4. Mass balance equation for the first billet in the next billet set is shown in Eq. (21). Equations (22)–(25) are to ensure that billet sizes selected are within limits for the extrusion press. Equations (26), (27) are to impose constraint on extrusion length for each billet set. The number of consecutive billets required, n_s, to meet customer order quantity can be calculated from (28). The total number of consecutive billet sets that can be packed into the first and next aluminum logs, h_1 and h_2, are calculated in (29)–(30). Equation (31) is used to determine how many aluminum logs, aside from the first, are used to meet or exceed customer order quantity. All decision variables are denoted in (32)–(33).

$$maximize \quad a(q_1 + D_2q_2 + q_3)(1 + h_1 + N_2h_2) \tag{17}$$

Subject to

$$B_1 = e + P + aw\{Lq_1 + L_s + h\} \tag{18}$$

$$B_2 = e + P + aw\left\{Lq_2 + \left(d - \frac{P}{aw}\right)\right\} \tag{19}$$

$$B_3 = e + P + aw\left\{L_s + t + Lq_3 + \left(d - \frac{P}{aw}\right)\right\} \tag{20}$$

$$B_1' = e + P + aw\left\{Lq_1 + \left(d - \frac{P}{aw}\right)\right\} \tag{21}$$

$$B_{min} \le B_1 \le B_{max} \tag{22}$$

$$B_{min} \le B_2 \le B_{max} \tag{23}$$

$$B_{min} \le B_3 \le B_{max} \tag{24}$$

$$B_{min} \le B_1' \le B_{max} \tag{25}$$

$$L_E = (L_s + Lq_1) + D_2(d + Lq_2) + (d + Lq_3 + t) \tag{26}$$

$$R_{min} \le L_E \le R_{max} \tag{27}$$

$$n_s = \left\lceil \frac{n - a(q_1 + D_2q_2 + q_3)}{a(q_1 + D_2q_2 + q_3)} \right\rceil \tag{28}$$

$$h_1 = \left\lfloor \frac{G - (B_1 + D_2B_2 + B_3)/b}{(B_1' + D_2B_2 + B_3)/b} \right\rfloor \tag{29}$$

$$h_2 = \left\lfloor \frac{G}{(B_1' + D_2B_2 + B_3)/b} \right\rfloor \tag{30}$$

$$N_2 = \left\lceil \frac{n_s - h_1}{h_2} \right\rceil \tag{31}$$

$$B_1, B_1', B_2, B_3 \ge 0 \tag{32}$$

$$D_2, q_1, q_2, q_3, n_s, h_1, h_2, N_2 \ge 0, integer \tag{33}$$

3.4 Solution Procedure

The following solution procedure is proposed to explore both extrusion setups and handle process variations.

Step 1. Solve the model in Sect. 3.2 by enumerating all independent variables, which are C_1, C_2, q_1 and q_2. Write all feasible solutions in a table and identify all optimal solutions.
Step 2. Solve the model in Sect. 3.3 by enumerating all independent variables, namely q_1, q_2, D_2 and q_3. Write all feasible solutions in another table and identify all optimal solutions.
Step 3. Select the better solution, i.e., higher optimal yield, from those optimal solutions obtained from step 1 and step 2.

Step 4. Perform actual extrusion using parameters identified in step 3. From the first billet, the remaining backend, e_{act}, can be measured. If e_{act} is significantly less than e, there exists a risk that ram may damage the die. On the other hand, if e_{act} is larger than e, more material scrap will ensue. The actual profile density w_{act} can be calculated from Eq. (2) or (18), depending on which extrusion setup is being applied, using e_{act}.
Step 5. Re-solve step 1 or 2 depending on the current extrusion setup. Select feasible solution to continue from the first billet or choose to start over with a new first billet. The methodology described is implemented using Visual Basic Application in Microsoft Excel.

4 Case Study

The model's capability will be demonstrated through a real-world example. The input data for the program is shown in Table 1. To make 2000 pieces of 1.9 m-long profile, the program identifies two optimal solutions one for multiple extrusions per billet and the other from multiple billets per extrusion cases. All enumerated solutions and all feasible solutions are not shown here. Instead, Tables 2 and 3 summarized the optimal solutions obtained from the enumeration program. Computation time is relatively small, mostly within a few minutes running on a personal computer, Intel Core-i7. Table 2 illustrates the results from multiple extrusions per billet. At the beginning, profile density w for the calculation is set at 0.572 kg/m, usually in the upper range. The original solution from the program is shown in the second column. Here the program identifies an optimal solution to be single extrusion per billet where first billet, 66.28 cm long, makes 17 pieces of product per die opening and each consecutive billet, 71.6 cm long, makes 20. With 2 die openings ($a = 2$), the first and consecutive billets make 34 and 40 pieces of product, respectively. The first billet along with 50 consecutive billets will make 2,034 pieces for customer demand. Since each aluminum log is 580 cm in length, 8 billets can be cut from an aluminum log. Thus, 7 logs are needed. If the entire log is cut, 2,234 pieces will be made with the resulting yield of 89.52%.

When the extrusion plan is put into practice, it is found that the first billet backend is larger than normal. Actual backend size can be put into Eq. (2) and solved for actual profile density w_{act}. Say, w_{act} is found to be 0.552 kg/m. Should the operator choose to continue the same extrusion plan, the actual yield will drop to 86.39%, shown in the third column. A reasonable approach is to re-run the program with actual profile density. Then, from the list of feasible solution, it is possible to select consecutive billet size that results in maximum overall yield. This is shown in the fourth column. Here, the first billet size was already used, consecutive billet size can be adjusted to make 21 pieces of product per profile. The number of logs required is 6, making 2008 pieces. The resulting yield is impressive, 90.59%. Should the actual profile density known beforehand, the optimal solution, shown in the rightmost column, will yield 90.86%.

Table 3 shows the results from multiple billets per extrusion case. Similar to the previous case, the original solution is compared with the result when there is no adjustment is made. Here, no second billet size is recommended by the program, hence, not shown in the table. Once again, yields drop more than 3% suggesting that simply ignoring to adjust the extrusion plan can have significant ramifications on the yield. Since the new

Table 1. Parameters used in case studies.

Aluminum log	$G = 387.44$ kg (length 5.8 m), $b = 0.668$ kg/cm
Customer order	$L = 1.9$ m, $n = 2000$ pcs
Extrusion press setup	$B_{min} = 23.38$ kg (35 cm), $B_{max} = 53.44$ kg (80 cm), $e = 1.75$ kg
Die	$a = 2$, $w = 0.572$ kg/m, $P = 1.25$ kg
Runout table	$R_{min} = 25$ m, $R_{max} = 47$ m, $L_s = 1.5$ m, $h = 0$ m, $t = 1$ m

Table 2. Extrusion plan from the enumeration program (multiple extrusions per billet)

	original solution	continue w/o adjusting	adjust 2nd billet size	new optimal solution
Profile density w (kg/m)	0.572	0.552	0.552	0.552
first billet size (cm)	66.28	66.28	66.28	73.53
pcs/profile x pushes/billet q_1xC_1	17x1	17x1	17x1	20x1
extruded length L_{E1} (m)	34.80	34.80	34.80	40.50
consecutive billet size (cm)	71.60	71.60	72.32	72.32
pcs/profile x pushes/billet q_2xC_2	20x1	20x1	21x1	21x1
extruded length (m) L_{E2}	40.50	40.50	42.40	42.40
number of consecutive billets n_{b2}	50	50	47	47
total no. of billet (set) per log	8	8	8	8
Total no. of log used	7	7	6	6
total no. of workpieces made (pcs)	2234	2234	2008	2014
total billet weight (kg)	2712.08	2712.08	2324.64	2324.64
yield (%)	89.52	86.39	90.59	90.86

Table 3. Extrusion plan from the enumeration program (multiple billets per extrusion)

	original solution	continue w/o adjusting	adjust 2nd billet size	new optimal solution
Profile density w (kg/m)	0.572	0.552	0.552	0.552
first billet size (cm)	42.48/38.04	42.48/38.04	42.48/36.79	41.14/36.79
pcs/profile x pushes/billet q_1xC_1	11	11	11	11
last billet size (cm)	42.32	42.32	44.06	44.06
pcs/billet q_3	11	11	12	12
extruded length (m) L_{E3}	44.30	44.30	46.20	46.20
number of sets n_s	45	45	43	43
total no. of billet (set) per log	7	7	7	7
Total no. of log used	7	7	7	7
total no. of workpieces made (pcs)	2156	2156	2254	2254
total billet weight (kg)	2712.08	2712.08	2712.08	2712.08
yield (%)	86.40	83.38	87.17	87.17

optimal solution, when w_{act} is 0.552, is very similar to that of the original solution, adjusting the second and continuing billets to achieve optimal solution is easier. The adjusted solution is very close to the optimal solution if the profile density is known in advance.

5 Conclusion

In this research, an alternative yield optimization methodology is proposed for online billet sizing in aluminum profile extrusion process. The model utilizes mass balance equations to consider two general ways to extrude aluminum profiles, namely multiple billets per extrusion and multiple extrusions per billet. Its objective function is simplified to maximize the total number of products made.

A computer program is developed to enumerate all feasible solutions. Its capabilities are demonstrated using a case study. Based on the case study, it has been shown that process adjustment is necessary to control material yield. The list of feasible solution allows process planner to adjust the extrusion plan by selecting the best continuing plan. More work can be done to identify hardware constraints that will further improve the yield.

References

1. Grand View Research. https://www.grandviewresearch.com/industry-analysis/aluminum-extrusion-market
2. Tabucanon, M.T., Treewannakul, T.: Scrap reduction in the extrusion process: the case of an aluminium production system. Appl. Math. Modelling **11**, 141–145 (1987). https://doi.org/10.1016/0307-904X(87)90158-2
3. Masri, K., Warburton, A.: Optimizing the yield of an extrusion process in the aluminum industry. In: Tamiz, M. (ed.) Multi-Objective Programming and Goal Programming. LNE, vol. 432, pp. 107–115. Springer, Heidelberg (1996). https://doi.org/10.1007/978-3-642-87561-8_9
4. Masri, K., Warburton, A.: Using optimization to improve the yield of an aluminum extrusion plant. J. Oper. Res. Soc. **49**(11), 1111–1116 (1998). https://doi.org/10.1057/palgrave.jors.2600616
5. Hajeeh, M.A.: Optimizing an aluminum extrusion process. J. Math. Stat. **9**(2), 77–83 (2013). https://doi.org/10.3844/jmssp.2013.77.83
6. Reggiani, B, Segatori, A., Donati, L., Tomesani, L.: Prediction of charge welds in hollow profiles extrusion by FEM simulations and experimental validation. Intl. J. Adv. Manuf. Technol. **69**, 1855–1872 (2013). https://doi.org/10.1007/s00170-013-5143-2
7. Oza, V.G., Gotowala, B.: Analysis and optimization of extrusion process using Hyperworks. Int. J. Sci. Res. Dev. **2**(8), 441–444 (2014)
8. Ferras, A.F., Almeida, F. De, Silva, E.C., Correia, A., Silva, F.J.G.: Scrap production of extruded aluminum alloys by direct extrusion. Procedia Manuf. **38** 1731–1740 (2019). https://doi.org/10.1016/j.promfg.2020.01.100
9. Sheppard, T.: Extrusion of Aluminum Alloys. Springer Science+Business Media Dordrecht (1999). https://doi.org/10.1007/978-1-4757-3001-2

Enhanced Hydrozoan Algorithm with Local Searches for Job-Shop Scheduling Problems

Daranat Tansui(✉) and Arit Thammano

Computational Intelligence Laboratory, Faculty of Information Technology, King Mongkut's Institute of Technology Ladkrabang, Bangkok 10520, Thailand
62606002@kmitl.ac.th, arit@it.kmitl.ac.th

Abstract. Job-Shop Scheduling Problem (JSSP) is one of the most difficult NP-Hard problems to solve. The goal of JSSP is to find a schedule of operations that minimizes the maximum completion time (or makespan) of n jobs with m machines. This work developed and evaluated an enhanced hydrozoan algorithm with local search strategies for efficiently solving JSSPs. The three incorporated local search strategies were one of the strengths of this algorithm, exploiting the local solution space thoroughly. Diverse exploration of global space was also enhanced by incorporation of a strong-parent crossover operator and three mutation operators. The evaluation was based on 18 widely used JSSP benchmark instances. The developed algorithm performed well, as intended; it provided the best solutions to all 18 instances that matched exactly with the best-known solutions reported in the literature. Our future work will be on attempting to apply the developed algorithm to flexible job shop problem, a more complex problem.

Keywords: Job-shop scheduling · Hydrozoan algorithm · Local search strategy

1 Introduction

Recently, job scheduling problems (JSSPs) have attracted increasing attention especially in the industrial sector. To solve them, engineering principles and mathematical models are required to achieve appropriate production plans with limited resources, such as machine restrictions, production route systems, and the time required to service customers. Shop scheduling is an interesting area of research in the field of scheduling and considered to be a combinatorial optimization problem [1, 2]. The main objective of a JSSP is to find the scheduling of operations that minimizes operation time or objective function. For industrial machine operations, the scheduling that provides a minimum operation time may minimize or maximize the efficiency of the machines that require the longest or the shortest time, respectively. Another way to explain machine operation scheduling is to say that a JSSP tries to find a schedule that minimizes makespan.

A makespan is the total time required to complete all n jobs for all m machines. Since each job is performed before or after another job is completed in the production

P. Meesad et al. (Eds.): IC2IT 2021, LNNS 251, pp. 244–256, 2021.
https://doi.org/10.1007/978-3-030-79757-7_24

process, finding an effective schedule is a challenge [3]. JSSP belongs to a class of NP-complete problem. Such problems are difficult problems because of their complexity [4]. The solution method is through exact algorithms such as linear programming or branch and bound algorithm. For small scale problems, some algorithms may be effective at finding the optimal solution, but for large scale problems, it is very difficult to calculate and find the optimal solution within a limited time. Most commonly used methods for solving JSSP in the industrial sector are heuristic and metaheuristic methods which are approximation methods for finding near optimal solutions in a relatively short time. Approximation methods are such as priority dispatching rules, shifting bottleneck procedure, meta-heuristic algorithms, and the like. At present, meta-heuristic algorithms such as GA, ant colony algorithm (ACO), particle swarm optimization algorithm (PSO), simulated annealing algorithm (SA), neural network, tabu search (TS) and artificial bee colony algorithm (ABC) have been widely used and shown good performance [5].

In a study, Janes et al. [6] attempted to develop an efficient scheduling method based on genetic algorithm for solving JSSPs. In the proposed genetic algorithm, the initial population was generated randomly, and some crossover and mutation operations are performed to them. Experimental results showed that the proposed method was able to find a satisfactory solution in a quite short time span, and it was sufficiently reliable for solving real-world JSSPs.Mekni and Faye [7] developed a modified invasive weed optimization (IWO) algorithm for optimization of multi-objective, flexible job shop scheduling problems. In their study, the smallest position value (SPV) method was used to convert continuous position values into discrete job sequences. Then, each solution in the population produced seeds based on its fitness. Experimental results showed that their algorithm was highly competitive with several state-of-the-art methods at that time.

This paper is organized as follows: in Sect. 2, JSSP is briefly presented; the background of the proposed algorithm is explained in Sect. 3; Sect. 4 shows the results; and Sect. 5 is the conclusion of the paper.

2 Theories and Background

2.1 Job Shop Scheduling Problems

Job shop scheduling problems (Brucker, 1994) [8] are problems of finding the optimal order for each job in a multiple-job operation of a machine with an objective to minimize the total time required to complete all jobs. The technical expression for this field is to minimize makespan, based on the priority of each job or each activity according to a specific demand, condition, or objective. The typical condition or parameter found in a job shop schedule problem is time as it is involved with job operations-all operations processed by each machine within a given time. Solving job shop scheduling problems becomes more complex when constraints on different machines (n) and jobs (m) are taken into consideration. For an $n \times m$ JSSP, the problem can be modeled by a set of m machines, denoted by $M = \{1, 2, \ldots, m\}$, to process a set of $n \times m$ operations, denoted by $O = \{0, 1, 2, \ldots, (n \times m) - 1\}$. The notations are as follows:

n: number of jobs
m: number of operations for one job

O_i: a job sequence for a complete job operation (i = 0, 1, 2, … (n × m) − 1)
t_i: processing time for a job operation i on a given machine
C_{max}: makespan

Each job consists of a different sequence of production on each machine, and the time required to complete each job depends on the constraints on each machine and job as follows [9]. A machine may not process more than one job at a time. job may not be processed by more than one machine at a time. The sequential arrangement of machine operations has a linear precedence structure. Every machine's processing time is known.

2.2 Hydrozoan Algorithm (HA)

Hydrozoan algorithm (HA) was inspired by the regeneration and transplantation processes in hydrozoan life cycle. Regeneration and transplantation consist of cloning, splitting, and mutation. Cloning transfers the good fitness of better parents to new offspring or solutions; splitting promotes generation of more offspring with better fitness; and mutation results in rich diversity of offspring. The personal, best solution achieved by each hydrozoa from previous iterations and the global best solution of the whole population are used to produce new and better solutions [10].

The standard HA was developed by Tansui and Thammno in 2017, initially designed for solving continuous optimization problems. Its main ideas can be summarized as follows. HA imitates the nature of three main stages of hydrozoan life cycle to make its search for better solutions more effective. The following operators are used in HA: 1) a type of mutation operator is used to imitate planula larvae swimming through water. 2) clonal selection is used to imitate asexual reproduction of polyps 3) another type of mutation operator and a crossover operator are used to mimic medusa behavior. The pseudocode of HA is shown in Algorithm 1 and described below.

Step 1: Specify an initial population of N hydrozoas and randomly generate the initial position of every hydrozoa, following Eqs. (1) and (2) below,

$$h_i(0) = [h_1^i, h_2^i, \ldots, h_D^i], \tag{1}$$

$$h_d^i(0) \in [lower, upper], \tag{2}$$

where $i = 1$ to N, and D is the dimensions of each hydrozoa.

Step 2: Input the initial position of each hydrozoa into the objective function and evaluate its fitness value for that position.

Step 3: In the local morphodynamic process, the growth value of each hydrozoa ($G_i(t)$) is determined from the activator value (Ac_i^{β}) and the inhibitor value (In_i^{α}) by Eq. (3)–(5).

$$G_i(t) \propto Ac_i^{\beta}, \tag{3}$$

$$G_i(t) \propto \frac{1}{In_i^{\alpha}}, \tag{4}$$

$$G_i(t) = \frac{Ac_i^{\beta}}{In_i^{\alpha}}, \tag{5}$$

where β is the significance factor of the activator and α is the significance factor of the inhibitor.

Step 4: Calculate the median value that represents the growth values. A median is the average of two middle numbers of a series of ordered values.

Step 5: The growth value of a hydrozoa in a particular swarm is calculated by Eq. (6). Hydrozoas are split into swarms according to $Split_i$ condition (Eq. 7).

Step 6: Choose and clone hydrozoas according to the $Split_i$ value.

$$Swarm_i = G_i - Median \tag{6}$$

$$Split_i = \begin{cases} = 0, \, if \;\; Swarm_i < 0, \\ = 1, \, if \;\; Swarm_i = 0, \\ = 1, \, if \;\; Swarm_i > 0, \;\; \& \; Saram_i = \min^+ \\ = 3, \, if \;\; Swarm_i > 0, \;\; \& \; Saram_i = \min^+ \\ = 2, \;\; otherwise \end{cases} \tag{7}$$

Step 7: Mutate the cloned hydrozoas. This kind of mutation is a small random adjustment of genes.

A gene h_d^i is selected and mutated by a low probability of mutation as follows:

$$RP = (\max - \min) * \varphi + \min, \tag{8}$$

$$h_d^i = h_d^i + (h_d^i \times RP), \tag{9}$$

where RP is a constant derived from 'max' and 'min', the lower and upper bounds of mutation rate specified as percentages by user, and ϕ is a random number in the range [0, 1].

Step 8: P_j is the probability that an individual hydrozoa i is selected, which is computed as follows,

$$P_j = f_{h_i} / \sum_{q=1}^{N} f_{h_q}, \tag{10}$$

where f_{h_i} is the fitness value of hydrozoa i^{th} at time t.

Step 9: Cross over the genes of the selected parents by an N-point crossover operator to obtain a variety of better offspring for the next generation.

Step 10: Then, a mutation operator operates on the offspring. The index of the gene to be mutated is randomly produced, and the indexed gene is mutated.

Step 11: Evaluate the fitness values of every offspring; Combine and sort all.

hydrozoas according to their fitness value; then, select a specified number of the best its.

Step 12: Check the stopping criterion. If the maximum number of iterations is reached, output the best solution.

3 The Proposed Algorithm

The original version of HA algorithm was developed for solving continuous optimization problems, but we used it to solve JSSP in this study. The genes of stronger parents were crossed over to create diverse and even stronger offspring. Furthermore, we integrated 3 local search operators into the original HA to improve the movement of hydrozoa in the search space. The details of the developed HA for JSSP are described below. In this study of JSSP, our goal was to develop an efficient algorithm for finding the best schedule of job operations of a JSSP benchmark instance that minimizes the makespan. Main details of this algorithm are described below.

3.1 Chromosome Representation

In Table 1, the developed permutation-based method was derived from the continuous HA version based on the smallest position value (SPV) rule [11], which is a random representation. According to the SPV rule, the smallest randomly generated position value was −2.26 corresponding to the (dimensional) index number 1. This position has an index number 3. The second smallest position value was 0.25, corresponding to the position with the index number 1. The third smallest position value was 0.57, corresponding to the position with index number 4. The fourth smallest value was 2.8, corresponding to the position with index number 5. The fifth smallest value was 4.3, corresponding to the position with index number 2. Finally, the sixth smallest was 5.1, corresponding to the position with index number 6. Therefore, the chromosome of this hydrozoa would be this permutation: (2, 5, 1, 3, 4, 6). In other word, the SPV rule sorts the indices of the elements in ascending order according to their position values. The permutation is a tentative solution under the mechanism of HA applied to a discrete problem. Consider the following FT06 (6x6) Job-Shop Scheduling Problem. The sequences of operations of every machine and their operation time are shown in Tables 2, 3, and 4.

Table 1. Permutation-based representation

Index number	Position value of hydra	Permutation/SPV
1	0.25	2
2	4.3	5
3	−2.26	1
4	0.57	3
5	2.8	4
6	5.1	6

Table 2 shows an example of job shop schedule. The first operation sequence is O_{21} = j2 (a job may have more than one operation); the first operation sequence is O_{11} = j1; the first operation sequence is O_{41} = j4, the first operation sequence is O_{61} = j6, the second operation sequence is O_{42} = j4, the first operation sequence is O_{31} = j3, and so on. The last operation sequence is O_{66} = j6 (Fig. 1).

Table 2. 6 × 6 Job shop schedule.

O_{21}	O_{11}	O_{41}	O_{61}	O_{42}	O_{31}
O_{43}	O_{32}	O_{22}	O_{33}	O_{62}	O_{51}
O_{63}	O_{23}	O_{34}	O_{24}	O_{12}	O_{52}
O_{13}	O_{44}	O_{14}	O_{64}	O_{35}	O_{53}
O_{45}	O_{54}	O_{65}	O_{36}	O_{25}	O_{15}
O_{55}	O_{56}	O_{16}	O_{46}	O_{26}	O_{66}

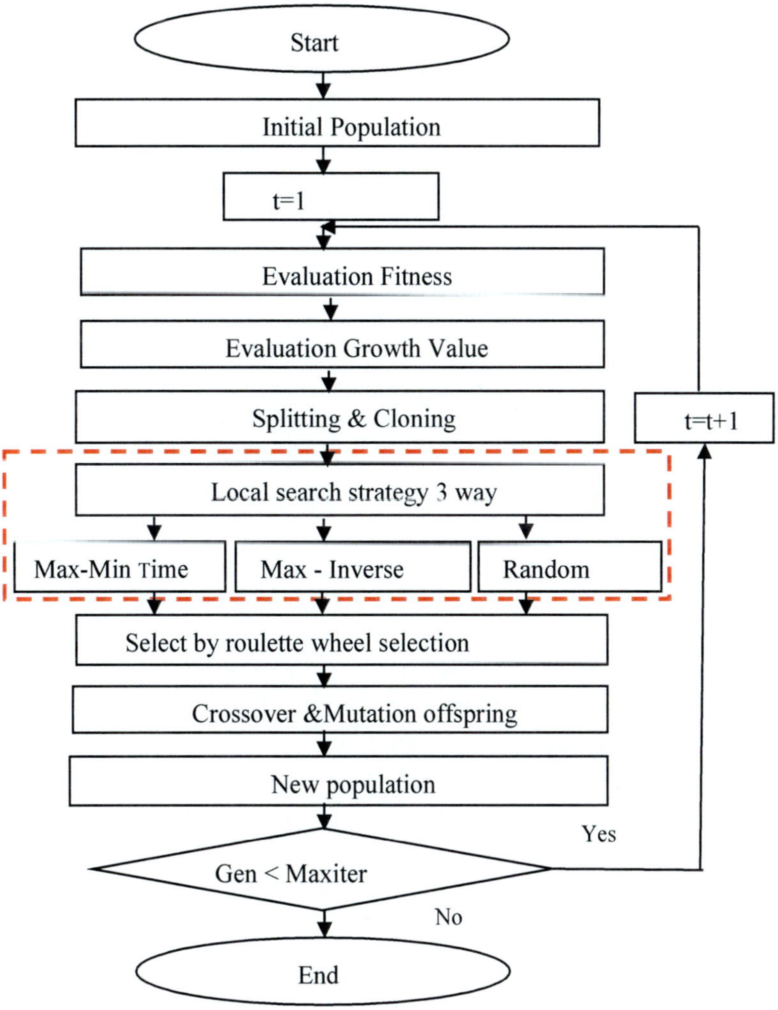

Fig. 1. Flowchart of hydrozoan algorithm for JSSP.

Table 3 shows routing and machining times of each operation in a job shop schedule for an FT06 standard JSSP benchmark dataset. The set of 6 jobs are processed by the set of 6 machines, listed in Table 2 . In Table 3, it can be observed that machine 2 performs the first operation of Job 2, the first job in the schedule, in 8 min.

Table 3. Routing and machining times for FT06 (6 × 6)

Job	Operation					
	1	2	3	4	5	6
1	3,1	1,3	2,6	4,7	6,3	5,6
2	2,8	3,5	5,10	6,10	1,10	4,4
3	3,5	4,4	6,8	1,9	2,1	5,7
4	2,5	1,5	3,5	4,3	5,8	6,9
5	3,9	2,3	5,5	6,4	1,3	4,1
6	2,3	4,3	6,9	1,10	5,4	3,1

Table 4 shows an example of how details of a schedule are encoded in the chromosome of an individual hydrozoa or solution. The first gene represents the first scheduled job sequence; the second gene represents the second scheduled job sequence, and so on. For example, a whole chromosome or solution consists of 36 genes for 36 operations (6jobs × 6machines): {214643432365621514116354563215514 26}.The first six genes of the chromosome, 214643, represent the following operation [O21, j2], [O11, j1], [O41, j4], [O61, j6], [O42, j4], [O31, j3].

Table 4. Schedule details encoded in HA chromosomes

Chromosome	2	1	4	6	4	3
Operation	O_{21}	O_{11}	O_{41}	O_{61}	O_{42}	O_{31}
Job	j2	j1	j4	j6	J4	j3
Machine, time	2,8	3,1	2,5	2,3	1,5	3,5

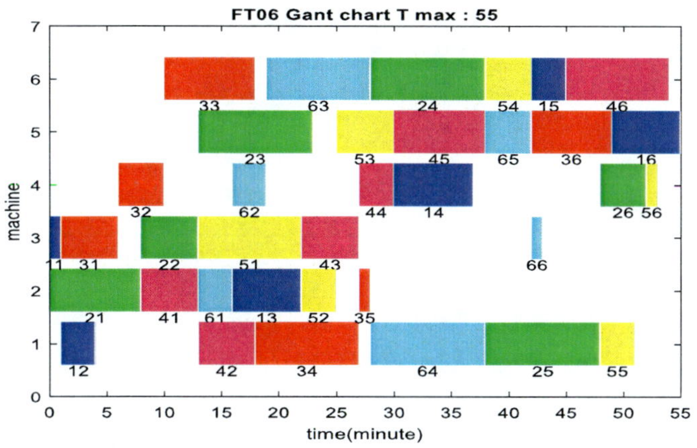

Fig. 2. GanttChart of Scheduled jobs for FT06.

Figure 2 shows a Gantt chart of scheduled jobs for FT 06. It can be observed that the minimum makespan for FT06 was 55.

3.2 Objective Function

The main goal of JSSP is to find the minimum makespan for a job schedule. The time for completing the whole schedule or the makespan is also the minimum finishing time for a set of *i* jobs. The sequence of all operations on all machines is arranged to minimize*Cmax* , Therefore, a makespan can be expressed as,

$$Cmax = max(C1, C2, \ldots Cj), j = 1, 2, \ldots, n \tag{11}$$

3.3 Local Search Strategy

The schedule obtained in the construction phase is still not optimal with respect to the local search neighborhood. Hence, a local search strategy can be applied to decrease the makespan. Three different local search strategies were incorporated into the proposed algorithm: two exchange local search techniques, where the exchange was based on operation times; and a local search strategy using a random operator.

The first exchange local search strategy used a **Max–Min Time operator** to, for example, as shown in Figure 3, swap job 6 with time index 1, the minimum time index, with job 3 with time index 10, the maximum time index.
The second exchange local search strategy used a **Max-inversion operator** to invert the sequence of jobs that followed the job with the maximum time index: 4123 => 3214 Fig. 4.
The third local search strategy used a **Random operator** to randomly select two job indexes, created previously for every job, and swap the jobs with these indexes. For example, shown in Fig. 5, job 4 is swapped with job 2.

	Min					Max				
Timeindex	**1**	3	6	7	3	**10**	8	5	9	7
chromJob	**6**	1	4	6	5	**3**	4	1	2	3
NewchromJob	**3**	1	4	6	5	**6**	4	1	2	3

Fig. 3. Max–Min Time operation.

						Max				
Timeindex	1	3	6	7	3	**10**	8	5	9	7
chromJob	6	1	4	6	5	3	**4**	**1**	**2**	**3**
NewchromJob	6	1	4	6	5	3	**3**	**2**	**1**	**4**

Fig. 4. Max– inversion operation.

			Random 1						Random 2	
index	1	2	**3**	4	5	6	7	8	**9**	10
chromJob	6	1	**4**	6	5	3	4	1	**2**	3
NewchromJob	6	1	**2**	6	5	6	4	1	**4**	3

Fig. 5. Random operation

3.4 Crossover Operation

Crossover is the main operation in the HA algorithm. This operation, to a large extent, is responsible for the global search capability of the algorithm. Selected pairs mate to generate an offspring. The crossover operator yields only one offspring that inherits more genes from the higher fitness parent than from the lower fitness parent. The number of genes that each parent contributes to the offspring is determined according to (12) and (13) (Fig. 6).

$$n_i = \left[\left(\frac{a-b}{a}\right) \times \left(\frac{d}{2}\right) + \left(\frac{d}{2}\right)\right] \tag{12}$$

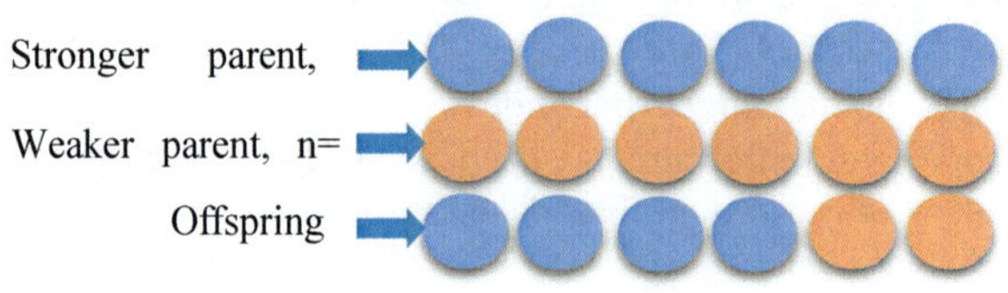

Fig. 6. Crossover operator.

$$offspring_i = \begin{cases} a = b -> \text{randomly change all genes} \\ a \neq b -> take\ n\ genes from\ the\ strong\ \ and \\ \qquad d - n\ genes from\ the\ weak \\ b = 0, -> take\ d\ genes from\ the\ strong \end{cases} \tag{13}$$

where d is the number of dimensions of the continuous search space and n = 1, 2, 3, …, D is the number of genes that the winner contributes. Therefore, a variety of better. offspring containing more genes from the winner are generated.

3.5 Mutation Operation

Some genes of each offspring may be mutated or not. We used three mutation operators The first move operator was **swapping operator** (Fig. 7) that randomly selected two job indexes and swapped the two jobs with the selected indexes. For example, as shown in Fig. 7, job 1 (with job index 1) and job 6 (with job index 5) are swapped. The second operator was **insertion operator** (Fig. 8) that randomly selected two job indexes. The job with the first selected index was inserted into the slot for the job with the second selected index, then the sequence of jobs in the slots between the selected jobs were shifted left. For example, as shown in Fig. 8, job 1 is inserted at the slot of job 6 and the sequence of jobs, 4656, was shifted left one slot. The third operator was **inversion operator** that did the insertion and shift operation like the insertion operator, but the sequence of jobs was inverted. For example, as shown in Fig. 9, job 1 is inserted at the slot for job 6, and the original sequence of jobs between the selected job, 4656, is inverted and shift to the left one slot to be 6564.

Insert mutation (i=1, j=5)

index	0	1	2	3	4	5	6	7	8	9
chromJob	3	1	4	6	5	6	4	3	2	6
NewchromJob	3	4	6	5	6	1	4	3	2	6

Fig. 7. Swap mutation.

Swap mutation (i=1, j=5)

index	0	1	2	3	4	5	6	7	8	9
chromJob	3	1	4	6	5	6	4	3	2	6
NewhromJob	3	6	4	6	5	1	4	3	2	6

Fig. 8. Insert mutation.

Inverse mutation (i=1, j=5)

index	0	1	2	3	4	5	6	7	8	9
chromJob	3	1	4	6	5	6	4	3	2	6
NewchromJob	3	6	5	6	4	1	4	3	2	6

Fig. 9. Inverse mutation.

4 Experimental Results

To evaluate the proposed algorithm, we used 18 instances from OR-Library (Beasley, 1990 [12]) as benchmark instances: FT06, LA01, LA02, LA03, LA04, LA05, LA06, LA07, LA08, LA09, LA10, LA11, LA12, LA13, LA14, LA15, LA17, and LA20. The evaluation was run on a Windows 10 PC with Intel Pentium Core i5 1.63 Hz processor and 12 GB RAM, and the algorithm was implemented in MATLAB. The simulation parameters are shown in Table 5.

Table 5. Parameter settings of the proposed algorithm

Parameter	Value	Meaning
N	200	Population size
α	1.1	Significance factor of the inhibitor
β	0.9	Significance factor of the activator
Pc	0.6, 0.7, 0.8, 0.9, 1	Crossover rate
Pm	0.15, 0.2	Mutation rate
MaxGeneration	5000	Maximum number of iterations

The performance of our proposed algorithm was compared to that of a traditional genetic algorithm (GA). The comparative results are shown in Table 6. For each benchmark instance, the maximum number of cycles, for terminating a run of the algorithm, was 5000, and the algorithm was run 50 times for each performance evaluation. The initial population in the proposed algorithm consisted of 200 chromosomes. Each run started with a different random initial population of 200 chromosomes and stopped when the specified maximum number of iterations was reached.

Performance was evaluated relative to best known reported solutions. Every output of the proposed algorithm was the same as the best-known reported solution for every benchmark instance tested on. The exact matching between the output and the corresponding best-known reported solution was achieved by manually tuning the crossover and mutation rates. The exact matching between the output and the corresponding best-known reported solution was achieved by manually tuning the crossover a mutation rates.

The tested crossover rate ranged from 0.6 to 1, while the tested mutation rate ranged from 0.1 to 0.2. We also compared the performances of the proposed algorithm with those of a traditional GA algorithm, listed in Table 6.

Table 6. Experimental Results

No	Instance name	Instance size (n × m)	Best known solution	Traditional genetic algorithm	Relative error (100%)	The proposed algorithm	Relative error (100%)
1	FT06	6 × 6	55	55	0.00	55	0.00
2	LA01	10 × 5	666	686	3.00	666	0.00
3	LA02	10 × 5	655	739	12.82	655	0.00
4	LA03	10 × 5	597	670	12.23	597	0.00
5	LA04	10 × 5	590	634	7.46	590	0.00
6	LA05	10 × 5	593	593	0.00	593	0.00
7	LA06	15 × 5	926	956	3.24	926	0.00
8	LA07	15 × 5	890	994	11.69	890	0.00
9	LA08	15 × 5	863	906	4.98	863	0.00
10	LA09	15 × 5	951	956	0.53	951	0.00
11	LA10	15 × 5	958	958	0.00	958	0.00
12	LA11	20 × 5	1222	1267	3.68	1222	0.00
13	LA12	20 × 5	1039	1071	3.08	1039	0.00
14	LA13	20 × 5	1150	1188	3.30	1150	0.00
15	LA14	20 × 5	1292	1292	0.00	1292	0.00
16	LA15	20 × 5	1207	1383	14.58	1207	0.00
17	LA17	10 × 10	784	906	15.56	784	0.00
18	LA20	10 × 10	902	1072	18.85	902	0.00

5 Conclusion

Aiming at minimizing the makespan of NP-hard JSSPs, we developed an enhanced hydrozoan algorithm with local search strategies and evaluated it on 18 standard JSSP benchmark instances. Compared to GA, the proposed algorithm provided a higher optimization performance, and since it could be tuned to match best-known reported solutions, the implementation of the proposed algorithm on common JSSPs was more practical. The success of the proposed algorithm rested on its three enhanced features: (1) Incorporation of three local search strategies to exploit the local search space more thoroughly; (2) Incorporation of a crossover operator on stronger parents to generate better offspring; and (3) Incorporation of three mutation operators to more widely explore the global search space.

References

1. Kaur, A., Khehra, B., Virk, I.: Makespan optimization in job shop scheduling problem using differential genetic algorithm. Int. J. Comput. Appl. **172**(10), 30–36 (2017)

2. Piroozfard, H., Wong K., Hassan, A.: A hybrid genetic algorithm with a knowledge-based operator for solving the job shop scheduling problems. J. Optim. 1–13 (2016). Author, F., Author, S., Author, T.: Book title. 2nd edn. Publisher, Location (1999)
3. Hasan, S., Sarker, R., Essam, D., Cornforth, D.: Memetic algorithms for solving job-shop scheduling problems. Memetic Comput. **1**(1), 69–83 (2008).LNCS Homepage. http://www.springer.com/lncs. Accessed 21 Nov 2016
4. Bhatt, N., Chauhan, N.: Genetic algorithm applications on job shop scheduling problem: a review. In: International Conference on Soft Computing Techniques and Implementation, pp. 7–14. ICSCTI, Faridabad, India doi: https://doi.org/10.1109/ICSCTI.2015.7489556
5. Peng, C., Wu, G., Liao T., Wang, H.: Research on multi-agent genetic algorithm based on tabu search for the job shop scheduling problem. PLOS ONE **14**(9), e0223182 (2019). https://doi.org/10.1371/journal.pone.0223182
6. Janes, G., Perinic, M., Jurkovic, Z.: An efficient genetic algorithm for job shop schedul-ing problems. Tehnicki vjesnik **24**(4), 1243–1247 (2017)
7. Mekni, S., Fayech, B.: A Modified invasive weed optimization algorithm for multi-objective flexible job shop scheduling problems. Comput. Sci. Inf. Technol. **4**, 51–60 (2014)
8. Brucker, P., Jurisch, B., Sievers, B.: A branch and bound algorithm for job-shop scheduling problem. Discrete Appl. Math. **49**, 105–127 (1994)
9. Tamilarasi, A., kumar, T.: An enhanced genetic algorithm with simulated annealing for job-shop scheduling. Int. J. Eng. Sci. Technol. **2**(1),144–151 (2010)
10. Tansui, D., Thammano, A.: Nature-inspired optimization method: hydrozoan algorithm for solving continuous problems. In: 18th IEEE/ACIS International Conference on Software Engineering, Artificial Intelligence, Networking and Parallel/Distributed Computing, pp. 23–28. IEEE Xplore (2017)
11. Kongkaew, W.: Bat algorithm in discrete optimization: a review of recent application Songklanakarin. J. Sci. Technol. **39**(5), 641–650 (2017)
12. Beasley, E.: OR-library: distributing test problems by electronic mail. J. Oper. Res. Soc. **41**(11), 1069–1072 (1990)

Model Predictive Control of a Fed-Batch Enzymatic Hydrolysis Reactor for Processing Lignocellulosic Feedstock

Sukunya Areeya(✉) and Malinee Sriariyanun

King Mongkut's University of Technology North Bangkok, Bangkok, Thailand
{sukunya.a,malinee.s}@tggs.kmutnb.ac.th

Abstract. Enzymatic hydrolysis process is abundantly applied in ethanol production because this process is operating under mild conditions, low energy requirement and high sugar yield obtained. However, the obstacles of enzymatic hydrolysis process are low rate of reaction and high cost of enzyme. This study was considering the maximization of glucose concentration at final time for enzyme economically used. Model Predictive Control (MPC) based on semi-mechanistic model with competitive inhibition of sugar product was formulated to calculate the manipulated variables in three case studies including: optimal dilution rate problem, optimal endoglucanase and exoglucanase loading problem, and optimal β-glucosidase loading problem. In batch processing time of 119 h, MPC response with dilution rate problem was increased the glucose concentration at final time by 16.79%, which is improving a yield of 84.46%. The closed-loop control has more performant than the open-loop control, while the plant-model mismatch occurred in the process because the error of process can compensate via feedback the measured variables on real-time.

Keywords: Enzymatic hydrolysis process · Fed-batch bioreactor · Lignocellulosic biomass · Model predictive control · Semi-mechanistic model

1 Introduction

Ethanol that is one of the most important biotechnology of alternative energy in term of quantity and market values has been abundantly studied to overcome the current energy demand and future potential energy crisis [1]. The typically process steps of ethanol production from lignocellulosic feedstock include four main steps: (a) the pretreatment process disrupting the crystalline structure of lignocellulose, which has strong chemical bonds, for accessible enzyme in hydrolysis process, (b) the hydrolysis process, which the several enzyme components are converting the pretreated lignocellulose to the reducing sugars such as glucose and xylose via chemical,

physicochemical and biological methods, (c) the fermentation process, which these reducing sugars are fermented to ethanol (d) distillation or dehydration that is ethanol purification to fuel grade specification [2–5]. The enzymatic hydrolysis process is important process in ethanol production because it decides the amount of glucose and ethanol.

P. Meesad et al. (Eds.): IC2IT 2021, LNNS 251, pp. 257–266, 2021.
https://doi.org/10.1007/978-3-030-79757-7_25

However, the main obstacles of enzymatic hydrolysis process are low rate of reaction and high cost of enzyme [6, 7].

The improvements of optimizing the enzymatic hydrolysis that can enhance economic feasibility are expensive and take a long time in a laboratory scale experiments. The model based optimization, which is appropriate strategies in the complex and non-linear process can develop in the short term solution and more appropriate than through heuristic method [3]. In the process optimization strategies, the open-loop control profile is solved by determined once through the offline optimization and unconsidered measurement feedback system. In previous research, the maximizing concentration of desired product at final batch time that is objective function with substrate feed flow rate as control variables (CVs) was implemented to increase a benefit and decrease a batch time. However, the objective function of maximizing concentration has enzyme requirement more than the objective function of minimizing batch time [1, 3]. Both semi-mechanistic multi-reaction model and epidemic-based model were successfully formulated and applied in the optimal control problem of the enzymatic hydrolysis process [6, 8–11], but the semi-mechanistic multi-reaction model is more extensive application than epidemic-based model [1, 3, 12]. To regulate level and output concentration of glucose, the cascade control that is advanced PID application with inlet feed flow rate and outlet flow rate as manipulated variables was formulated in completely stirred tank reactor (CSTR) for enzymatic hydrolysis process of soluble potato starch via aspergillus niger glucoamylase [13].

Model Predictive Control (MPC) strategy which has been abundantly used in chemical industries was applied to fed-batch enzymatic hydrolysis reactor of lignocellulosic feedstock [2]. The closed-loop optimization problem based on a semi-mechanistic model with sugar products as a competitive inhibitor was presented to implement the objective function, which is minimization of an error and a change in manipulated input, with reactor volume as process constraint at final time.

The performance of MPC was given better response by comparing with A classical proportional integral derivative controller and Generic model control (GMC). MPC based on dynamic enzyme-cost FBA model (deFBA) was also applied to the fed-batch bioreactor of *Escherichia coli* for ethanol production [14]. An open-loop and a closed-loop control were implementing a bilevel optimization problem including the substrate feeding and process-level regulation for maximal productivity. The compensate of modelling errors of a closed-loop control are evaluated by comparing with the performance of an open-loop control.

The purpose of this study is applying The MPC strategy to maximize glucose concentration at final time, which is objective function of optimal control problem, for enzymatic hydrolysis of lignocellulosic feedstock in fed-batch reactor. The three problem formulations as manipulated variables, which are including: optimal dilution rate problem, optimal endoglucanase and exoglucanase loading problem, and optimal β-glucosidase loading problem are formulated to implement the objective function. Furthermore, the result of this study will be compared with open-loop system, which also referred to non-feedback system.

2 Nomenclature

E_{iB}	bound enzyme concentration of the ith enzyme (g/Kg cellulose)
E_{iF}	free enzyme concentration of the ith enzyme (g/Kg cellulose)
E_{imax}	the maximum mass of enzyme that can adsorb into a unit mass of cellulose of the ith enzyme (g protein/g cellulose)
K_{3M}	cellobiose saturation constants (g/kg)
K_{iad}	the dissociation constant for the enzyme adsorption of the ith enzyme (g/Kg)
K_{jIG}	inhibition constants for glucose of the jth reaction (g/Kg)
K_{jIG2}	inhibition constants for cellobiose of the jth reaction (g/Kg)
K_{jIX}	inhibition constants for xylose of the jth reaction (g/Kg)
k_{jr}	rate constants of the jth reaction (Kg/g·h)
CB	concentrations of cellobiose (g/Kg)
E_1	enzyme concentration of endoglucanase and exoglucanase (g/Kg cellulose)
E_2	enzyme concentration of β-glucosidase (g/Kg cellulose)
G	concentrations of glucose (g/Kg)
S	concentrations of cellulose (g/Kg)
D	dilution rate (h^{-1})
V	reactor volume (l)

3 Descriptions of Enzymatic Hydrolysis Process

Both batch and fed-batch reactor are broadly used in enzymatic hydrolysis process for lignocellulosic feedstock, nevertheless, batch reactor has high solids loading and high substrate inhibition of lignin through binding or physical interference of by-products. The potential advantage of a fed-batch reactor is the maintenance of a low solid loading during the reaction. Therefore, fed-batch reactor can operate under regulating of solid content and end-product inhibition [12]. In the fed-batch enzymatic hydrolysis reactor, the pretreated lignocellulose is stirred by agitators sited in cylindrical vessel reactor. The dilution substrate is fed into reactor during reaction, oftentimes enzyme also consists in the feed, with no output until completely batch operation. The outer reactor is covered by jacket to maintain the reactor temperature at optimum condition.

Enzymatic hydrolysis is synergistic action of three type of enzymes, which are including endoglucanase (EG), exoglucanase (CBH), and β-glucosidase (Cellobiases) [5]. Usually, these enzymes are together called cellulase or cellulolytic enzymes [4]. The steps of enzymatic hydrolysis consist of three steps. The first step is a heterogeneous reaction, which endoglucanase in liquid phase binds to low-crystallinity regions of cellulose fiber to create free chain-ends. The second step is exoglucanase decomposing the sugar chain to remove cellobiose, which is dimer of glucose from free chain-ends. At last step, β-glucosidase degrades the cellobiose molecules into glucose [2–4].

To simplify the model development, Kadam et al. [10] has concept to organize the six of sugar productions including glucose, cellobiose, galactose, mannose, xylose, and arabinose into three group of sugar, viz. cellobiose, xylose that is substitute of pentose (xylose and arabinose), and glucose that is substitute of hexoses (glucose, galactose, and mannose).

In optimal control problem of fed-batch enzymatic hydrolysis, the optimal dilution rate, which is a ratio of feed flow rate and reactor volume is mostly computed to obtain the objective function such as maximizing concentration of desired product,

minimizing batch time and maximizing profit. The assumption of this study is the product components measured on real-time and the reactor volume at final time as the constraint of optimal control problem. The objective function in this study is maximizing concentration of desired product.

4 Kinetic Modelling of Enzymatic Hydrolysis Process

4.1 Hydrolysis Rate Equation

Owing to a fed-batch operation causing the change in volume variable, Hodge et al. [15] is modifying the semi-mechanistic model [1] to evaluate the concentrations of substrates and product in fed-batch reactor. The assumption of these models is perfect mixing. In other assumption, due to the jacket can great maintain the reactor temperature at 45 °C, the effect of temperature variability and energy balance equation are neglected. Modified model equations are as follows.

$$\frac{dS}{dt} = D(S_{feed} - S) - r_1 - r_2 \tag{1}$$

$$\frac{dCB}{dt} = D(CB_{feed} - CB) + 1.056r_1 - r_3 \tag{2}$$

$$\frac{dG}{dt} = D(G_{feed} - G) + 1.111r_2 + 1.053r_3 \tag{3}$$

$$\frac{dE_1}{dt} = D(E_{1feed} - E_1) \tag{4}$$

$$\frac{dE_2}{dt} = D(E_{2feed} - E_2) \tag{5}$$

$$\frac{dV}{dt} = DV \tag{6}$$

The Michaelis-Menten kinetics with competitive inhibition of sugar products were appropriated to simulate the enzymatic hydrolysis of cellulose. The three hydrolysis reactions are including: (1) the heterogeneous reaction of cellulose breakdown to cellobiose (r_1) with competitive inhibition of cellobiose, glucose and xylose that has synergistic action of endoglucanase and exoglucanase together represented by E_1, (2) the heterogeneous reaction of cellulose breakdown to glucose (r_2) with competitive inhibition of cellobiose, glucose and xylose that has synergistic action of cellulolytic enzymes, and (3) the homogeneous reaction of hydrolyzing cellobiose breakdown to glucose (r_3) by β-glucosidase represented by E_2 with competitive inhibition of glucose and xylose. These hydrolysis reactions are given by Eq. (7), (8), and (9) respectively.

$$\frac{dS}{dt} = D(S_{feed} - S) - r_1 - r_2 \tag{7}$$

$$\frac{dCB}{dt} = D(CB_{feed} - CB) + 1.056r_1 - r_3 \quad (8)$$

$$\frac{dG}{dt} = D(G_{feed} - G) + 1.111r_2 + 1.053r_3 \quad (9)$$

4.2 Enzyme Adsorption

The Langmuir adsorption isotherm is used to describe the amount of enzyme bounding to substrate surface, given by Eq. (10). The total enzymatic concentration at any time t ($E_{iT}(t)$), is sum of free enzymatic concentration (E_{iF}) and bound enzymatic concentration (E_{iB}), given by Eq. (11).

$$E_{iB} = \frac{E_{i\max} K_{iad} E_{iF} S}{1 + K_{iad} E_{iF}} \quad (10)$$

$$E_{iT}(t) = E_{iF}(t) - E_{iB}(t) \quad (11)$$

5 Model Predictive Control of Enzymatic Hydrolysis Process

An open-loop control profile also called a non-feedback control profile is determined by an optimal control strategy through once offline optimization. A Closed-loop optimal control strategies are applied in Model predictive control (MPC). MPC is model based control to find optimal manipulated variables for implemented objective function and can apply in Multiple-Input Multiple-Output (MIMO) systems. Moreover, it can combine feedforward and feedback action, which indicate the ability of MPC to forecast the future state behavior from input variable deviations [16, 17]. To consider the structure of MPC, the control horizon (M) which contains a present input u(t) is defined and calculated to reach the objective function that must cover throughout all a set of predicted output (P) [2]. Tuning parameters of MPC that are P and M were approximated by FOPDT models [18, 19]. The MPC structure is shown as follows [1].

$$\max_{u(t)}(G(t_f)) \quad (12)$$

$$\dot{X} = g(X(t), u(t), t); \; X(t_0) = X_0 \quad (13)$$

$$V(t + t_f) = V_{\max} \quad (14)$$

$$u_{\min} \le u(t) \le u_{\max} \quad (15)$$

In this study, we aimed to maximize the objective function that is the concentration of desired product at final time ($G(t_f)$), which is shown in Eq. (12). The accurate dynamic models, which represent in terms of state variables, input variables and manipulated variable, are taken to MPC structure given by Eq. (13). The constraint at final time, which this study was volume of the reactor is given by Eq. (14). The manipulated variable (u(t)) is optimized under conditions of inequality process constraint given by Eq. (15). MPC was formulated to implement this aim through three problem formulations of different the optimal manipulated variables in order to improve the fed-batch operational performance of the enzymatic hydrolysis reactor.

Table 1. The operational parameters and constraint of the manipulated variable for the formulations of control problems [1, 3].

Parameter		MVs for each scenario				Initial condition (t_0)	The initial values of MVs	u_{min}	u_{max}
		Base case	DP	EIP	EIIP				
1	D	✓	✓			0.00345	0.00345	0	2.64
2	E_1			✓		45	0	0	20.54
3	E_2				✓	4.16	0	0	20.54

The first control scenario of the three problem formulations is the optimal dilution rate problem (DP) that the dilution rate as manipulated variable is calculated to implement the maximizing glucose concentration [1]. The second control scenario is the optimal endoglucanase and exoglucanase loading problem (EIP). The endoglucanase and exoglucanase concentrations (E_1) in feed flowrate are defined as the manipulated variable [20]. The last of problem formulations is the optimal.

β-glucosidase loading problem (EIIP), which β-glucosidase is a manipulated variable [20]. The operating conditions and constraint of the manipulated variable for the formulations of control problems are defined by Table 1.

Table 2. Hydrolysis kinetic parameters for lignocellulosic feedstock [11, 13].

Parameter	Value	Unit
E_{1max}, E_{2max}	0.06, 0.01	[g protein/g cellulose]
k_{1r}, k_{2r}	22.3, 7.18	[kg/g h]
k_{3r}	285.5	[h^{-1}]
K_{1ad}, K_{2ad}	0.4, 0.1	[g protein/g cellulose]
K_{1IG2}, K_{2IG2}	0.015, 132	[g/kg]
K_{1IG}, K_{2IG}, K_{3IG}	0.1, 0.04, 3.9	[g/kg]
K_{1X}, K_{2X}, K_{3X}	0.1, 0.2, 201.0	[g/kg]
K3M	24.3	[g/kg]

6 Results and Discussion

To maximize the glucose concentration at final time through MPC strategy, the forward Euler method that is numerical procedure was used to solve kinetic model of enzymatic hydrolysis process represented in system of ordinary differential equations (ODEs), which was explained in Sect. 4. The results of MPC response were solved in MATLAB. In this study, M and P were approximated as 56 and 14, respectively. The kinetic data of these kinetic models reported by Kadam et al. [10] were used in this study, as shown in Table 2. The initial and other related parameter values were reported by Vegi et al. [1].

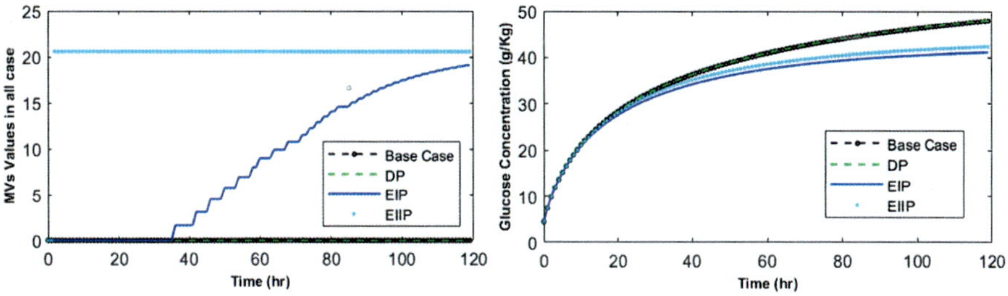

Fig. 1. MPC response for maximization of glucose concentration at final time in difference case of control problem formulations.

The volume reactor at final time (V_{max}) of 10,000 L was considered for all case of control problems studied [1]. This study was not considered the impact of particle size on process performance. The initial conditions of cellulose, glucose, cellobiose, and xylose were 59.2, 4.5, 3.2, and 30 g/Kg, respectively. The initial condition of volume reactor was given as 6,610 L. An operating batch time was defined to 119 h.

The amount of enzyme, which is one of the most important factor in process economic was accounted for all case of control studied. The results of MPC response to implement the maximizing glucose concentration at final time were studied and compared in the three problem formulations, moreover, these results were compared with open-loop system. The results of Enzymatic hydrolysis control were discussed as follows.

The DP, which was implementing the maximizing glucose concentration at final time slightly went up by 4.93% of the desired product yield (Table 3) by comparing with the performance of an open-loop control (Base Case) that was probably caused by the limitation of the optimization under conditions of inequality process constraint. Both the EIP and the EIIP was poor performance in control by comparing with base case, which were Declined in 14.23% and 11.71%, respectively, while enzyme consumption of the EIP and the EIIP were higher than base case in 18.76% of E_1 loading and 489.16% of E_2 loading, respectively. However, the closed-loop control has more performant than the open-loop control, while the plant-model mismatch occurred in the process because the error of process can compensate via feedback the measured variables on real-time. MPC response for maximization of glucose concentration at final time in difference case of control problem formulations is given in Fig. 1.

Table 3. Comparison of MPC results for maximization of glucose concentration at final time (119 h) in different manipulated variables (MVs).

Problem formulations	$G\,(t_f)$ [g/Kg]	Yield [%]	Total enzyme addition in the reactor (g protein/ batch)	
			E_I [g/Kg]	E_I [g/Kg]
1. Open-loop system (Base case)	47.88	80.87	45	4.16
2. Optimal dilution rate problem (DP)	50.00	84.46	45	4.16
3. Optimal E1 loading problem (EIP)	41.06	69.36	53.44	4.16
4. Optimal E2 loading problem (EIIP)	42.27	71.40	45	24.53
Plant-model mismatch in an error of −20% in the k_{3r}				
1. Open-loop system (Base case)	40.97	69.21	45	4.16
2. Optimal dilution rate problem (DP)	47.85	80.83	45	4.16
3. Optimal E1 loading problem (EIP)	40.74	68.82	55.28	4.16
4. Optimal E2 loading problem (EIIP)	42.17	71.23	45	24.50

To evaluate the performance of the close-loop controller, the plant-model mismatch was considered in term of an error of -20% in the rate constants of homogeneous reaction for hydrolyzing cellobiose to glucose (k_{3r}), which has most effective in the performance of controller for enzymatic hydrolysis of lignocellulosic feedstock in fed-batch reactor [2]. To consider the percent yield of the desired product (Table 3), the DP can be reached highest yield and obtained 16.79% increase by comparing with the performance of base case. The EIIP was also slightly higher than base case of 2.92%. However, not only

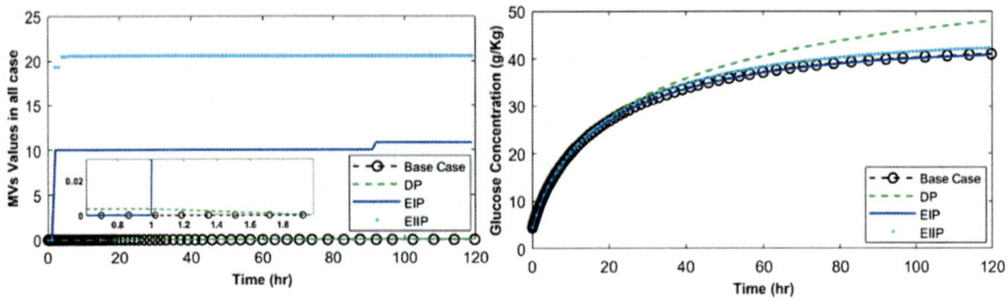

Fig. 2. MPC response for a −20% in k_{3r} for maximization of glucose concentration at final time in difference case of control problem formulations.

was the EIP less than base case in 14.90%, but the EIP was also using 22.84% more E_1 loading than base case. The consumption of the EIIP was numerously a 488.94% increase in E_2 loading when compared with base case (Fig. 2).

7 Conclusion

The optimal manipulated variables in difference case of control problem formulations including: optimal dilution rate problem, optimal endoglucanase and exoglucanase loading problem, and optimal β-glucosidase loading problem was formulated to maximize the glucose concentration at final time. The optimal dilution rate problem can improve yield of process by 16.79% for batch time of 119 h. The closed-loop control has more performant than the open-loop control, while the plant-model mismatch occurred in the process because the error of process can compensate via feedback the measured variables on real-time.

Acknowledgements. The authors would like to thank King Mongkut's University of Technology, North Bangkok (Research University Grant No. KMUTNB-BasicR-64–37) for financial support of this work.

References

1. Vegi, S., Shastri. Y.: Optimal control of dilute acid pretreatment and enzymatic hydrolysis for processing lignocellulosic feedstock. J. Process Control **56**, 100–111 (2017)
2. Areeya, S., Tawai, A., Sriariyanun, M.: Model based control for fed-batch enzymatic hydrolysis reactor of lignocellulosic biomass. In: 6th International Conference on Control, Automation and Robotics, pp. 536–540. IEEE Xplore, Singapore (2020)
3. Fenila, F., Shastri, Y.: Optimal control of enzymatic hydrolysis of lignocellulosic biomass. Resour.-Eff. Technol. **2**, 96–104 (2016)
4. Taherzadeh, J.M., Karimi, K.: Enzyme based hydrolysis processes for ethanol from lignocellulosic materials: a review. BioResources **2**, 707–738 (2007)
5. Prunescu, M.R., Sin, G.: Dynamic modeling and validation of a lignocellulosic enzymatic hydrolysis process – a demonstration scale study. Biores. Technol. **150**, 393–403 (2013)
6. Angarita, D.J., Souza, A.B.R., Cruz, G.J.A., Biscaia, C.E., Secchi, R.A.: Kinetic modeling for enzymatic hydrolysis of pretreated sugarcane straw. Biochem. Eng. J. **104**, 10–19 (2015)
7. Gupta, R., Kumar, S., Gomes, J., Kuhad, C.R.: Kinetic study of batch and fed-batch enzymatic saccharification of pretreated substrate and subsequent fermentation to ethanol. Biotechnol. Biofuels **5**(1), 1–10 (2012)
8. Tai, C., Arellano, G.M., Keshwani, R.D.: Epidemic based modeling of enzymatic hydrolysis of lignocellulose biomass. Biotechnol. Prog. **30**(5), 1021–1028 (2014)
9. Sin, G., Meyer, S.A., Gernaey, V.K.: Are mechanistic cellulose hydrolysis models reliable for use in biofuel process design? - Identifiability and sensitivity analysis. In: 7th International Conference on the Foundations of Computer-Aided Process Design, pp. 93–106. Taylor & Francis Group, United States (2009)
10. Kadam, L.K., Rydholm, C.E., McMillan, D.J.: Development and validation of a kinetic model for enzymatic saccharification of lignocellulosic biomass. Biotechnol. Prog. **20**(3), 698–705 (2004)

11. Ye, X., Chen, Y.: Kinetics study of enzymatic hydrolysis of paulownia by dilute acid, and ultrasonic assisted alkali pretreatments. Biotechnol. Bioproc. Eng. **20**, 242–248 (2015)
12. Tai, C., Keshwani, R.D., Voltan, S.D., Kuhar, S.P., Engel, J.A.: Optimal control strategy for fed-batch enzymatic hydrolysis of lignocellulosic biomass based on epidemic modeling. Biotechnol. Bioeng. **112**(7), 1376–1382 (2015)
13. Fonseca, R., Dantas, S.S.T., Silva, D.V.F.: Simulation and control of enzymatic hydrolysis reactor of soluble potato starch by aspergillus niger glucoamylase for fermentation process feed. COBEQ. Brazil **1**, 12448–12455 (2015)
14. Jabarivelisdeh, B., Findeisen, R., Waldherr, S.: Model predictive control of a fed-batch bioreactor based on dynamic metabolic-genetic network models. IFAC-PapersOnLine **51**, 34–37 (2018)
15. Hodge, B.D., Karim, N.M., Schell, J.D., McMillan, D.J.: Model-based fed- batch for high-solids enzymatic cellulose hydrolysis. Appl. Biochem. Biotechnol. **152**(1), 88–107 (2009)
16. Romagnoli, A.J., Palazoglu, A.: Introduction to process control. 2nd ed. CRC Press (2012).
17. Kittisupakorn, P.: Model based control for batch chemical process, 1st edn. Chulalongkorn University Printing House, Thailand (2008)
18. Shridhar, R., Cooper, D.J.: A tuning strategy for unconstrained SISO model predictive control. Ind. Eng. Chem. Res **36**(3), 729–746 (1997)
19. Jha, S.K., Karthika, S., Radhakrishnan, T.K.: Modelling and control of crystallization process. Resour.-Eff. Technol. **3**(1), 94–100 (2017)
20. Rodríguez, M.R., Capron, M., Huusom, K.J., Sin, G.: Controlled fed-batch operation for improving cellulose hydrolysis in 2G bioethanol production. In: 20th European Symposium on Computer Aided Process Engineering, Italy (2010)

Split-Match-Merge Method with Multi-matcher Algorithm to Handle Data Interoperability Problems in Small Farm Management Information System

Henriyadi[1](✉), Vatcharaporn Esichaikul[2], and Chutiporn Anutariya[2]

[1] Information Management, Asian Institute of Technology, Bangkok, Thailand
st118501@ait.ac.th

[2] ICT Department, Asian Institute of Technology, Bangkok, Thailand

Abstract. Farm Management Information System (FMIS) is an application to help farmers manage their farm. Small Farm Management Information System (sFMIS) is FMIS application targeted at small scale farmers. sFMIS only provide functionality that required by farmers. Besides, sFMIS utilize as much as possible free open data sources to reduce the cost. However, utilization of external data sources to support the sFMIS is not trivial, complicated, and faces many heterogeneity problems. This study proposed a split-match-merge method with a multi-matcher algorithm to handle those problems. Split-match-merge method aimed to make the matching process faster than a conventional method. A multi-matcher algorithm used to improve the quality in matching. The multi-matchers algorithm is a combination of several algorithms in hybrid and composite models. The hybrid model consists of exact-matcher, synonym matcher and composite similarity-matcher. Composite matcher contains Levenshtein, Jaro-Winkler, LCCS, Cosine-similarity, and Sorensen-dice algorithms. This research conducts evaluations by comparing split-match-merge methods with traditional methods and calculating the precision and recall multi-matcher algorithm. There are 282 records extracted from ontoFMIS.rdf and 430 records pulled from the data source database used for evaluation. Split-match-merged reduces the matching process from 74 to four minutes. The split-match-merge also reduce the number of matched tuples from 180 into 123 matched tuples. It means that 57 tuples indicate an orphan-tuple, a tuple that assigns as matched but has a schema heterogeneity problem. Moreover, the precision of the multi-matcher algorithm is 0.946 and recall is 0.88. This score indicated the algorithm promising in handling data interoperability problems.

Keywords: FMIS · sFMIS · Small Farm Management Information System · Split-match-merge · Multi-matchers algorithm

1 Introduction

Crop production is a consecutive production process that commences from pre-planting, planting, harvesting, post-harvesting until selling the product. Crop productions involve

P. Meesad et al. (Eds.): IC2IT 2021, LNNS 251, pp. 267–278, 2021.
https://doi.org/10.1007/978-3-030-79757-7_26

natural, human-made, and human resources. To manage the limited land resources, labour, and capital require farm management; therefore, crop production can run effectively and efficiently [1]. Besides manually, farmers can apply farm management electronically. Farm Management Information System (FMIS) is a common term used for farm management electronically. Sorensen [2] define FMIS as an information system for gathering, handling, putting away and disseminating farm data as information expected to convey. Furthermore, Fountas et al., [3] identify the six main activities or functions of the FMIS, namely (a) to manage and monitor the farm production process, (b) to manage and monitor the availability of resources, (c) to manage human resources that involved in the farm production process, (d) to manage the availability of material, (e) to manage finance, and (f) take care of external regulation. Nowadays, FMIS more sophisticated through integration and implementation of new technologies [4].

Existing FMIS generally characterized as complex applications requiring high operational costs, specialized operators, and segmented to large-scale farm. In contrast, about 98% of farmers globally are small scale or family farmers [5]. Providing an FMIS application at an affordable price for small-scale farmers is challenging [6]. Therefore, this paper proposes an FMIS application at an affordable price called a small Farm Management Information System (sFMIS). There are two approaches use in sFMIS, namely (1) only provides functionalities that indeed required by small farmers in a specific area and (2) utilize as much as possible free external data sources.

However, using free open data source through a data interoperability framework in sFMIS is not trivial, complicated and face many problems. Bergman [7] identify about 40 distinct potential sources of data interoperability problems. Furthermore, this paper focuses on handling three main issues in data interoperability comprehensively, namely:

- *Heterogeneity schema.* Heterogeneity schema exists when different data models or different schemas within the same data model.
- *The granularity of data.* The granularity of data happens when there is a different level of detail of data presentation.
- *Mismatch the entity naming or data unit.* Mismatch entity naming or data unit happen when two domains have different entity name or data unit even though both entities are similar.

2 Previous Works and Problem Statement

There are many previous researchers proposed method in handling data interoperability problems through schema and entity matching method. Some researchers already propose methods to overcome the heterogeneities of schema [8]. The most use and discuss in scientific article [9] are AgreementMaker [10], COMA++ [11, 12], Cupid [13], Falcon-AO [14], and S-Match [15]. Concerning handling granularity of data, there are many approaches already proposed, such as neutrality [16], adaptive mining technique [17], and Clustering analysis [18]. Some researchers also proposed approaches to overcome the mismatch with entity naming and data unit issues [19–24]. However, there is not find yet previous research explicitly states that their approach could handle the three main data interoperability problem comprehensively.

Furthermore, no research paper calculates the algorithm performance in handling big-size data sets. This aspect is essential because schema and entity matching between two data sets generally involve big size records. It is implying the number of iterations as a multiplication number of records of two tuples also big. Suppose one schema consists of 500 tuples and another schema consisting of 1000 tuples, the total number iteration of matching is equal with $500 \times 1000 = 500{,}000$ iterations with possibility matching only about $1/500 = 0.2\%$. Therefore, besides the quality of matching, should consider the time to finalize the matching process. How to handle all heterogeneities problems in schema level with high accuracy in reasonable execution time? Therefore, this study proposed a *split-match-merge* method with a *multi-matcher algorithm* to comprehensively tackle semantic heterogeneities problems in the schema level. This method aims to tackle data interoperability problems comprehensively with high matching accuracy in reasonable execution time.

3 Material and Methods

This study uses two materials, namely (1) OntoFMIS ontology as a reference schema and (2) temporary database that contain the results extraction process of the external data as the external data source database. OntoFMIS is a proposed ontology concerning small farm management information system. OntoFMIS combines create-own ontology and existing ontologies such as FoaF, DCMI, schema.org, AGROCOV, good relation, soil ontology, crop ontology, and weather. The temporary URI for OntoFMIS available at http://124.81.126.50/ontoFMIS/ontofmis.rdf. Moreover, the external data source database consists of some non-related tables. Each table consists of the schema and data resulting from all external data sources' extraction process. The temporary database is available at http://124.81.126.50/ontoFMIS/eds.sql.

Moreover, the research methodology for this study present in Fig. 1. Short description of the research methodology as follows:

1. The first step is extracting both materials into a set of tuples. This phase consists of two activities, (a) extraction external data source database structure into a list of data source tuple schema and (b) extraction OntoFMIS ontology into a set of reference tuple schema. Data source tuple schema considers as 'ds', and reference schema considers as 'rs'.
2. The following step is finding concept matching between 'rs' and 'ds', which matches the *table_name* of 'ds' with *class* or *subclass* of 'rs'. The result of this step is a concept matching table. This matching process employs multi-matcher algorithm.
3. The next step is applying split-match-merge method. This step consists of three consecutive processes: split, match, and merge. This phase's result is a list of tuples in reference schema 'rs' that correlates with one or more tuples in the data source schema 'ds'.
4. Multi-matcher algorithm. The multi-matcher algorithm is an algorithm to involve in the string-matching process. This algorithm involves some algorithms, namely: word preprocessing, string matching, string similarity, and datatype matching.

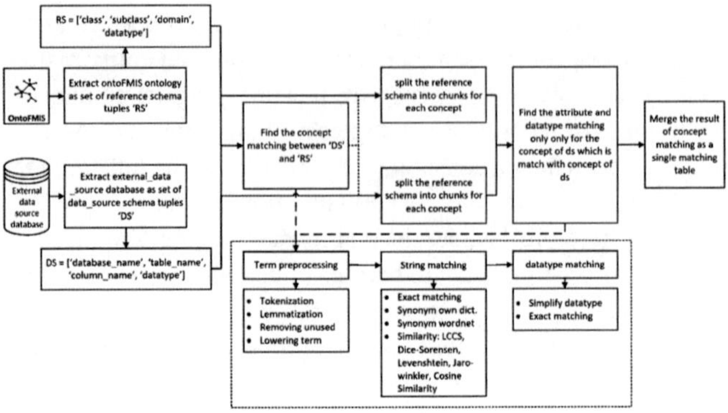

Fig. 1. Research methodology

4 Results and Discussion

4.1 Extraction Database and Ontology into a Schema

Extract the External Data Source Database into Data Source Schema (ds). Extraction external data source database is a process to extract schema of the external data source database into set tuples of the data source schema (ds). This schema consists of four columns, namely database_name, table_name, column_name, and datatype. Algorithm 1 presents the detail of extracting external data source database.

```
# Algorithm 1. extracting database into ds schema
Input: external data source MySQL database
Output: list of external data sources tuples
Package imported: pandas, mysql.connectors
Step  Description
1     established connection to MySQL database server
      #use mysql.connectors
2     read TABLE_SCHEMA, TABLE_NAME, COLUMN_NAME,
      DATA_TYPE from "INFORMATION_SCHEMA.COLUMNS"
      from the external data source database
3     extract schemas (db_name, table_name, column_name,
      datatype)
4     save as data_source_schema.csv
```

Extract OntoFMIS Ontology as Reference Schema (rs). Extract OntoFMIS ontology as a reference schema (rs) is an extraction process from OntoFMIS ontology to set tuples as a reference schema. The schema consists of four columns, namely class, subclass, domain, and datatype. The algorithm to extract OntoFMIS ontology into reference schema present in Algorithm 2.

```
# Algorithm 2. Extracting ontology as reference schema
Input: ontoFMIS.rdf ontology
Output: list of reference_schema tuples
Package imported: pandas, rdflib
Step Description
1    find URL of ontoFMIS.rdf
2    open the rdf URL location #use rdflib package
3    read the rdf graph as a set of s, p, o
4    remove unused string #do schema cleansing
5    remove non-text characters #do schema cleansing
6    convert RDF triple (s, p, o) into set of tuples
7    save the result as reference_schema.csv
```

4.2 Concept Matching

The following process is finding concept matching between *class/subclass* of '*rs*' and *table_name* of '*ds*'. The output is a matching table of *class/subclass* of 'rs' that matches with *table_name* of 'ds'. The detail of concept matching present in Algorithm 3.

```
# Algorithm 3. Concept matching
Input: data source schema and reference schema
Output: concept matching table
Package imported: pandas, multi-matcher, word_preprocess
Step Description
1    open and read reference_schema.csv as rs
2    open and read data source_schema.csv as ds
5    for each row in ds:
6      get value table_name
7      for each row in rs
8        get value class of rs
9        get value subclass of rs
10       if string matching (class, table_name) is match
11         result= class is upper level of table_name
12         save the result
13       elif string matching (subclass, table_name) is match
14       result = subclass is same level with table_name
15       save the result
16   save all result into concept_matching.csv
```

4.3 Split, Match and Merge Method

The split-match-merge method, as present in Fig. 2 is the modification of split-apply-merge method proposed by Wickham [25]. The modification from the original method is related to "split" and "apply" phase. The comparation between split-apply-merge method with the split-match-merge method shown in Table 1.

Table 1. Comparation between split-apply-merge method with the split-match-merge method.

No	Step	Original split-apply-merge	Modify split-match-merge
1	Split	Splitting the single list of tuples into several subsets of tuples	Splitting two long lists of tuples into subset tuples based on some criteria
2	Apply/match	Focus on mathematical or Boolean process	Change from apply to match. Focuses on string matching
3	Merge	Merge the result	Merge the result

Split. Split is the first phase. Split is splitting the long list of tuples data into small chunk subset of tuples. This splitting use list of class/subclass matching that already executes previously. Algorithm 4 present the detail of the split-phase of reference schema. The similar algorithm also applies for data source schema.

```
# Algorithm 4. Split schema into set chunks of concept
Input: concept matching schema and reference schema
Output: set of concept chunks for reference schema
Package imported: pandas
Step Description
1      open and read reference schema as rs
2      open and read concept matching as cm
3      group of rs based on class
4      for i in cm.index:
5        get value rs concept
6      for key, item in rs_grouped:
7        rs_group = rs_grouped.get_group(key)
8        for g in rs_group.index:
9          get_value of class
10         get_value of subclass
11         if class == subclass and subclass == cm value):
12           Get value of class, subclass, attribute, range
13          elif class == cm_value:
14           Get value of class, subclass, attribute, range
15          elif subclass == cm_value:
16           Get value of class, subclass, attribute, range
17     merge all result
18     save the result into rs_concept_chunk.csv
```

Match and Merge. Matching and merging is an integrated process. The matching process only executes which *class/subclass* of 'rs' matches with a *table_name* of 'ds'. While Merge is process merges all the matching results into one file as a matching table for the ETL process. Algorithm 5 present the matching and merging process.

```
# Algorithm 5. Matching and Merging
Input: chunk of data source schema, chunk of reference
schema, and concept matching table
Output: table of concept matching
Step  Description
1   read and convert concept matching.csv as a dictionary
2   read list chunk of concept reference_schema as rsc
3   read list chunk of concept data_source_schema as dsc
4   for d, r in (dsc.index and rsc.index):
5    get_value of rs_class
6    get_value of ds_class
7   for key, value in class_dict.items():
8   if ds_concept in key and value == rs_concept:
9      get_value of ds_attribute
10      get_value of rs_attribute
11      if string_matching(ds_attribute,
          rs_attribute)) == 'match':
12        save the result as 'match'
13      else:
14        save the result as 'do not match.'
15    merge all result # merge all match attributes
16    save as concept_matching.csv
17    count attribute matches for each match concept
      #concepts match if at least two chunks are matches
18    if total number of attribute match > 2
19      reference concept and data source concept 'match'
20    else
21      reference concept and data source 'do not match'
```

4.4 Multi-matcher Algorithm Based on Decision Tree

Multi matcher algorithm applies in both concept matching and string matching. Multi-matcher is a combination of hybrid and composite from many matcher algorithms. Decision tree used in this research because the matching process is a hierarchical process that uses a specific algorithm for each branch. The result for each branch of decision tree process only provides two answers: (a) "Yes" if both strings match, and (b) "No" if both strings do not match. Matcher algorithms used in this paper as follows:

- Exact matcher: a matcher to find whether two words are exactly matches.
- Word synonym: a matcher to check whether two strings are synonym. Three sources used as of dictionary, namely (1) Wordnet in English, (2) Wordnet in the Indonesian Language, and (3) create own dictionary.
- Similarity matcher: a matcher to find similarity of two terms. This study uses a hybrid model as a combination of five string-similarity algorithms, namely: (1) Levenshtein[1], (2) Jaro-Winkler[2], (3) Cosine similarity[3], (4) Longest Common substring (LCS)[4], and (5) Sorensen-dice[5]. The result for each algorithm will be normalized and calculate the average similarity with acceptance score is 0.9.

Algorithm 6 present the detailed algorithm for word preprocessing. Whereas Algorithm 7 present the detailed algorithm for string matching.

```
#Algorithm 6. Word preprocessing
Input: data source schema and reference schema
Output: "root" of the word for each term
Package imported: pandas, re, nltk.corpus, nltk.stem,
                  nltk.tokenize
Step  Description
1     for each list of tuples
2       read string
3       split string into word or term
4       convert each word into lowercase
5       remove stop-word # use stop_words
6       tokenize word # use word_tokenize
7       lemmatize word # use WordNet_Lemmatizer
8       save the result
9     continue for next process
```

[1] https://en.wikipedia.org/wiki/Levenshtein_distance.
[2] https://en.wikipedia.org/wiki/Jaro%E2%80%93Winkler_distance.
[3] https://en.wikipedia.org/wiki/Cosine_similarity.
[4] https://en.wikipedia.org/wiki/Longest_common_substring_problem.
[5] https://en.wikipedia.org/wiki/S%C3%B8rensen%E2%80%93Dice_coefficient.

#Algorithm 7. String matching

```
Input: word1, word2
Output: term matching
Package imported: pandas, nltk.corpus, textdistance
Step  Description
1     read own_dictionary.csv
2     converts own_dictionary.csv into dictionary
3     word 1 == word from reference schema
4     word2 == word from data source schema
5     if word1 == Word2
6       result = 'match'
7       go to next matching process
8     else
9     for key, value in List_of_abbrv():
10      if word1 in value and value == word2:
11        result = 'match'
12        Go to next matching process
13     else:
14    synonyms = []
15    for syn in wordnet.synsets(word2):
16      for line in syn.lemmas():
17        synonyms.append(line.name())
18    synonyms_idn = []
19      for syn in wordnet.synsets(word2, lang='ind'):
20        for i in syn.lemmas():
21          synonyms_idn.append(i.name())
22          if word1 in synonyms: #synonym in English
23          result = 'match'
24          go to next matching process
25      else:
26      if word1 in synonyms_idn: #synonym in Bahasa
27        result = 'match'
28        go to next matching process
29  else: #calculate the similarity based on some method
30    lev=levenshtein.normalized_similarity(word1, word2)
31    jaro=jaro_winkler.normalized_similarity(word1, word2)
32    cos=cosine.normalized_similarity(word1, word2)
33    lccs=lccs.normalized_similarity(word1, word2)
34    dice=dicesorensen.normalized_similarity(word1, word2)
35    sim_avg = (lev+jaro+cos+lccs+dice)/5
36    if sim_avg >= 0.9:
37      result = 'match'
38      go to next matching process
```

4.5 Evaluation of the Algorithm and Method

There are two approaches in evaluation the algorithm and method, namely (1) calculating the precision, recall, and accuracy of the result of the multi-matcher algorithm process and (2) comparing between *split-match-merge* method with a non-splitting method based on four indicators.

Evaluation of the Multi-matcher Algorithm. Precision, recall, and accuracy analysis is a method to evaluate the performance of the multi-matcher algorithm. The evaluation conducted through executing the multi-matcher algorithm between term from reference schema and term from data source schema. There are 218 terms extracted form reference schema and 139 terms extracted from data source schema. Table 2 present the result of the calculation the True Positive (TP), False Positive (FP), True Negative (TN), and False Negative (FN). The multi-matcher algorithm found 93 term matches between the term from 'rs' and term from 'ds', and 30,209 terms do not match. From 93 terms indicate as match terms, use manual found that five terms exactly not match. Whereas 'not match' result, from 30,209 that resulted not match found 12 terms that should be matching.

Table 2. Calculation of TP, TN, FP, FN.

		Manual checking	
		Positive	Negative
Multi-matcher algorithm	Positive	TP = 88	FP = 5
	Negative	FN = 12	TN = 30,193

Furthermore, the accuracy, precision and recall the proposed algorithm as follows:

$$\begin{aligned}
\text{Accuracy} &= (\text{TP} + \text{TN})/(\text{TP} + \text{FP} + \text{FN} + \text{TN}) \\
&= (88 + 30193)/(88 + 5 + 30193 + 12) \\
&= 0.99 \\
\text{Precision} &= \text{TP}/(\text{TP} + \text{FP}) \\
&= 88/(88 + 5) \\
&= 0.946 \\
\text{Recall} &= \text{TP}/(\text{TP}+\text{FN}) \\
&= 88/(88 + 12) \\
&= 0{,}88
\end{aligned}$$

The result of calculation accuracy, precision, and recall indicates the algorithm is promising to use to handle the data interoperability problems for sFMIS.

Evaluation Split-Match-Merge Method. The evaluation of the method used four indicators: (1) time to execute, (2) non-duplicate of iteration, (3) number of tuples matched and (4) percentage matched tuples. Reference schema that extracted from ontoFMIS.rdf consists of 282 sets of tuples and data source schema extracted from a data source database consisting of 430 sets of tuples. Data presented at Table 3 indicate that split-match-merge method gives a significant improvement in execution time, about 4 min against 74 min. Furthermore, the split-match-merge approach also reduces the number of iterations from 3647 iterations to 2787. The method also found 57 orphan tuples. Orphan tuples are a concept that only has one attribute matches to another concept.

Table 3. Calculation result for each indicator for each method.

No	Method	Indicators			
		Time to execute	Iterations*	Matched tuples	% matched
1	Non-splitting	1:14:37 (h:m:s)	3647	180	4.49%
2	Split-match-merge	0:03:52 (h:m:s)	2787	123	4.41%

* Iteration is the process of comparing each tuple from reference schema (rs) with data source schema (ds). Reducing the number of iteration processes means reducing the time to complete the matching between rs with ds.

5 Conclusion

Small Farm Management Information System (sFMIS) is an FMIS application that targeted for small scale farmers. sFMIS provide the functionality required by farmers and use external open data sources as much as possible to reduce the operational cost. However, utilization of external data sources to support the sFMIS is not trivial, complicated and face many heterogeneity problems. Three main problems that may arise in schema level are schema heterogeneity, granularity data, and mismatch entity and datatype. The calculation precision and recall result indicate that multi-matcher algorithm promising in handling string matching between reference schema and data source schema. Compared with the traditional method, the split-match-merge method significantly reduces the time to execute the matching process. Moreover, utilizing create-own matching dictionary could help increase the quality of schema matching.

References

1. Kay, R., Edwards, W., Duffy, P.: Farm Management, 8th edn. McGraw-Hill Education, New York (2015)
2. Sørensen, C.G., et al.: Conceptual model of a future farm management information system. Comput. Electron. Agric. **72**, 37–47 (2010). https://doi.org/10.1016/j.compag.2010.02.003
3. Fountas, S., Sorensen, C.G., Tsiropoulos, Z., Cavalaris, C., Liakos, V., Gemtos, T.: Farm machinery management information system. Comput. Electron. Agric. **110**, 131–138 (2015). https://doi.org/10.1016/j.compag.2014.11.011
4. Nikkilä, R., Seilonen, I., Koskinen, K.: Software architecture for farm management information systems in precision agriculture. Comput. Electron. Agric. **70**, 328–336 (2010). https://doi.org/10.1016/j.compag.2009.08.013
5. Graeub, B.E., Chappell, M.J., Wittman, H., Ledermann, S., Kerr, R.B., Gemmill-Herren, B.: The state of family farms in the world. World Dev. **87**, 1–15 (2016). https://doi.org/10.1016/j.worlddev.2015.05.012
6. Murakami, E., Saraiva, A.M., Ribeiro, L.C.M., Cugnasca, C.E., Hirakawa, A.R., Correa, P.L.P.: An infrastructure for the development of distributed service-oriented information systems for precision agriculture. Comput. Electron. Agric. **58**, 37–48 (2007). https://doi.org/10.1016/j.compag.2006.12.010

7. Bergman, M.: Sources and classification of semantic heterogeneities. AI3, Adapt. Inf., pp.1–6 (2006)
8. Euzenat, J., Shvaiko, P.: Ontology Matching, 2nd edn. Springer, Berlin (2013). https://doi.org/10.1007/978-3-540-49612-0
9. Otero-Cerdeira, L., Rodríguez-Martínez, F.J., Gómez-Rodríguez, A.: Ontology matching: a literature review. Expert Syst. Appl. **42**, 949–971 (2015). https://doi.org/10.1016/j.eswa.2014.08.032
10. Cruz, I.F.: AgreementMaker: efficient matching for large real-world schemas and ontologies. In: PVLDB, pp. 1586–1589 (2009). https://doi.org/10.14778/1687553.1687598
11. Do, H.-H., Rahm, E.: COMA: a system for flexible combination of schema matching approaches. In: Proceedings of the 28th International Conference on Very Large Data Bases, pp. 610–621 (2002). https://doi.org/10.1016/B978-155860869-6/50060-3
12. Aumueller, D., Do, H.-H., Massmann, S., Rahm, E.: COMA++ - schema and ontology matching with COMA. In: Proceedings of the 2005 ACM SIGMOD International Conference on Management of Data (SIGMOD 2005), p. 906 (2005). https://doi.org/10.1145/1066157.1066283
13. Madhavan, J., Bernstein, P.A., Rahm, E., Bernstein, P.A.: Generic schema matching with cupid. In: VLDB, vol. 10, pp. 49–58 (2001). https://doi.org/10.1007/s007780100057
14. Hu, W., Qu, Y.: Falcon-AO: a practical ontology matching system. Web Semant. **6**, 237–239 (2008). https://doi.org/10.1016/j.websem.2008.02.006
15. Shvaiko, P., Giunchiglia, F., Yatskevich, M.: Semantic matching with S-match. In: Semantic Web Information Management: A Model Based Perspective (2010). https://doi.org/10.1007/978-3-642-04329-1
16. Schulz, S., Boeker, M., Stenzhorn, H.: How granularity issues concern biomedical ontology integration. Stud. Health Technol. Inform. **136**, 863–868 (2008). PMID no. 18487840
17. Gaber, M.M., Krishnaswamy, S., Zaslavsky, A.: Adaptive mining techniques for data streams using algorithm output granularity. In: AusDM (2003)
18. Ding, S., Du, M., Zhu, H.: Survey on granularity clustering. Cogn. Neurodyn. **9**(6), 561–572 (2015). https://doi.org/10.1007/s11571-015-9351-3
19. Checiu, L., Ionescu, D.: A new algorithm for mapping XML schema to XML schema. In: ICCC-CONTI 2010 - International Joint Conference on Computational Cybernetics and Technical Informatics, pp. 625–630 (2010). https://doi.org/10.1109/ICCCYB.2010.5491337
20. Xu, Z., Li, Z.: A schema mapping technique for XML-based semantic geodata translation. In: 2009 17th International Conference on Geoinformatics (2009). https://doi.org/10.1109/GEOINFORMATICS.2009.5293506
21. Wang, B., Guo, B.: An algorithm for indirect schema mapping composition. In: Proceedings of the 1st International Workshop on Education Technology and Computer Science (ETCS 2009), vol. 3, pp. 723–726 (2009). https://doi.org/10.1109/ETCS.2009.696
22. Sekhavat, Y.A., Parsons, J.: SESM: semantic enrichment of schema mappings. In: Proceedings of the International Confereence on Data Engineering, pp. 7–12 (2013). https://doi.org/10.1109/ICDEW.2013.6547415
23. Mecca, G., Papotti, P., Santoro, D.: A short history of schema mapping systems. In: Proceedings of the 20th Italian Symposium on Advanced Database Systems (SEBD 2012), pp. 99–106 (2012)
24. Hai, D.H.: Schema matching and mapping-based data integration. Dep. Comput. Sci. 222 (2006)
25. Wickham, H.: The split-apply-combine strategy for data analysis. J. Stat. Softw. **40**, 1–29 (2011)

Intelligent Triage Assistant

Wannarat Duangdee and Pattarachai Lalitrojwong(✉)

Faculty of Information Technology, King Mongkut's Institute of Technology Ladkrabang, Bangkok, Thailand

61606005@kmitl.ac.th, pattarachai@it.kmitl.ac.th

Abstract. The triage of outpatient department is a vital process to screen patients. In addition, assessing the emergency case by the outpatient department can determine the quality of service. In emergency case, the patients have to be treated as soon as they arrive. If the triage of this case is delayed and the patients have major symptoms, it may develop serious complications leading to the cause of death or disability. Accordingly, assessing and making triage decisions, and assigning the level of patient acuity require being accurate, rapid and trustworthy in order to take care and treat them in time. Due to the fact that nowadays there is only one nurse on duty for the triage, he or she cannot well handle a bunch of patients arriving for treatment. According to Hospital and Healthcare Standards, 4th edition, the Healthcare Accreditation Institute (Public Organization) allows the use of technology to improve hospital services. This research aims to develop an intelligent triage assistant. To help the process by making triage decisions for all the patients promptly, assigning an acuity level and screening and assigning the patients to an area based on acuity. If unsuitable results happen, the triage nurse can be on duty promptly. Our intelligent triage assistant has been implemented using Visual Studio C#, Microsoft SQL Server for the database, and QnA Maker for the knowledge base. The application has been tested by outpatient nurses. They are quite satisfied with the primary outcome and recommend further improvements for future work.

Keywords: Triage · Outpatient · Chatbot · Expert system · Knowledge base

1 Introduction

The triage of patients is the basic standard for patient service. The emergency crisis can happen any time from the first admitted patient coming and waiting for treatment. That may bring to the cause of emergency dead or permanent disability [1]. According to this reason, the outpatient department nurses have to be able to assess for making decisions rapidly and assign acuity patients accurately in order to satisfy the services and achieve the strategy of Access to Health Care Services.

Since Phraphuttabat hospital has insufficient nursing personnel for triage service, other medical personnel, such as medical records officers, hospital porters, and so on, have to assist the triage nurses to handle an overwhelming number of patients. That affects emergency crisis and risk in the outpatient department when the patients are waiting for treatment, as shown in the incident reports (year 2017–2019), even if they are supposed

P. Meesad et al. (Eds.): IC2IT 2021, LNNS 251, pp. 279–287, 2021.
https://doi.org/10.1007/978-3-030-79757-7_27

to be sent to the accident and emergency department. As a result, it is suggested by The Healthcare Accreditation Institute (Public Organization) that the hospital requires a corrective process to achieve the criteria assessment of the Hospital and Healthcare Standards.

Based on this circumstance, Phraphuttabat hospital has strived constantly to improve the triage system to support the comprehensive assessment of the patient's condition to become faster and more standardized. Therefore, the patients can receive timely and appropriate treatment. It can be stated that triage is significantly crucial to handle patient's risk and emergency crisis that can happen to the patients at any moment from start and while waiting for treatment.

This research aims to develop an intelligent triage assistant. To help the process by making triage decisions for all the patients promptly, assigning an acuity level and screening and assigning the patients to an area based on acuity. Our intelligent triage assistant has been implemented using Visual Studio C#, Microsoft SQL Server for the database, and QnA Maker for the knowledge base.

2 Literature Reviews

2.1 Expert Systems

An expert system is a computer system emulating knowledge of a specific problem such as medical profession or science to perform decision-making ability like a human expert [2]. It can provide a second opinion to a human expert or break a tie in case of disagreements by multiple human experts to increase confidence that correct decision was made. This research requires its fundamental knowledge, including knowledge representation, justification and machine learning.

2.2 Chatbots

A chatbot is a software application created to conduct chat conversation for various purposes like answering questions, providing assistance, providing entertainment, and so on, depending on the usage [3]. It is designed to simulate the way a human would behave as a conversational partner.

Rule-based bots can answer questions based on a predefined set of rules that are embedded into them. AI bots are self-learning bots that are programmed with Natural Language Processing (NLP) and machine learning. They are able to communicate in complex ways with different users [4]. Building such rule-based bots is much simpler and faster than building AI bots. However, they cannot self-learn and hard to maintain in the long run.

2.3 Triage Process

When a patient arrives at the triage front desk, the triage nurse mainly operates the following steps: 1) Giving a "critical first look" 2) Screening for infectious diseases 3) Interviewing and assessing the presenting complaint and vital signs 4) Making the triage decision and assign an acuity level 5) Assigning the patient to an area based on acuity 6) Provide symptom relief for patients assigned to the waiting area, and 7) Re-assessing & Re-triage waiting area patients to make the screening system effective.

Therefore, the essential triage nursing skills are as follows: 1) public relations skill 2) interviewing skill 3) critical thinking, and 4) communication skill. To prevent or reduce the incidence of risk in patients, the patients should be evaluated properly and treated promptly, without serious complications.

It is recommended that interviewing and assessing the presenting complaint and vital signs should be done directly to the patients if possible. Besides, patients encountering the incident brought by the courier or the medical personnel should take less than five minutes to be interviewed [5].

2.4 Previous Related Research

MESTRIMAN [6] is an expert system providing proper medical assistance in catastrophes and medical emergencies. It is a subsystem of SIAC, a computer system used for basic life support and advanced medical aid. MESTRIMAN deals with the principles of triage and performs the clinical management of patients in catastrophes. The knowledge acquisition process consists of direct observation of the expert at work and a set of interviews between knowledge engineers and clinical experts. Subsequently knowledge representation is performed for patient classification in the form of IF-THEN rules. This system can be viewed as an advisor for the personnel involved in the management of catastrophes to take shorter time and gain more reliability. It is undoubtedly useful in particular when they lack experts.

Chang, P. et al. [7] extended the support systems for the Emergency Medical Services (EMS) to cover pre-hospital emergency medical services and onsite evaluation forms for the mass gatherings. Twenty-three nurses and six physicians in the medical center were invited to examine the new systems and answer the questionnaire. Most of the subjects agreed that the systems were easy to use and useful for mass gatherings.

IntelTriage [8] was proposed as a smart triage and hospital monitoring system. It dynamically assigns priorities to patients in an emergency department and monitors their vital signs and location during their stay in the clinic through wearable biosensors. The integration of customized machine learning algorithms and statistical models will provide dynamic monitoring of patients' health, using time series analysis and real-time changes detection, aiming at optimal triage of cases and reprioritization on the fly.

3 Current Triage Process

Triage is a critical process that determines the quality of service to patients who are having serious symptoms or requiring emergency attention upon the arrival to the hospital. If the analysis of this risk group is inaccurate or delayed, then the patients may develop serious complications leading to the cause of death or disability. According to this case, it could affect litigation, claims, increased maintenance costs, and impact on service satisfaction.

Based on hospital data in 2017–2019, it was reported that the number of critical patients waiting for treatment at the outpatient department was higher than the accident and emergency department [1]. This causes a negative effect on the access and treatment of the patients. According to the Healthcare Accreditation Institute (public organization) standard, patients have to be screened urgently in a short period of time by competent triage nurses and appropriate equipment. This is the top priority for treatment in order to prevent potential crises and emergencies that can arise to the patients at any time like sudden death and permanent disability.

Patients who need urgent and emergency care need immediate attention by competent staff and appropriate equipment. However, the current triage process cannot reach the patient promptly. The implementation of the operational system takes time to prepare based on the availability of resources such as technology, staff, troubleshooting knowledge and skills of specialists. Due to insufficient triage nurses, additional staff members come to help screening patients. Unfortunately, they lack of professional knowledge and observation skills. Patients may be sent to the wrong area. This problem results in a waste of time for the patients or it may have a negative effect on them. Furthermore, it can be shown that current actual work does not meet the standard.

As a result, triage procedures at the outpatient department need to be adjusted and improved to meet the standard. This research proposes an automated intelligent triage assistant, a knowledge based application helping the process by making triage decisions for all the patients promptly, assigning an acuity level and screening and assigning the patients to an area based on acuity. It is expected that the risk of death, disability, and complications with patients waiting for treatment must be reduced.

4 System Work Flow

As shown in Fig. 1, when a patient comes to a triage entry point, a user who can be either a triage nurse or any staff member enters a patient's ID card number or patient ID number. Then the triage assistant connects to the hospital database via API to retrieve patient's medical history. The key data to be used in the screening process are drug allergy, congenital disease, and patients' rights to medical care.

The user then enters the patient's blood pressure, heart rate and body temperature data to allow our assistant produce the results and display the severity type. The triage of the hospital exploits a three-level system dividing patients into the groups "emergent" (cannot safely wait until a space in the clinical area becomes available), "urgent" (can safely wait a short amount of time until a space in the clinical area becomes available), and "non-urgent" (can safely wait a long time until a space in the clinical area becomes available). For emergent and urgent patients, the assistant indicates patient status to allow a triage nurse to respond immediately because they must be seen promptly by a physician. Therefore the patients will be sent to the emergency room straight away. On the other hand, for non-urgent patients, the assistant via its chatbot asks the user for three additional key symptoms of the patients. Based on its knowledge base, the automated intelligent triage assistant makes triage decisions and guides the user on the appropriate clinical area or examination room to be assigned for the patients. We can see that by this intelligent application the duration of the triage process can be reduced significantly. It

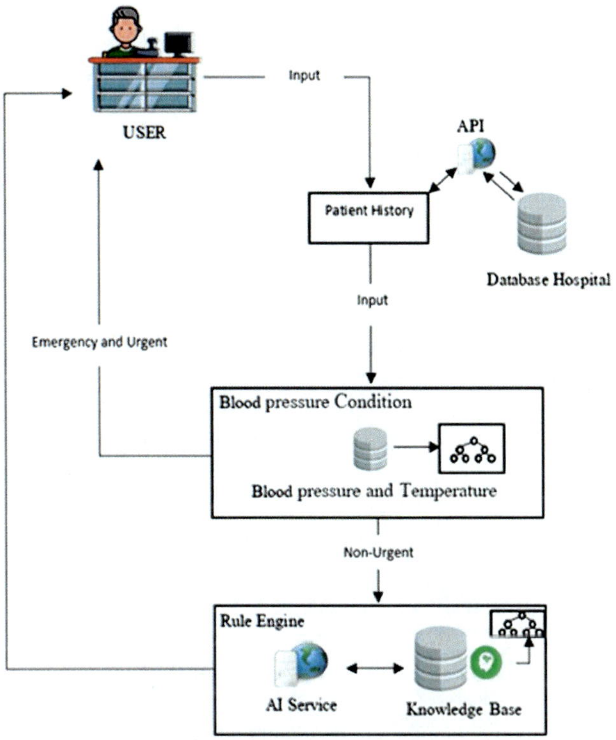

Fig. 1. Triage process work flow

serves as an assistant to inexperienced nurses and non-clinical staff to perform triage more quickly, more accurately, and more effectively.

5 Developing a Knowledge Base for the Intelligent Triage Assistant

In this stage we begin with gathering information about the patient's initial illness. There are two main knowledge sources we collect from: hospital's medical records and human expert's experiential knowledge. We gathered patient symptoms over the past year in order to categorize symptoms for each examination room in order to establish an association between the illnesses and that examination room. Besides, we had an interview with triage nurses to acquire their professional knowledge from experience.

After acquiring knowledge from both sources, we analyzed triage information based on patient symptoms, associated it with the examination rooms, and represented it in a decision tree format that can be used to develop the intelligent triage assistant knowledge base as partially shown in Fig. 2. Therefore, the intelligent triage assistant has a rule-based chatbot utilizing its knowledge base to perform classification. Patients will be asked for three key symptoms. Then it can make an inference to screen them and assign them to appropriate examination rooms. This facilitates the triage process more quickly so the patients can be treated readily.

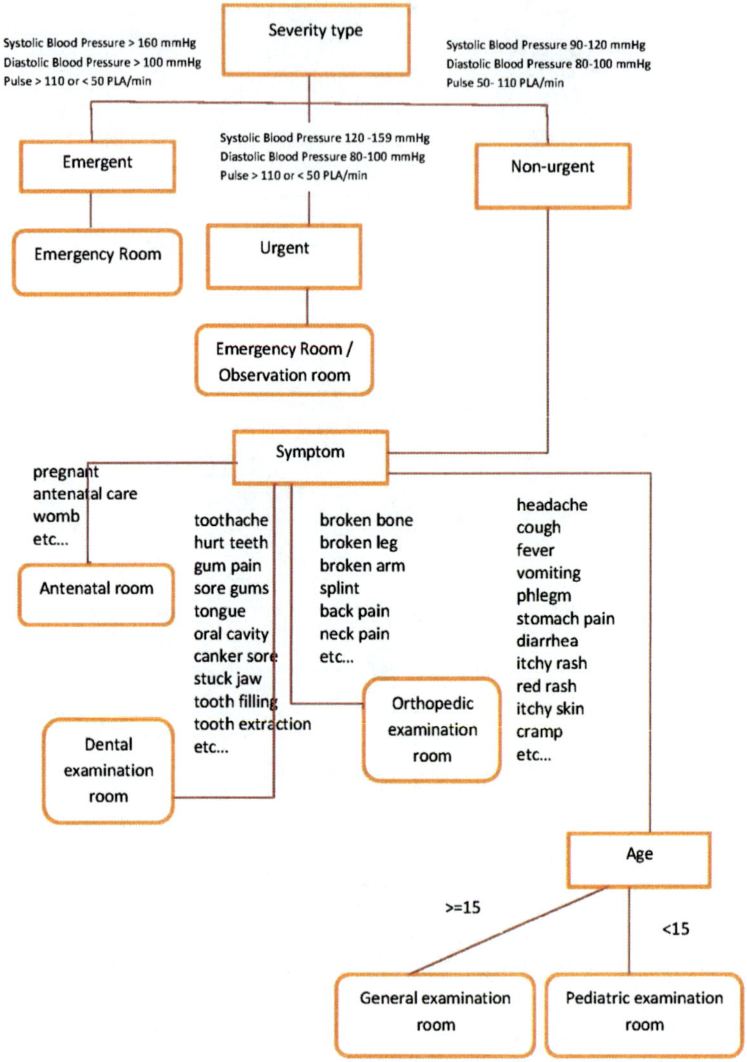

Fig. 2. Partial decision tree for screening patients by initial symptoms.

6 Implementation of Automated Intelligent Triage Assistant

In our work, we use MS Visual Studio 2017 cooperating with MS SQL Server to implement the application. A patient database had been created. In addition, Microsoft Azure had been utilized for handling a cloud computing service. Intelligent cloud is a computing service which is embedded with public cloud and artificial intelligence. Azure supports many AI services like machine learning, bot services, cognitive services, and web services. We employed QnA Maker [9] to develop a chatbot as a triage expert system to make an inference decision screening patients to appropriate examination room. QnA

Maker, a cloud-based Natural Language Processing (NLP) service, is commonly used to build conversational client applications, which include social media applications, chat bots, and speech-enabled desktop applications. It is used to find the most appropriate answer for any input from custom knowledge base of information.

After logging in, the first screen can be illustrated in Fig. 3. On the left area, the user can search for patient data by ID card number or patient ID number. The Triage Assistant application sends a request to access and retrieve patient records to the hospital database management system. All necessary patient data will be displayed including drug allergy, congenital disease, and patients' rights to medical care. So the user can verify the identity of the patient at the time of admission and use these data as part of making decisions.

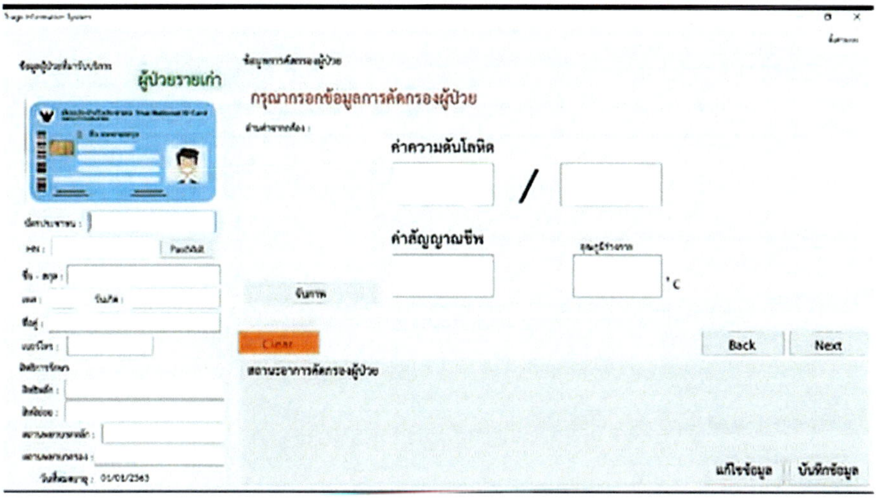

Fig. 3. First screen of the Intelligent Triage Assistant

Next step, the user fills in the patient's blood pressure, heart rate and body temperature on the right area. At practice, these can be automatically obtained from the medical equipment.

The subsequent screen as shown in Fig. 4. internally performed by a chatbot asks for the main symptoms of the patient. The user can fill out only 1–3 symptoms. Then the assistant does triage and determines the examination room for the patient to have a medical treatment in the next procedure (Fig. 5).

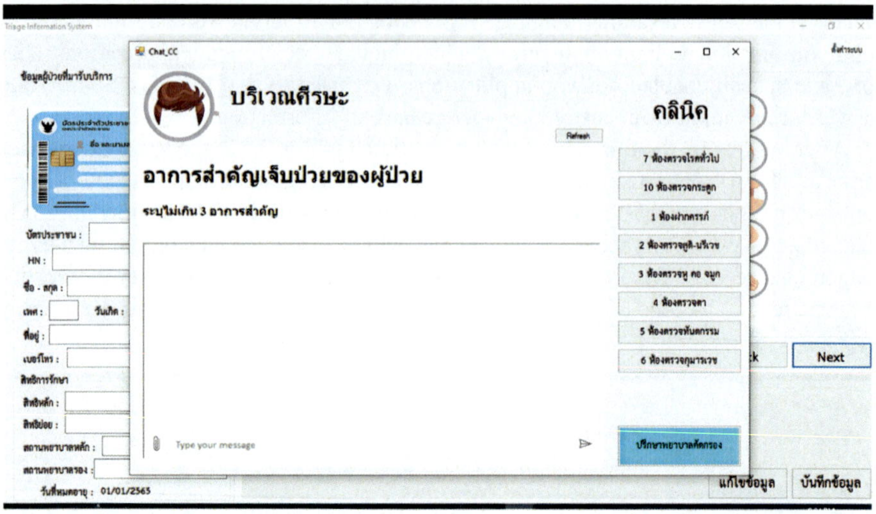

Fig. 4. Second screen asking for patient's key symptoms

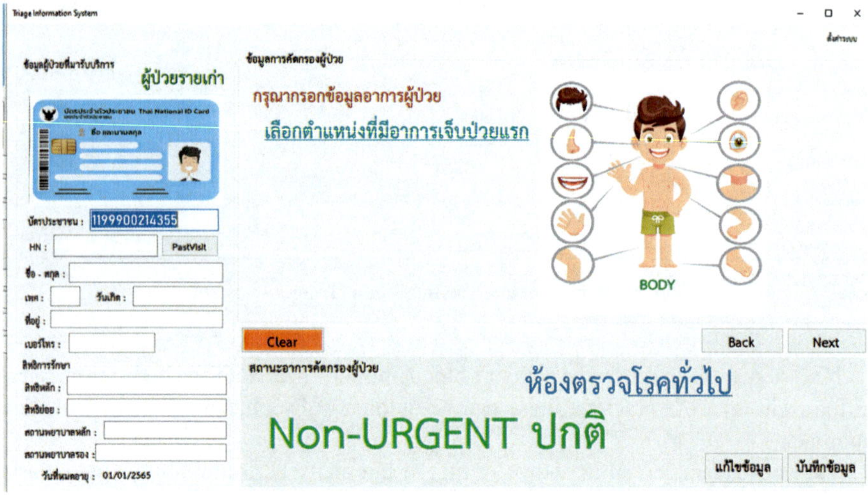

Fig. 5. Third screen displaying the examination room assigned for the patient

7 Conclusions

The intelligent triage assistant was developed as an application written in C Sharp (C#) using Dot Net Framework (.NET Framework) technology and using Microsoft SQL Server as a database management system to store patient data. This can solve the problem of losing paper-form triage documents and reduce data redundancy. Besides, we use Microsoft Azure as Microsoft's cloud-based operating system for running programs or storing online data on high-speed internet. We also employ QnA Maker to implement a chatbot to answer questions according to the system's knowledge base.

The application has been tested and verified by outpatient nurses. They are quite satisfied with the primary outcome and recommend further improvements for future work. Based on the indicators on the risk assessment list, this application can facilitate the triage to handle an overwhelming number of clients more quickly and more conveniently.

References

1. Healthcare Accreditation Institute (Public Organization) (2009). https://www.ha.or.th/Backend/fileupload/Quality%20Tools/Attach/HAOverallScoring2010%20Ver4.pdf. Accessed 12 Jan 2021
2. Expert system. https://en.wikipedia.org/wiki/Expert_system. Accessed 12 Jan 2021
3. Sawangwattananutat, N.: What is Chatbot? How important is it? (2017). http://blogs.cipher.co.th/what-is-chatbot. Accessed 12 Jan 2021
4. Savina, N.: Five different types of chatbot. In: Voice UI. https://medium.com/@hellonastasyasavina. Accessed 12 Jan 2021
5. Sukswang, S.: Triage nurse: beyond main process through practice. J. Health Sci. Sch. **5**(2), 1–14 (2018)
6. Pazos Sierra, A., Blanco Ferro, A., Maojo Garcia, V., Martín Sanchez, F., Barreiro Sorrivas, J.: MESTRIMAN: an expert system for medical triage and clinical management of patients in catastrophes. Expert Syst. Appl. **6**(4), 449–457 (1993)
7. Chang, P., Hsu, Y.S., Tzeng, Y.M., Sang, Y.Y., Hou, I.C., Kao, W.F.: The development of intelligent, triage-based, mass-gathering emergency medical service PDA support systems. J. Nurs. Res. **12**(3), 227–236 (2004)
8. Billis, A., et al.: Towards the definition of an intelligent triage and continuous monitoring system for hospital emergency departments and clinics. In: Ohno-Machado, L., Séroussi, B. (eds.) MEDINFO 2019: Health and Wellbeing e-Networks for All (2019). https://doi.org/10.3233/SHTI190574
9. Microsoft: What is QnA Maker? (2020). https://docs.microsoft.com/en-us/azure/cogni-tive-services/qnamaker/overview/overview. Accessed 12 Jan 2021

Development of Physical and Mental Support System Using Communication Robot for Isolated COVID-19 Patients

Hiro Kuwahara[1], Aiko Mieno[2], Tae Yamaguchi[2], and Kazuo Hemmi[3](✉)

[1] Graduate School of Regional Design and Creation, University of Nagasaki, Nagasaki, Japan
mc220501@sun.ac.jp
[2] Department of Nursing Science, University of Nagasaki, Nagasaki, Japan
{aikoie,tae0813}@sun.ac.jp
[3] Department of Information Systems, University of Nagasaki, Nagasaki, Japan
hemmi@sun.ac.jp

Abstract. The symptoms of the COVID-19 infection are statistically commonly observed in the order of fever, cough, and shortness of breath. These symptoms can exacerbate over time and require continuous measurements of body temperature and percutaneous arterial oxygen saturation. In addition, it has been suggested that patients in isolation become psychologically anxious, and therefore, communication is important to promote mental health. A system that automatically supports both physical management and mental health of COVID-19 patients has not been developed yet; therefore, our research group developed a system that supports physical management and promotes mental health through a communication robot called Unibo. Regarding the feasibility of the proposed system, it was verified that the developed system could be implemented in terms of both physical management support and the promotion of mental health. The results showed that the subject's impression of winsomeness towards Unibo worked in a psychologically positive manner.

Keywords: COVID-19 · Communication robot · Body temperature measurement · SpO_2 measurement · Human interface

1 Introduction

According to a survey conducted by the World Health Organization, the number of COVID-19 patients is still increasing worldwide, and this is still a global problem [1]. Patients suspected or confirmed to have a mild COVID-19 infection are recommended to be quarantined in medical institutions or at home to contain the virus [2]. They are required to self-observe their symptoms in an isolated environment. S. K. Brooks et al. describe that when a patient infected with COVID-19 is quarantined, the patient feels strong anxiety and loneliness due to quarantine, and conversation is important for both promoting mental health to reduce loneliness and boredom [3]. According to a study by Ishii et al. [4], symptoms of COVID-19 infection are first fever (73.0%), next cough

P. Meesad et al. (Eds.): IC2IT 2021, LNNS 251, pp. 288–297, 2021.
https://doi.org/10.1007/978-3-030-79757-7_28

(48.3%), and then shortness of breath (28.0%). In addition, 32.5% of people need oxygen inhalation, accounting for one-third of the total. Therefore, early detection of fever and exacerbation of respiratory symptoms is very important. It is known that COVID-19 causes more rapid deterioration of respiratory status than expected, even if mild at first [5]. These results indicate that continuous physical condition management and early detection, and treatment of respiratory disorders are important for COVID-19 patients.

Consequently, it can be inferred that the issue is how to promote mental health and support physical management during the isolation of COVID-19 patients. Ćosić et al. have discussed the potential of digital technologies such as AI to address the shortage of mental health professionals in pandemics [6], and Libin et al. have reported that humans cause positive emotions such as joy and show interest in their involvement with interactive robots that have the appearance of living things [7]. In addition, it is effective to measure percutaneous arterial oxygen saturation concentration (we call it "SpO_2") using a pulse oximeter for early detection of the exacerbation of respiratory symptoms; however, a previous study in which a robot was used to measure body temperature and SpO_2 [8] does not include the function of automatic conversation for the patient. Therefore, in response to this issue, our research group has developed a system that supports the measurement of body temperature and SpO_2 through a conversation function using the communication robot Unibo (we call it "Unibo") of Unirobot Corp. First, this study describes the developed system, and then the feasibility verification results.

2 Communication Robot Unibo

In this study, we develop the proposed system using Unibo. The appearance of Unibo is shown in Fig. 1. It is a multifunctional communication robot with a deformed child appearance. Wireless communication with the outside can be performed using Wi-fi and Bluetooth 4.0. The speaker's voice is picked up using the built-in microphone above the Unibo's head. This picked up voice is recognized, interpreted, and synthesized by the cloud server (we call this "conversation API"). Responses are provided through the abdominal speaker. The conversation API can support Japanese, English, Chinese, and Korean languages. The LCD screen on the face is a full HD: 1920 * 1200 touch screen. The soles of the feet have touch sensors, each of which can determine the presence or absence of contact. The neck and arms are movable; the neck is operated at 140° left and right, and 44° up and down using a 2-axis motor, and both arms are operated at 90° back and forth using a uniaxial motor.

3 System Configuration

In this study, we develop a system that supports physical management and promotes mental health using communication robot.

For the physical management support in this system, Unibo gives instructions to the subject through conversation, and the subject understands their own symptoms by measuring body temperature and SpO_2, and through Unibo's interview, and then Unibo contacts the necessary persons, such as medical staff, through e-mail. The system configuration diagram is shown in Fig. 2. The system consists of Unibo as the core, which

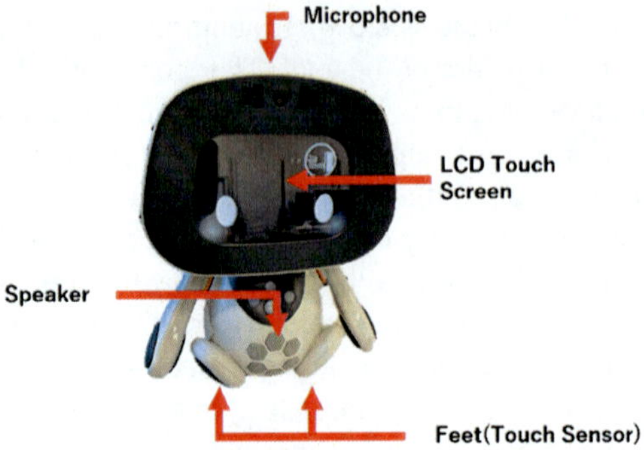

Fig. 1. Appearance of Unibo.

is always connected to the Internet. Unibo and Unirobot's cloud server are connected via the Internet and communicate conversation API, and the software update of Unibo is performed as appropriate. A thermistor-type electronic thermometer is connected for the axilla (estimated 30 s) (UT-201BLE manufactured by A & D Co., Ltd.), and a fixed fingertip pulse oximeter (PLS-01BT manufactured by CUSTOM) is connected to Unibo using Bluetooth 4.0. The results of each measuring device are automatically collected wirelessly by Unibo, and the patient's subjective symptoms are notified to the necessary parties through e-mail.

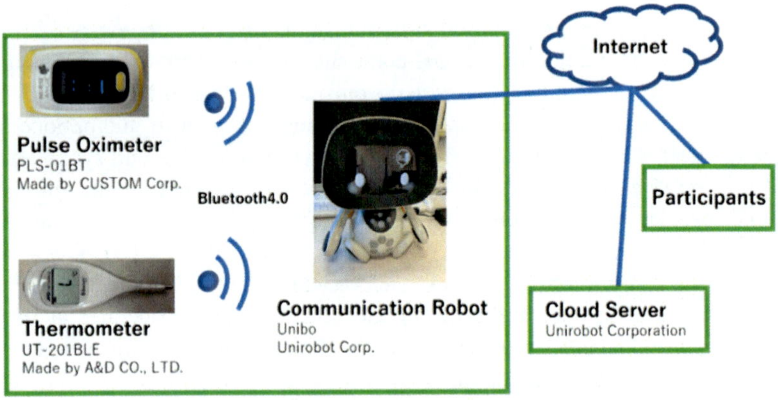

Fig. 2. System configuration diagram.

The measured values of SpO_2 were programmed to branch the conversation of Unibo according to the acquired values. The value of SpO_2 branch is set according to the medical standard [5] of the Ministry of Health, Labor and Welfare. If $SpO_2 \geq 96\%$, it is mild, if $96\% > SpO_2 > 93\%$, it is Moderate I (patient does not suffer respiratory failure), and if $SpO_2 \leq 93\%$, it is Moderate II (patient suffers respiratory failure).

Figure 3 shows the activity diagram of the physical condition management support. The physical management support is launched using the timer function whenever Unibo

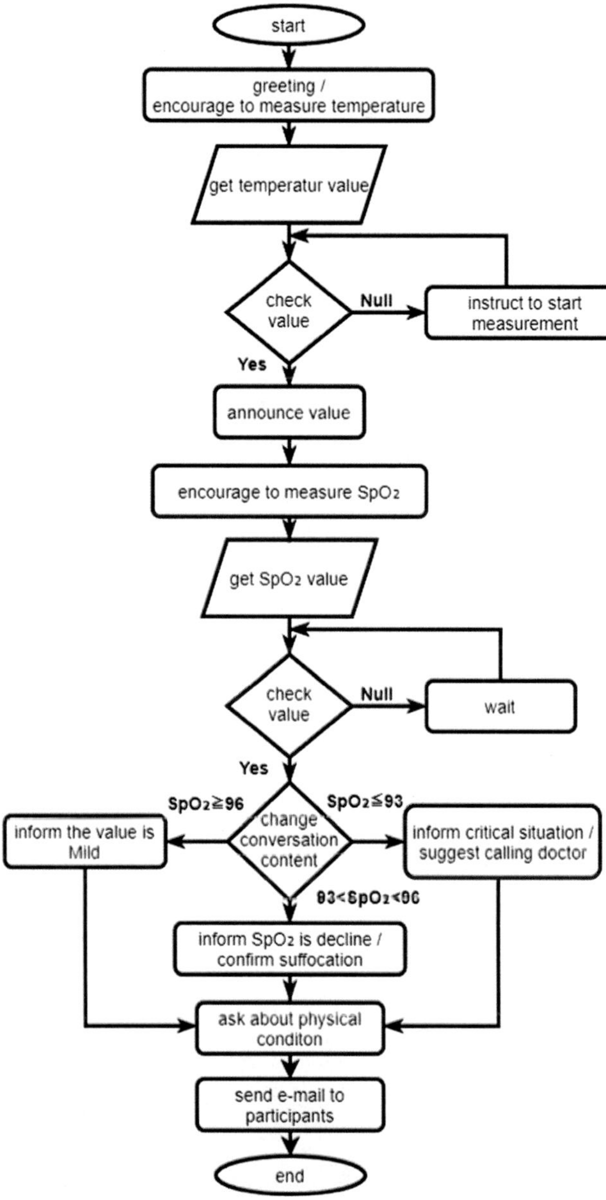

Fig. 3. Physical management support system activity diagram.

speaks. Next, Unibo gives verbal instructions as appropriate. The measurement result is read aloud by Unibo and transmitted to the subject. Interviews are conducted by Unibo's utterance and text display on the screen. The subject responds by touching the touch sensor on the sole of each Unibo's foot or ○ × display shown on the screen according to the subjective symptom. The Centers for Disease Control and Prevention has listed the symptoms of COVID-19 at the time of illness [9], and based on this, the questions in our interview are whether dysgeusia, cough, runny nose, headache, malaise, suffocation, digestive organs, and other symptoms have worsened or not. The physical management support system is terminated by sending the collected data to the necessary parties. In the email sentence, the name of the subject, the date and time of the measurement, the acquired body temperature/SpO_2 value, the result of the interview, and the time taken for measurement are itemized. Concrete values are entered for the date and time of the measurement, the acquired body temperature and SpO_2 values, and yes or no for the results of the interview.

In this study, the promotion of mental health in an isolated environment is carried out through conversation with Unibo. Unibo continues to pick up external sounds, except for physical management supporting system running time, and uses the conversation API to respond appropriately to the voice entered. Responses may include chats, information retrieval from the web, and games such as shiritori and quizzes. Unibo's face display can be transformed into 10 different facial expressions: joy, trust, fear, surprise, sadness, disgust, anger, expectation, sleep, and a straight face. Unibo can move its neck and hands while talking; therefore, it can add gestures as per conversation. Onyeulo and Gandhi describe the usefulness of robot facial expressions and gestures in robot–human communication [10]. In this study, we use the Unibo's functions described above to express conversations and emotions, and to realize interactive communication between robots and humans.

4 Feasibility Verification Method

To test the feasibility of the proposed system, a pilot test was conducted by recruiting the students at University A as subjects. Figure 4 shows the execution environment and state of the pilot test. Unibo is placed near the center of the desk with a width of 110 cm, a depth of 80 cm, and a height of 70 cm, and a thermometer and a pulse oximeter required for the experiment are placed in front of Unibo. The subject is seated in front of Unibo, the left and the back of the subject are covered with two partitions 157 cm long, 90 cm

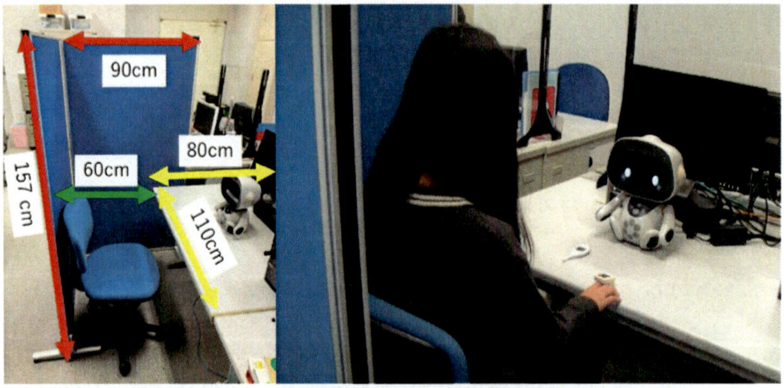

Fig. 4. Experiment executed environment and experimental scenery.

wide, and 3 cm wide, and the experimental scenery was photographed from the back right using a recording camera. The distance between the back partition and the desk is 60 cm, and subjects were instructed to stay seated in this gap chair during the experiment.

The feasibility verification test was divided into three phases and conducted for approximately 20 min. Each phase was further divided into three phases: testing Unibo's physical management support by subject, the mental promoting test through conversation with Unibo, and collecting the subject's impression of the system after the experiment using a questionnaire. The phase of testing the physical management support was performed by the subject testing through the system content described in the previous section, while following Unibo's instructions. In the conversation test phase, the subject spoke independently to Unibo for 5 min, and Unibo returned appropriate conversations. In the conversation test, an example of talking was shown to the subject for the purpose of encouraging him/her to talk. Table 1 shows the example displayed at the time.

Table 1. Conversation example.

Let's play word chain game!	Sing please	Tell me today's news
What can you do for me?	Hello!	How is the weather today?
Tell me your recommended movies	Who are you?	What's your favor?
Give me a quiz		

After the experiment was conducted, the questionnaire for the subjects was created by writing together the selection items and the free description field, printing them on both sides of A4 size paper, and distributing them to the subjects. The contents of the questionnaire are shown in Sect. 5.

5 Result

Table 2 shows the characteristics of the subjects obtained from the questionnaire. There were 33 subjects, including both male and female (6 males and 27 females). The age range in this study was 17 to 24 years, and the average age $\pm$ SD was 20 $\pm$ 1.2 years. All the subjects had experience using mobile terminals such as smartphones and tablets. Three subjects (9%) had experience using communication robot, and 30 (91%) had no experience using them. In addition, no one had motor dysfunction of the upper limbs and fingers.

Table 2. Characteristics of subjects.

$n = 33$	
Age, y	
Mean ± SD	20.5 ± 1.2
Range	17–24
Sex	
Male	6 (18%)
Female	27 (82%)
Experience using mobile terminals (smartphones and tablets)	
Yes/No	33 (100%)/0 (–)
Experience using communication robot	
Yes/No	3 (9%)/30 (91%)
Are there any motor dysfunctions (especially dominant arms) in the upper limbs (and fingers)?	
No	33 (100%)
Have dysfunctions but is able to operate the system	0 (–)
Disable to operate the system	0 (–)

Table 3 shows the results of the questionnaire obtained from the subjects in the experiment to verify the feasibility. Regarding the subject's impression of the system, 23 (70%) answered it was extremely good impression and 10 (30%) answered it was moderately good impression. Following this result, the overall impression of this system was generally good. Regarding the operability of both thermometer and pulse oximeter equipment, more than 25 subjects (more than 85%) answered extremely easy. Regarding the touch operation, 32 subjects (97%) answered that there was no problem with operability, but 1 (3%) subject evaluated that there was a problem with touch operability. In the questionnaire that asked about the enjoyment of conversation with Unibo, 21 subjects (61%) answered extremely enjoyed, 11 (33%) answered moderately enjoyed, and 1 (3%) answered moderately not enjoyed. Regarding the questionnaire items asking whether the loneliness and anxiety from the isolation can be reduced using this system, 20 subjects (61%) answered that they could do it, and 12 subjects (36%) answered that they could do it moderately. One person (3%) answered that they could not do it.

Table 3. Results of testing physical management support and promotion of mental health.

$n = 33$	
How do you feel about this system?	
Extremely good impression	23 (70%)
Moderately good impression	10 (30%)
Moderately bad impression	0 (–)
Moderately bad impression	0 (–)
Was the operation easy during body temperature measurement?	
Extremely easy	28 (85%)
Moderately easy	5 (15%)
Moderately hard	0 (–)
Extremely hard	0 (–)
Was the operation easy when measuring SpO_2?	
Extremely easy	29 (88%)
Moderately easy	4 (12%)
Moderately hard	0 (–)
Extremely hard	0 (–)
Did the touch operation problem?	
Yes/No	1 (3%)/32 (97%)
Did you enjoy the conversation with Unibo?	
Extremely enjoyed	21 (64%)
Moderately enjoyed	11 (33%)
Moderately not enjoyed	1 (3%)
Extremely Not Enjoyed	0 (–)
Do you think this system can reduce the loneliness and anxiety of isolated people?	
Able	20 (61%)
Moderately able	12 (36%)
Moderately unable	1 (3%)
Unable	0 (–)

6 Conclusion and Future Works

In this study, we have described the outline of a system developed by our research group to support physical management and promote mental health of isolated COVID-19 patients, and verified its feasibility. As a result of feasibility verification, 70% of the subjects answered that the impression of the system was extremely good. Regarding the feasibility validation test of physical management support, 85% of the subjects answered that the each device was very easy to operate, indicating that it is feasible. Regarding the promotion of mental health, 61% of the subjects answered that this system was able to alleviate the anxiety and loneliness of an isolated person, and 36% said that it was moderately capable, and 3% subject was answered it was moderately unable. For the impression of the system in this study, we counted the number of appearances of adjectives in the subjects' free opinions. It was found that 15 (45%) subjects described Unibo as "kawaii," and "kawaii" was the most appeared adjective. "kawaii" is "a word that has emotional and social value in Japan [11], and the feeling of "kawaii" is known to have a positive psychological effect [12]. In this study, it is suggested that the subjects stimulated by "kawaii" through conversation with Unibo made their minds positive. However, in response to the question whether the impression of the system and the conversation with Unibo were enjoyable, the evident critical opinions were that the response of Unibo was sometimes slow and sometimes silent, even though subjects had spoken to Unibo. Onyeulo and Gandhi responded that it is important for robots to understand various linguistic and non-verbal cues that they give to humans in human–robot conversations, and to understand and implement them along with a natural pause between speakers [10]. By improving these speech and listening functions of Unibo, it is possible for Unibo to further support the promotion of mental health, and we will consider this in the future studies.

Acknowledgment. This work was supported by JSPS KAKENHI Grant Number JP19K10830.

References

1. World Health Organization: Weekly epidemiological update - 15 December 2020. https://www.who.int/publications/m/item/weekly-epidemiological-update. Accessed 29 Dec 2020
2. World Health Organization: Clinical management of COVID-19 Interium guidance 27 May 2020 (2020)
3. Brooks, S.K., Webster, R.K., Smith, L.E., Woodland, L., Wessely, S., Greenberg, N., et al.: The psychological impact of quarantine and how to reduce it: rapid review of the evidence. Lancet **395**(10227), 912–920 (2020). https://doi.org/10.1016/S0140-6736(20)30460-8
4. Ishii, M., Terai, H., Kabata, H., Masaki, K., Chubachi, S., Tateno, H., et al.: Clinical characteristics of 345 patients with coronavirus disease 2019 in Japan: a multicenter retrospective study. Elsevier Public Health Emergency Collection. Published (2020). https://doi.org/10.1016/j.jinf.2020.08.052
5. Ministry of Health, Labour and Welfare JAPAN: Clinical Management of Patients with COVID-19: A guide for front-line healthcare workers version 2.1 (2020)
6. Ćosić, K., Popović, S., Šarlija, M., Kesedžić, I.: Impact of human disasters and COVID-19 pandemic on mental health: potential of digital psychiatry. Psychiatr. Danub. **32**(1), 25–31 (2020)
7. Libin, A.V., Libin, E.V.: Person-robot interactions from the robopsychologists' point of view: the robotic psychology and robotherapy approach. Proc. IEEE **92**(11), 1789–1803 (2004). https://doi.org/10.1109/JPROC.2004.835366
8. Huang, H., Chai, P., Ehmke, C., Merewether, G., Dadabhoy, F., Feng, A., et al.: Agile mobile robotic platform for contactless vital signs monitoring. TechRxiv. https://doi.org/10.36227/techrxiv.12811982.v1. Accessed 29 Dec 2020
9. CDC Homepage. https://www.cdc.gov/coronavirus/2019-ncov/symptoms-testing/symptoms.html. Accessed 29 Dec 2020

10. Onyeulo, E.B., Gandhi, V.: What makes a social robot good at interacting with humans? MDPI Open Access Journals. https://doi.org/10.3390/info11010043. Accessed 29 Dec 2020
11. Nittono, H.: The two-layer model of 'kawaii': a behavioural science framework for understanding Kawaii and cuteness. East Asian J. Popular Culture **2**(1), 79–951 (2016). https://doi.org/10.1386/eapc.2.1.79_1. Accessed 29 Dec 2020
12. Nittono, H., Ihara, N.: Psychophysiological responses to Kawaii pictures with or without baby schem. SAGE Journals (2017). https://doi.org/10.1177/2158244017709321. Accessed 29 Dec 2020

A Cloud Based Knowledge Sharing System for Traditional Medicines

Madiha Naz[1], Talha Waheed[1], Muhammad Aslam[1](✉), and A. M. Martinez-Enriquez[2]

[1] Department of Computer Science and Engineering, UET Lahore, Lahore, Pakistan
{ftwaheed,maslam}@uet.edu.pk
[2] Department of Computer Science, CINVESTAV-IPN, Mexico City, Mexico
ammartin@cinvestav.mx

Abstract. Use of traditional medicines to resolve healthcare issues in developing countries is widespread and increasing. Despite of this, knowledge of traditional medicines is getting lost. One of the major reasons is absence of appropriate knowledge preservation and sharing system. The aim of this paper is to design a comprehensive cloud based knowledge sharing system for traditional medicines community. This has been done by doing a comparative study of existing systems and techniques. It provides an additional feature of disease diagnosis on the basis of symptoms and guides about their treatment. It has the ability to provide data manipulation functions through cloud. Data is store on cloud knowledge base in form of ontology. Validation process of system is done by quantitative analysis of general and expert users response on technology acceptance model based questioner. This questioner shows 60–80% users give positive response. System usability features based comparison is also done to show system effectiveness. Comparison shows almost all features of comprehensive system are available in the developed system.

Keywords: Traditional medicines · Knowledge sharing system · Cloud · Cloud computing · High performance computing · Ontology

1 Introduction

Use of herbs and plants as medicines is ancient. Around 70–80% population of world still depends on traditional medicine (TM) for their primary health concerns [1]. Large volume of TMs knowledge is either available in oral traditions or in printed form that is present in local languages. Its concepts and terminologies are diverse and non-standard. This knowledge is passed on from generation to generation. Now it's in danger of getting lost because of the increasing drift of picking different professions. That mandates the development of a system for sharing and preservation of this knowledge. World Health Organization (WHO) advised its member countries to preserve and integrate knowledge of TM [2]. Some digitalization and automation efforts are available in the domain of TMs [3–5]. However, such work has been in the initial phase for TMs system.

Preservation and sharing of TM knowledge by knowledge sharing system (KSS) is the sub-discipline of knowledge management (KM). KM is required in health related

P. Meesad et al. (Eds.): IC2IT 2021, LNNS 251, pp. 298–309, 2021.
https://doi.org/10.1007/978-3-030-79757-7_29

disciplines to facilitate information digitalization and globalization. KM increases the collaboration, cooperation, intelligibility and simplicity in health care systems.

In order to preserve and share TMs knowledge, a Cloud Base high performance (HP) KSS for TM (CK-System4TM) needs to be developed. Cloud computing is an internet-based computing that allows sharing of resources and provides services to its users with the help of internet [6]. It saves time, power, space and money. Moreover its emergence has improved capabilities of knowledge sharing which are more effective, cost proficient and reliable. Cloud computing provides on demand resources according to user requirements. It removes need for installation and running of an application on own systems. Besides, it diminishes burden of staff, system monitoring and maintenance. Cloud computing mainly divided into three categories on the basis of services they provide [7]. These are Software as a Service (Saas), Platform as a Service (PaaS) and Infrastructure as a Service (IaaS). IaaS provides infrastructure for knowledge storage and sharing. So use of cloud as IaaS for the development and deployment of CK-System4TM is effective. To find out specific requirement and vital features of CK-System4TM, a comparative study of similar knowledge-based systems is required in other domains of TM, so the achieved results can be used to develop a novel CK-System4TM.

The CK-System4TM comprehends information about the overall system architecture, entities, entity's roles, processes, components and communication network. It provides shared platform for knowledge engineers, general users and domain experts to understand, use, apply, transfer, preserve and most importantly share their knowledge of TM practices. Proposed CK-System4TM helps not only to preserve and share this vital TM knowledge but also facilitates in upcoming knowledge-based applications. This is due to ontology based, TM knowledge base (KB) and data manipulation functionality through cloud, by using HPC.

The rest of the paper has been structured as follows: Section II focuses on the detail study of knowledge management system (KMS) and highlights the need for the development of CK-System4TM. Section III discusses design and architecture of CK-System4TM. Section IV provides validation whereas Section V holds the conclusion.

2 Literature Survey

A brief survey of existing efforts of KMS in the field of TM is presented. A KMS is designed on basis of critical factors and KM processes. System design and development is decided according to critical factors while processes are decided according to system activities.

Li et al., 2018 [8] designed a knowledge sharing system for mould alliances. System has application, blockchain network, knowledge resource and enterprise four layers. Private cloud based system is proposed. Knowledge searching is enhanced by using k Nearest Neighbor method.

Anshari et al., 2016 [6] proposed a knowledge system with features of cloud computing, big data and social networks. Knowledge is updated with the help of semantic web and cloud computing. Cloud computing used to provide on demand services while web 2.0 provides knowledge generation. Big data provides integration of knowledge.

Han-Lin et al., 2016 [3] provided a synthesis that is appropriate in support of KMS development for TChM. TChM is judged against western medicines in multiple aspects.

According to TChM beliefs five basic elements “earth, fire, metal, water and wood” had influence on human body. The KMS for TChM has KM processes such as knowledge collection, organization, retention, analyze, design reengineering and evaluation.

Lin et al., 2015 [4] discussed a case study of KMS for TChM. An appropriate and helpful inquiry system is designed, taking into account the attributes of this TM system. Use of IT in medicines field is very effective and valuable. System is designed which provide facilities such as market recognition, automation and future studies of TChM.

Yu et al., 2015 [5] discussed use of semantic web for assimilation of TChM and western medicine as both are different with respect to their language, notion and treatment techniques. Proposed semantic based system consists of resources, knowledge base and RDF base graphs. Ontology based semantics KB is developed and integration of TChM. The system facilitates discovery, reusability and distribution of vital knowledge.

Amjad et al., 2014 [9] discussed a KMS for traditional Unani medicines using ontologies. Ontology is developed and organization & classification of herbs is done in a KB. Proposed expert system have interface for user interaction. System inputs symptoms and produces output in the form of disease diagnosis and treatment.

Mustaffa et al., 2012 [10] proposed an online KB system via semantic; for four different types of TMs. KB is formulized for the preservation and linkage of herbal knowledge through ontology. KB is developed considering seven stage technique. This system provides many facilities such as searching, validation and management. Decision making feature is also present in the system to provide assistance.

Cloud computing has three delivery models which are deployed using deployment models. Delivery models are Software as a Service (SaaS), Platform as a Service (PaaS) and Infrastructure as a Service (IaaS) while deployment models are private, public, community and hybrid. SaaS delivery model provides services related to the applications, PaaS provides services related to business tools and platforms whereas IaaS provides services related to resources. SaaS model can be hosted on top of IaaS and PaaS infrastructure while SaaS and PaaS can be hosted on top of IaaS and PaaS infrastructure while SaaS and PaaS can be hosted directly on cloud infrastructure. IaaS provides an infrastructure for knowledge storage and sharing. IaaS service model is based on virtualization model. In order to provide scalability and multi tenancy characteristics. Use of cloud as IaaS with private deployment model, for the development and deployment of CK-System4TM is more effective and proficient.

The review of KMS shows that a comprehensive knowledge sharing system should have some critical factors and processes. Research shows that different. KMS have different critical factors and KM processes. Heisig, 2009 [11] identified the most generic set of four critical factors (people, processes, technology and strategy) and five KM processes (create, identify, apply, store and share). For effectively implementation KMS must address these four aspects. These four critical factors and five processes addressed by reviewed authors for development of their systems are summarize in Table 1. Different systems compose of different processes. Tick (✓) represents presence while cross (✗) represents absence of these factors and processes. On basis of functionality different processes are mapped.

These tabulated comparisons of KMS processes and critical factors of Table 1 leads to the conclusion that a comprehensive CK-System4TM must address these five KM

Table 1. Knowledge management critical factors and processes comparison.

Papers	People	Processes					Technology	Strategy
		Create	Store	Identify	Use	Share		
HTKMS [12]	✔	Not ontology based	Only by admin	✔	✘	✘	CBR, E-R diagram, MySQL	✘
ATHMed [13]	✔	✔	Only by admin	✔	✘	✔	Owl, xml,J2EE, ML	✔
Indigenous Medicine in Sri Lanka [14]	✔	✘	Manual	✘	Man ual	Man-ual	✘	✘
Herbal Medicine Ontology [15]	✔	Use Sesame server directly	Through sesame server only	✘	✔	✔	Rdf, IIS, DBMS	✘
Malaysian Indigenous Herbs Knowledge [16]	✔	✔	Only by admin	✘	✔	✘	Owl, KB, PHP, MySQL, XML	✘
Chinese Medicine [3]	✔	✔	Only by admin	✘	✔	✘	Business process modeling, UML	✘
Traditional Chinese Medicine [5]	✔	✘	Only by admin	✘	✘	✔	Semantic, SQL	✘
Chinese Medicine [4]	✔	✔	Only by admin	✘	✔	✘	Business process modeling, UML	✘
Traditional Unani Medicines [9]	✔	✔	Only by admin	✘	✔	✘	Jena API, RDF	✘
HMKB [17]	✔	✔	Only by admin	✘	✔	✘	✘	✔
KMFTI MODEL [18]	✔	✔	Only by admin	✔	✘	✔	Not selected	✔
Herbal Medicine Knowledge [10]	✔	✔	Only by admin	✔	✘	✘	Semantic, visual semantic query	✘
BioBridge Systems Medicine [19]	✔	✔	Only by admin	✘	✘	✔	Java, semantic, RDBMS	✘
Traditional Chinese Medicine [20]	✔	✔	Only by admin	✘	✔	✔	Rdf, XML	✔

processes considering four critical factors. Survey also concluded that there is a requirement for CK-System4TM as very less work is available in domain of knowledge sharing for TMs.

3 Cloud Based Knowledge Sharing System Design

A CK-System4TM is designed mainly on the basis of knowledge processes, and KM critical factors. Critical factors provide foundation in design and development of CK-System4TM. While KM processes are used to provide knowledge sharing and other activities required by system users. These processes are decided on basis of user requirements. Knowledge is stored in KB using private cloud as IaaS.

Developed CK-System4TM address four critical factors (people, processes, technology and strategy) as shown in Fig. 1. People represent different kind of users of the CK-System4TM. First of all KM processes are finalized keeping in mind the type of knowledge and user requirements. Then appropriate technology is chosen to develop the system prototype. While strategy of CK-System4TM provides an overall governance and management activity for the other three factors.

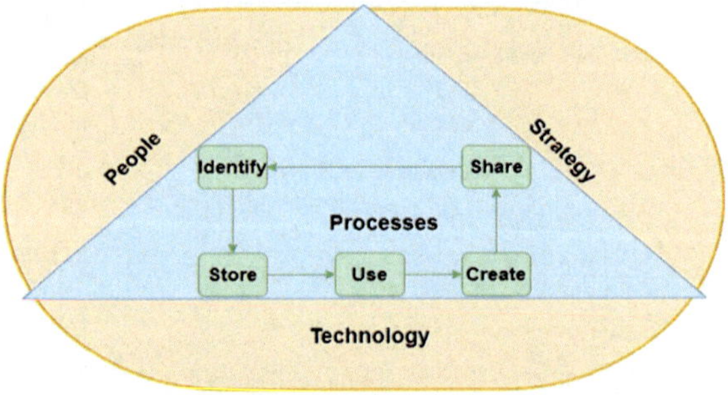

Fig. 1. Knowledge sharing system critical factors and processes.

CK-System4TM has five knowledge processes, identify, store, use, create and share, in order to provide an inclusive and thorough system. These processes are determined from available KMS processes (above Table 1) and KSS requirements. These processes work in steady progression to fulfill tasks (as shown in Fig. 2). However, the focus of our developed CK-System4TM is on knowledge sharing while all other KM processes act as supporting ones.

Figure 2 shows the hierarchical view of CK-System4TM. The prototype system is designed for three user types which are domain expert, knowledge engineer and general user. Domain Experts have expertise in different areas of TM knowledge, while Knowledge Engineers have expertise in knowledge modeling. General users are people which use the CK-System4TM. Different technologies such as HTML, SPARQL and JAVA have been used in the development of CK-System4TM.

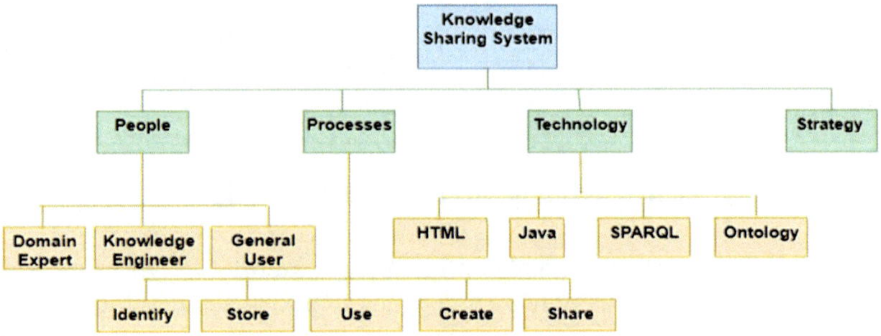

Fig. 2. Hierarchical view of CK-System4TM modules

4 System Architecture of Developed Cloud Based Knowledge Sharing System

The proposed layered architecture of CK-System4TM is shown in Fig. 3. This cloud based model help user to access CK-System4TM anywhere anytime. It consists of 4 layers; i.e. UI layer, Knowledge Processes layer, Service layer and Data Access layer. First of all different types of users can access the CK-System4TM to perform their activities as per their pre-specified roles, using UI layer. Domain Experts can modify KB using UI after authentication. Knowledge Engineers create or update TM ontology. General users perform required actions using their UI interface. Second layer is of knowledge process layer that consists of five knowledge processes (create, identify, share, use and store). These actions are performed using modules of services layer which store or retrieve required data from cloud based KB using Data Access Layer. Third layer consists of different services modules, like security, collaboration, Search & Navigation, inference reasoning, Data and Web services. In data access layer jena adapter is used to access or to store knowledge in the cloud based KB. Overall system is deployed using private deployment model for IaaS.

In CK-System4TM knowledge is represented in the form of ontology. Ontology defines a set of representational primitives in IT and CS context with which we can model a domain of knowledge or discourse [9].

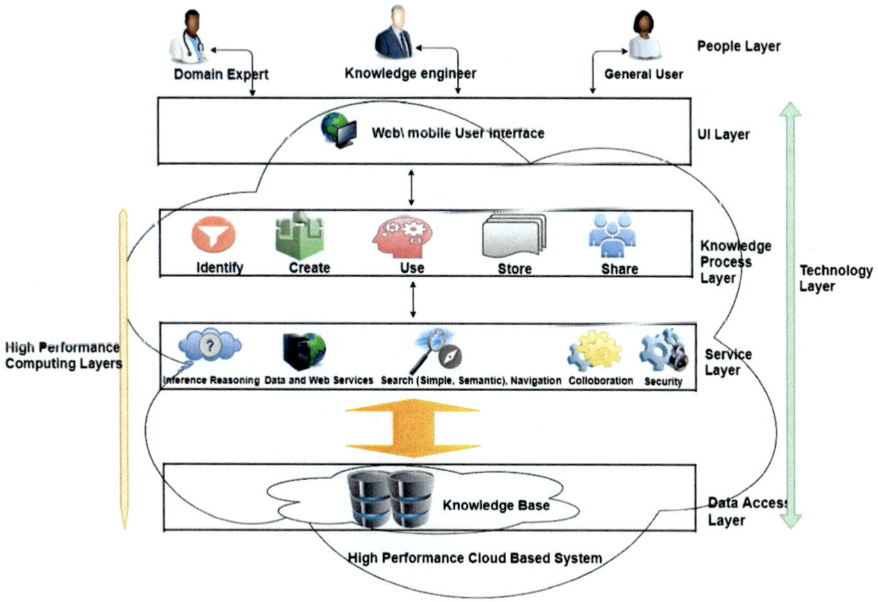

Fig. 3. CK-System4TM architecture design

4.1 TM Ontology

An ontology consists of classes (concepts of any domain), their attributes, individuals and their relationships between different classes and individuals. Individuals are the ground level components (or instances).Ontology is responsible for categorizing or grouping individuals. Core classes (concepts) of TM domain ontology consist of Treatment, Illness, Illness Symptoms and Sources individuals. Relationship graph of available data is shown in Fig. 4.

On the basis of proposed graph, ontology for CK-System4TM is implemented protégé as shown in Fig. 5.

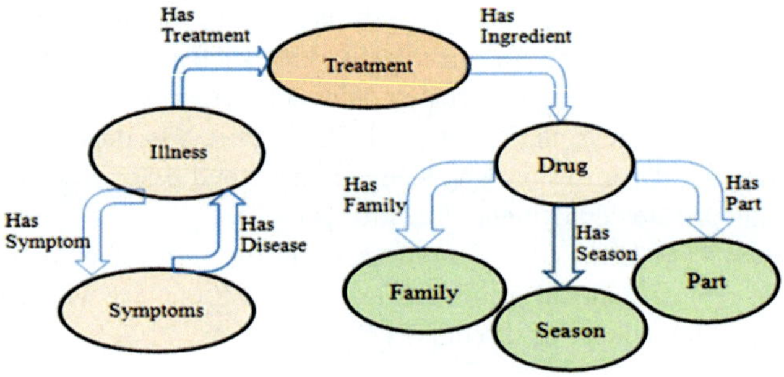

Fig. 4. Data relationship graph

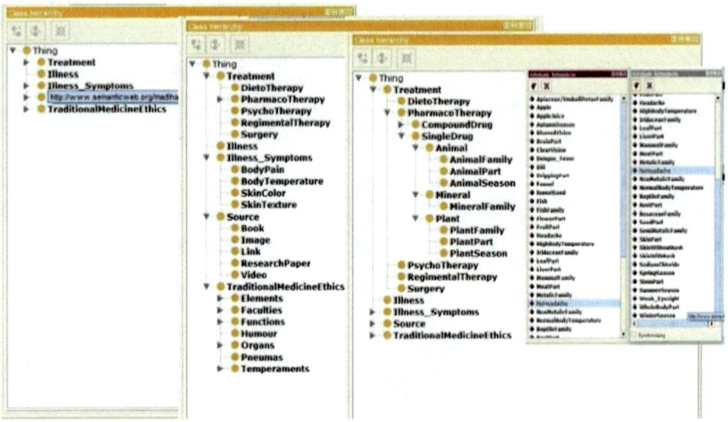

Fig. 5. CK-System4TM ontology

Object and Data Properties. Object properties mutually link classes (concepts) to each other. It also refers the individuals to their specific classes and subclasses. The term 'Data property' defines the attributes of individuals and also provides their values.

SPARQL Queries. SPARQL query language helps too view different scenarios. Queries are used to search, retrieve, insert, update and delete data from KB. Some sample SPARQL queries for data retrieval are shown in Fig. 6. Insert data module and query are shown in Fig. 7.

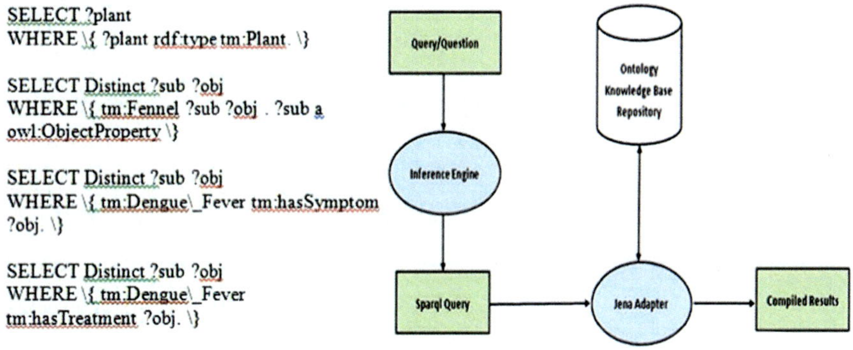

Fig. 6. Queries and model of CK-System4TM search

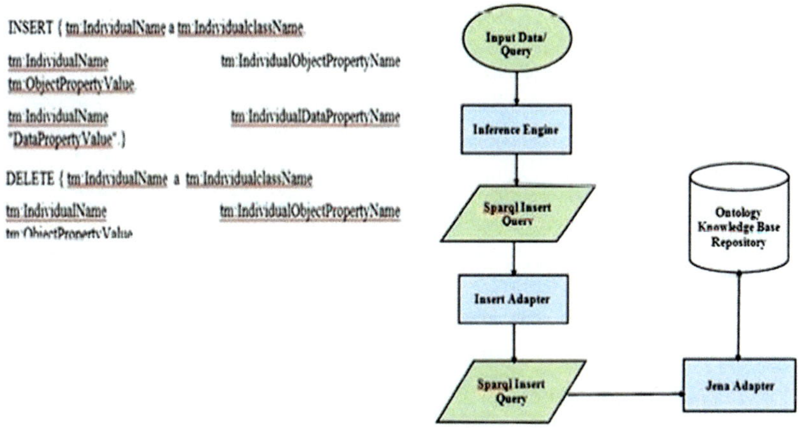

Fig. 7. Queries and model of CK-System4TM insert

CK-System4TM is developed using NetBeans IDE and Jena adapter. Data manipulation is done by passing query to inference engine. Inference engine inspects input and formulate query with valid data objects. This query then passed to adapter for execution. Query execution is done with the help of Jena by communicating with KB.

4.2 Feature-based Comparison of CK-System4TM and Existing Systems

A comparison is done between existing systems for TM domain and developed CK-System4TM on the basis of usability features. Table 2 shows that four TM systems are compared with CK-System4TM on basis of ten different usability features. These are generated by inspecting relevant literature and brainstorming. Some features deals with knowledge availability and effective response of system to its users. Similarly other features concerned with KM factors, processes and personalization.

Free accessibility means users can freely access available knowledge without any cost. Content sharing means system provides platform to experts for sharing knowledge after authentication. Decision Support is the facility of the system that it can suggest medicine after examines patient symptoms. Expert personalized data used for authentication. System is developed inline with KM critical factors and KM process. Navigation and easy search for drugs is available in the system. Module based system development leads to flexibility of change.

Table 2. CK-system4tm usability features comparison

Features	TM Systems				
	HerbMed	HTKMS	HMKB	Tropicos	CK-System4TM
Free Accessibility	✓	✓	✓	✗	✓
Content Sharing	✗	✗	✓	✗	✓
Decision Support	✗	✗	✗	✗	✓
Expert Personalized Data	✗	✗	✗	✗	✓
KM Critical Factors Consideration	✗	✗	✗	✗	✓
KM Process Used	✗	✗	✗	✗	✓
Navigation	✓	✓	✓	✓	✓
Search Response	✗	✓	✓	✗	✓
System Flexibility	✗	✗	✓	✗	✓
Text Search	✓	✓	✓	✓	✓

4.3 Validation of CK-System4TM by Technology Acceptance Model

Technology acceptance model (TAM) has two parameters 'Usefulness' and 'Ease-of-Use [21]. To validate CK-System4TM a TAM questionnaire based survey is performed from general users and domain experts on the basis of above two parameters. 15 questions have been given to the general users and 20 questions have been given to domain expert to collect their feedback about this CK-System4TM Fig. 8. Feedback form of general users is filled by 100 random persons who belong to different parts of the country while the Domain Expert feedback form have been filled by 30 different professional experts having a remarkable experience in TM field. General users show positive response about the CK-System4TM and express their interest in this prototype. They find it easy to use with its advanced and promising features. Domain expert have found this prototype user friendly and effective, while mentioning some limitations of UI and functionality. Statistical data analysis of both questionnaire response is shown in Fig. 9.

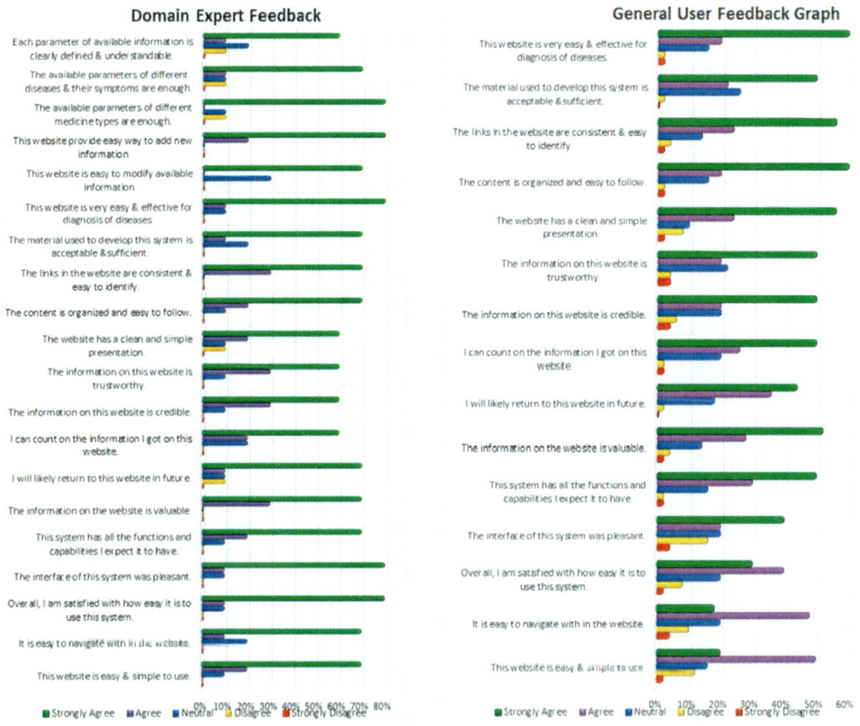

Fig. 8. Feedback from general users and domain expert

Domain Expert Feedback

	Strongly Disagree	Disagree	Neutral	Agree	Strongly Agree
This website is easy & simple to use.	0%	0%	10%	20%	70%
It is easy to navigate with in the website.	0%	0%	20%	10%	70%
Overall, I am satisfied with how easy it is to use this system.	0%	0%	10%	10%	80%
The interface of this system was pleasant.	0%	0%	10%	10%	80%
This system has all the functions and capabilities I expect it to have.	0%	0%	10%	20%	70%
The information on the website is valuable.	0%	0%	0%	30%	70%
I will likely return to this website in future.	0%	10%	10%	10%	70%
I can count on the information I got on this website.	0%	0%	20%	20%	60%
The information on this website is credible.	0%	0%	10%	30%	60%
The information on this website is trustworthy.	0%	0%	10%	30%	60%
The website has a clean and simple presentation.	0%	10%	10%	20%	60%
The content is organized and easy to follow.	0%	0%	10%	20%	70%
The links in the website are consistent & easy to identify.	0%	0%	0%	30%	70%
The material used to develop this system is acceptable &sufficient.	0%	0%	20%	10%	70%
This website is very easy & effective for diagnosis of diseases.	0%	0%	10%	10%	80%
This website is easy to modify available information	0%	0%	30%	0%	70%
This website provide easy way to add new information	0%	0%	0%	20%	80%
The available parameters of different medicine types are enough.	0%	10%	10%	0%	80%
The available parameters of different diseases & their symptoms are enc	0%	10%	10%	10%	70%
Each parameter of available information is clearly defined & understand	0%	10%	20%	10%	60%

General User Feedback

	Strongly Disagree	Disagree	Neutral	Agree	Strongly Agree
This website is easy & simple to use.	2%	12%	16%	50%	20%
It is easy to navigate with in the website.	4%	10%	20%	48%	18%
Overall, I am satisfied with how easy it is to use this system.	2%	8%	20%	40%	30%
The interface of this system was pleasant.	4%	16%	20%	20%	40%
This system has all the functions and capabilities I expect it to have.	2%	2%	16%	30%	50%
The information on the website is valuable.	2%	4%	14%	28%	52%
I will likely return to this website in future.	0%	2%	18%	36%	44%
I can count on the information I got on this website.	2%	2%	20%	26%	50%
The information on this website is credible.	4%	6%	20%	20%	50%
The information on this website is trustworthy.	4%	4%	22%	20%	50%
The website has a clean and simple presentation.	2%	8%	10%	24%	56%
The content is organized and easy to follow.	2%	2%	16%	20%	60%
The links in the website are consistent & easy to identify.	2%	4%	14%	24%	56%
The material used to develop this system is acceptable &sufficient.	0%	2%	26%	22%	50%
This website is very easy & effective for diagnosis of diseases.	2%	2%	16%	20%	60%

Fig. 9. Statistical data of feedback from general users and domain expert

5 Conclusion

The primary focus of the research is to design a Cloud based Knowledge Sharing System for Traditional Medicines. A thorough study of processes and critical factors of existing KMS leads to design and development of CK-System4TM. System is designed by comparing existing systems, identifying their critical factors and knowledge processes they address. The four critical factors are identified; i.e. people, processes, technology and strategy. These factors are determined in order of first knowledge type of the system on the premise of people for whom the system is developed. Afterwards KM processes are finalized while technology is decided in the last to develop the system. CK-System4TM strategy provides an overall management activity for the other three factors. CK-System4TM address five knowledge processes which are identify, store, use, create and share.

A cloud based system is designed and developed. SPARQL query is used for data operations. In this system, different types of users can access CK-System4TM using their specified roles. CK-System4TM offers the facility of searching and other data manipulation operations through cloud. The prototype has been validated by comparing usability features of the existing TM systems. Comparison shows almost all features of comprehensive system are available in the developed system. TAM based questioner is designed for validation. Survey has been conducted from general users and from experts about CK-System4TM. This questioner shows 70–80% users strongly agree with the system. After deployment, CK-System4TM can be used by the general users and the domain experts.

References

1. Gao, P.F., Watanabe, K.: Introduction of the world health organization project of the international classication of traditional medicine. Zhong Xi Yi Jie He Xue Bao = J. Chin. Integr. Med. **9**(11), 1161–1164 (2011)
2. WHO traditional medicine strategy: 2014–2023. http://www.who.int/iris/handle/10665/92455. Accessed 20 Nov 2020
3. Han-Lin, C., Yang, A.W.H., Pittayachawan, S., Wickramasinghe, N.: An analysis on the utilisation of health information technology to support clinical operation of Chinese medicine. In: Complementary and Alternative Medicine: Breakthroughs in Research and Practice, pp. 572–592. IGI Global (2019)
4. Lin, C.H., Wei, A., Yang, H., Pittayachawan, S., Vogel, D., Wickramasinghe, N.: Inquiring knowledge management systems a Chinese medicine perspective. In: 2015 48th Hawaii International Conference on System Sciences, USA, pp. 3682–3690. IEEE (2015)
5. Yu, T., Liu, J., Yang, S., Li, J., Jia, L.: Semantic web for knowledge integration between traditional Chinese medicine and biomedicine. In: 2015 7th International Conference on Information Technology in Medicine and Education (ITME), China, pp. 229–233. IEEE (2015)
6. Anshari, M., Alas, Y., Guan, L.S.: Developing online learning resources: Big data, social networks, and cloud computing to support pervasive knowledge. Educ. Inf. Technol. **21**(6), 1663–1677 (2015). https://doi.org/10.1007/s10639-015-9407-3
7. Rawat, S., Chowdhary, R., Bansal, A.: Data integrity of cloud data storages (CDSS) in cloud. Int. J. **3**(3) (2013)
8. Li, Z., Liu, L., Barenji, A.V., Wang, W.: Cloud-based manufacturing blockchain: secure knowledge sharing for injection mould redesign. Procedia CIRP **72**(1), 961–966 (2018)
9. Amjad, S., Waheed, T., Enriquez, A.M.M., Aslam, M., Syed, A.: An ontology based knowledge preservation model for traditional unani medicines. In: Ślęzak, D., Tan, A.-H., Peters, J.F., Schwabe, L. (eds.) BIH 2014. LNCS (LNAI), vol. 8609, pp. 278–289. Springer, Cham (2014). https://doi.org/10.1007/978-3-319-09891-3_26

10. Mustaffa, S., Ishak, R.Z., Lukose, D.: Ontology model for herbal medicine knowledge repository. In: Lukose, D., Ahmad, A.R., Suliman, A. (eds.) KTW 2011. CCIS, vol. 295, pp. 293–302. Springer, Heidelberg (2012). https://doi.org/10.1007/978-3-642-32826-8_30
11. Heisig, P.: Harmonisation of knowledge management comparing 160 km frameworks around the globe. J. Knowl. Manag. **13**(4), 4–31 (2009)
12. Fausat, O.B., Olayinka, T.A.: Design and implementation of herbal therapy knowledge management system (HTKMS). Sci. J. Public Health **7**(2), 44 (2019)
13. Kolog, E., Devine, S., Sutinen, E., Sääksjärvi, I.: Knowledge-based service for African traditional herbal medicine: a hybrid approach (2019)
14. Padmasiri, G.: Management of indigenous knowledge in Srilanka, with special reference to indigenous medicine. Inf. Dev. **34**(5), 475–488 (2018)
15. Alkhatib, B., Briman, D.: Building a herbal medicine ontology aligned with symptoms and diseases ontologies. J. Digital Inf. Manag. **16**(3), 114 (2018)
16. Sahri, Z., Aziz, M.E.S.A., Ramli, N., Canda, R.: The design and implementation of Malaysian indigenous herbs knowledge management system based on ontology model. In: Knowledge Management International Conference (KMICe), pp. 439–445 (2016)
17. Lin, C.H., Pittayachawan, S., Yang, A.W.H., Wickramasinghe, N.: Using is/it to support the delivery of Chinese medicine: a Chinese medicine clinic management system. Int. J. Biomed. Eng. Technol. **16**(3), 223–243 (2014)
18. Elashaheb, M., Beyh, S., Aouad, G.: The implementation of the knowledge management framework within the telecommunication industry: the KMFTI model (2005)
19. Maier, D., et al.: Knowledge management for systems biology a general and visually driven framework applied to translational medicine. BMC Syst. Biol. **5**(1), 38 (2011)
20. Yu, B., Dongfeng, C., Yanju, D.: An ontology based knowledge management system architecture for traditional Chinese medicine. In: 2010 IEEE International Conference on Bioinformatics and Biomedicine Workshops (BIBMW), China, pp. 640–643. IEEE (2010)
21. Hou, C., Plaisent, M., Ming-Hsun, J., Bernard, P.: Technology acceptance model of consumer satisfaction with the physical and spiritual health with stress-relief and health industry as an example. Int. J. Comput. Sci. Electron. Eng. **1**(2), 278–282 (2013)

An Analytical Review of Data Access Control Schemes in Mobile Cloud Computing

Pattavee Sanchol and Somchart Fugkeaw(✉)

School of Information, Computer, and Communication Technology, Sirindhorn International Institute of Technology, Thammasat University, Khlong Luang 12120, Pathum Thani, Thailand
somchart@siit.tu.ac.th

Abstract. Owing to the cryptographic-based access control property, ciphertext policy attribute-based encryption (CP-ABE) has been widely adopted as an effective solution for data access control in cloud computing. Nevertheless, CP-ABE cryptographic construct is based on bilinear paring rendering high computation cost. Therefore, using CP-ABE in MCC environment is non-trivial because mobile devices are resource-constraint (such as limited battery, limited computing power) and they are not suitable to run such high complexity of crypto operation. This research issue is unmatured and requires in-depth investigation. In this paper, we investigate the research problems and challenges of data access control in mobile cloud computing (MCC). Specifically, we conduct the comprehensive review of existing solutions based on CP-ABE. The review is done through the analysis and discussion of recent solutions. We also provide the comparative analysis of the selected works. Finally, we point out possible research issues and challenges.

Keywords: Cloud computing · Mobile cloud · CP-ABE · Access control

1 Introduction

With the popularity of cloud computing technology, many organizations have employed cloud as a strategic service deployment. To leverage cloud services, it is inevitable that data are processed and even stored in the cloud system. To this end, security and privacy management of outsourced data is paramount of importance. Most cloud platforms provide basic security mechanisms such as firewall, anti-virus, authentication models. However, fine-grained access control and encryption requirements are not available as an integrative solution. To this end, data owners are required to encrypt their data before they are uploaded to the cloud. Also, the access control system (authentication and authorization) should be implemented to work with the encryption mechanism. Managing both access control and multiple users' keys is non-trivial, for example the data owner or the developer need to have a database for controlling users' access rights and cryptographic mechanism for supporting key management to correspond to the access control rules.

Nowadays mobile devices have not been only used for presenting the information, but they are also extensively used for processing several computation tasks and storing

P. Meesad et al. (Eds.): IC2IT 2021, LNNS 251, pp. 310–321, 2021.
https://doi.org/10.1007/978-3-030-79757-7_30

data. Also, they can leverage data access service from cloud computing to enable high accessibility and effective use of service provisioning. Such combination renders an emerging cloud platform called mobile cloud computing (MCC).

Existing commercial cloud applications generally adopt encryption techniques and access control model to support the security and privacy of the outsourced data. For the fine-grained access control in cloud computing, ciphertext-policy attribute-based encryption (CP-ABE) is generally regarded as an effective approach. This is because it possesses the encryption feature and access control enforcement. With this approach, the access is granted to the individual user level based on their privilege characterized by their attributes. In CP-ABE, the data is encrypted with an access policy constituting from a set of attributes and logical operators (OR, AND, KofN). To decrypt the file, the user needs to have a secret key containing a set of attributes that satisfies the policy. For example, a policy is expressed as <Admin or (Head and Level > 9). Here, this policy only allows user who has a key attribute < Admin > or a key attribute < Head > and attribute < Level > 9 > to decrypt the file encrypted with this policy.

CP-ABE is based on bilinear-mapping theory that generally deals with pairing and exponentiation operation. These mathematical operations are computationally expensive. Even though CP-ABE is powerful mechanism for supporting fine-grained access control for data outsourcing scenario, but it is not designed to be efficiently implemented in MCC environment. Therefore, implementing CP-ABE in resource-constraint devices are not practical.

Existing solutions [1–7] are based on offloading partial CP-ABE cost to the proxy or outsourcing server while allow the lightweight decryption operation to be run by the mobile devices. Even though they share the mutual resolution in enabling small computation cost to be done at mobile devices, there are some differences in their proposed cryptographic protocol and security primitives.

In this paper, we provide the analytical review of existing works dedicated for data access control in MCC environment. Specifically, we focus on the investigation of decryption outsourcing schemes as it is the desirable feature in MCC where mobile devices are mostly used to support data access. To enable more generalization on how CP-ABE is integrated into the mobile cloud solution, we describe and give the example of the system process in the diagram together the discussion of security issue of the overall process. In addition, we conduct the comparative analysis of related works and points out pros and cons of the papers reviewed. We also point out possible research issues.

2 CP-ABE Background

This section provides the details of Ciphertext-Policy Attribute-Based Encryption (CP-ABE). CP-ABE is a kind of attribute-based encryption [8] and it was proposed in 2007 by Bethencourt et al. [9]. In [9], the important definitions for CP-ABE construct and security analysis are given. They also published and shared the CP-ABE library [5].

Basically, the construction of CP-ABE is based on the bilinear maps. A description of the formal definition of bilinear maps is shown below.

Bilinear Maps

Let G_0 and G_1 be two multiplicative cyclic groups of prime order p and e be a bilinear map e: $G_0 \times G0 \rightarrow G_1$. Let g be a generator of G_0. Let H: $\{0, 1\}^* \rightarrow G_0$ be a hash function that the security model is in random oracle. The bilinear map *e* has the following properties:

- Bilinearity: for all $u, v \in G_1$ and $a, b \in Z_p$, $e\,(u^a, v^b) = e\,(u, v)^{ab}$
- Non-degeneracy: $e\,(g, g) \neq 1$.

There are four major CP-ABE algorithms [5] as described below.

Setup. The setup algorithm takes no input other than the implicit security parameter. It outputs the public parameters *PK* and a master key *MK*.

Key Generation $(\boldsymbol{MK}, \boldsymbol{S})$**.** The algorithm takes as input the master key *MK* and a set of attributes S that describe the key. It outputs a user decryption key *UDK*. Technically, key generation is based on Bilinear Maps.

Encrypt $(\boldsymbol{PK}, \boldsymbol{M}, \boldsymbol{A})$**.** The encryption algorithm takes as input the public parameters *PK*, a message *M*, and an access structure A (an access tree *T*) over the universe of attributes. The algorithm encrypts *M* and produces a ciphertext *CT*.

Decrypt $(\boldsymbol{PK}, \boldsymbol{CT}, \boldsymbol{SK})$**.** The algorithm takes as input the public parameters *PK*, a ciphertext *CT*, and a user secret key *SK*, which is a key containing a set S of attributes. If the set S of attributes satisfies the access structure *A* then the algorithm will decrypt the ciphertext and return a message *M*.

3 Data Outsourcing Scenario in Mobile Cloud Computing

3.1 Overview

In MCC, mobile devices are used to access data stored in the cloud or even process some tasks provided by the cloud service. This enables high degree of accessibility and service delivery. To deal with a cloud-based solution, mobile devices need to possess the cryptographic processing capability. For example, they should encrypt or decrypt the data stored in the cloud. Nevertheless, CP-ABE is not designed to be implemented in the resource-constraint devices. Enabling CP-ABE to be feasibly integrated in the crypto primitive of MCC is thus a real challenge.

To date, most works have introduced lightweight cryptography based on CP-ABE to address the limited processing power of mobile devices. Basically, the key idea is to reduce the computation cost and optimize storage cost of mobile device based on the computation outsourcing strategy. The outsourcing can be done through key generation [4], encryption [6], and decryption [1–4, 7, 10]. To this end, CP-ABE operations are partially or fully outsourced to the proxy or the outsourcing server while the mobile clients deal with lighter computation cost of another symmetric encryption. The general access control model in MCC is illustrated in Fig. 1.

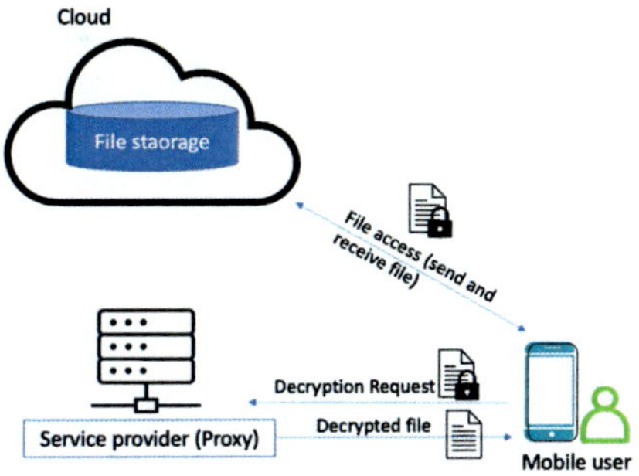

Fig. 1. Data access model in MCC.

From Fig. 1, data owners outsource their data in the encrypted format to the cloud. Mobile users can access data by getting the ciphertext and decrypting the file with their secret key. Basically, the CP-ABE related costs are outsourced to be done by the proxy that can be located in the cloud. However, the additional encryption layer is required in order to allow mobile users to decrypt the intermediate ciphertext to get the plaintext from their mobile devices.

Based on our review, there are two general solutions including outsourcing CP-ABE encryption or decryption and outsourcing CP-ABE with symmetric encryption to a proxy. Both solutions aim at applying CP-ABE in MCC that enables encrypted file stored on the cloud server to be decrypted by mobile devices with cost-effective manner.

Here, we describe these two outsourcing solutions in details and provide a review of key research works for each solution.

3.2 Outsourcing CP-ABE Encryption/Decryption to a Proxy Server

In this solution, CP-ABE decryption is partially outsourced to be processed by the proxy. Then, it returns intermediate ciphertext and it is sent to mobile users for final decryption. Based on this outsourcing strategy, three approaches are selectively chosen, and their details are described as follows.

a. *Outsourcing the decryption of ABE ciphertexts*
   In 2011, Green et al. [1] proposed a new method for outsourcing decryption of ABE ciphertexts by modifying the traditional CP-ABE key generation function to produce two keys. First, a short El Gamal [11] secret key is generated as the first private key component kept by the user. Another key is called a "transformation key" (TK) that is shared with the service provider and can be publicly distributed. This scheme consists of the following five algorithms.

Setup (λ, U)**.** The setup algorithm takes security parameter and attribute universe description as input. It outputs the public parameters PK and a master key MK.

Encrypt (PK, M, I_{enc})**.** The encryption algorithm takes as input the public parameters PK, a message M, and an access structure (aka. attribute set) I_{enc}. It outputs the ciphertext CT.

KeyGen$_{\text{out}}$ (MK, I_{key}). The key generation algorithm takes as input the master key MK and an attribute set (resp. access structure) I_{key} and outputs a private key SK and a transformation key TK.

Transform (TK, CT). The ciphertext transformation algorithm takes as input a transformation key TK for I_{key} and a ciphertext CT that was encrypted under $I_{enc.}$ It outputs the partially decrypted ciphertext CT' if $S \in A$ and it returns $\perp$ otherwise.

Decrypt$_{\text{out}}$ (SK, CT'). The decryption algorithm takes as input a private key SK for I_{key} and a partially decrypted ciphertext CT' that was originally encrypted under I_{enc}. It outputs the message M if $S \in A$ and it returns $\perp$ otherwise.

At high level, the outsourced key generation function is initially run based on bilinear maps. Then it chooses a blinding factor exponent $Z \in Z_p$ (for groups of prime order p) and raises all elements to z^{-1} (mod p). This produces the transformation key TK, while the blinding factor z can serve as the secret key. This scheme is regarded as a CP-ABE based outsourcing decryption scheme that has been adopted by several works. Specifically, this scheme allows the proxy server to run bilinear paring which is the major cost of CP-ABE approach and leave the simple exponentiation to the user. Actually, the decryption cost at user is significantly reduced. The authors also conducted the experiment to test in the mobile devices and it was proven that the proposed scheme was efficiently implementable. For the security point of view, this scheme was securely proven in both CPA and replay able CCA security model. However, the proposed scheme was not designed to support outsourcing CP-ABE encryption. Also, the trust of semi-trusted proxy is not given.

b. *Attribute-based encryption with encryption and decryption outsourcing*
   In 2014, Muhammad Asum et al. [2] proposed a CP-ABE encryption and decryption outsourcing scheme based on the extended solution of [1]. Essentially, the proposed scheme was designed to empower the key generation function algorithm to yield a new SK_{pdec} for partial decryption and SK_{pdnc} for partial encryption. Also, they proposed encryption and decryption algorithm to be partially executed by service providers or proxies. This scheme requires two proxies to perform encryption and decryption. The proposed scheme is constructed from the following algorithms.

Setup (λ). The setup algorithm takes security parameter and attribute universe description as input. It outputs the public parameters PK and a master key MK.

Key Generation$_{\mathbf{d}}$ (MK, PK, A)**.** This algorithm takes as input the master secret key MK, Public key PK and a set of attributes A. It outputs a user's secret *key* SK_A that contains a set of user attributes A. SK_A is composed of two parts, (1) SK_{pdec} as a decryption key used by a proxy, and (2) SK_{udec} as the final decryption key used by the user.

Key Generation$_{\mathbf{e}}$ (MK, PK)**.** This function is used to generate a unique secret key to deliver to the proxy server. The algorithm takes as input the master secret key MK and public parameters PK. The output of this algorithm is SK_{penc}. The SK_{penc} was used by encryption proxy for constructing the access policy in the encryption process.

Encryption (PK, M, T)**.** The encryption algorithm takes as input the public parameters PK, a message M, and an access tree T over a universe of attributes. To deal with encryption process, the host or data owner needs to produces a partial ciphertext, that consists of encrypted message, encrypted secret key, and the set of attributes.

The latter component is sent to the proxy to compute a cryptographic policy.

Policy Creation $\left(CT', SK_{penc}\right)$**.** This function is run by the proxy responsible for encryption as of the encryption time. It takes as input the partial ciphertext CT' and the proxy encryption secret key SK_{penc}. Then, it creates the cryptographic access policy related to the access tree T. Finally, it outputs the ciphertext CT.

Policy Verification $\left(SK_{pdec}, CT\right)$**.** This algorithm is run by the decryption proxy. It takes as input the proxy decryption key SK_{pdec} related to SK_A and the ciphertext CT. The output of this algorithm is a partially decrypted ciphertext CT' (El Gamal style ciphertext [5]) if the attributes in satisfies access tree T.

Decryption $\left(SK_{udec}, CT'\right)$**.** This algorithm is run by the user. It takes as input the partially decrypted ciphertext CT' and a user's secret key SK_{udec}. The output of this stage is the message M.

Actually, this scheme supports both encryption and partial decryption outsourcing. The scheme was proven to be secure in generic group model. However, this approach relies on two proxies to perform encryption and decryption tasks. The cost in maintaining two proxies and the complexity still exists at both data owner and users.

In 2020, Shulan Wang et al. [4] proposed an access control for supporting mobile healthcare network. In this solution, cost of CP-ABE operation is generally reduced because the decryption cost on mobile device is based on the random decryption on the element of CP-ABE ciphertext. Basically, the model construct is designed to offload heavy operations such as key generation, data encryption and partial ciphertext decryption are executed by several service providers. For example, key generation function is outsourced to the Key Generation Provider (OKGSP), the encryption function is outsourced to the Encryption Provider (OESP). and also outsourced the decryption function to the decryption Provider (ODSP). In this scheme, mobile users are required to decrypt the random value embedded in the CP-ABE ciphertext. This scheme consists of four phases and eight functions as follows:

Phase 1. System initialization: this phase includes an algorithm AC.Setup.

AC.Setup $(1^{\kappa}) \rightarrow (PK, MSK, SK_{OKGSP})$**.** AC runs the probabilistic operation which inputs security parameter κ and returns public key PK, master secret key MSK, and OKGSP's secret key SK_{OKGSP}.

Phase 2. User's key generation: this phase consists of two functions AC.KeyGen and OKGSP.KeyGen.

AC.KeyGen $(PK, MSK) \rightarrow (SK_1, OSK_{OKGSP})$**.** AC executes the local algorithm. It inputs PK and MSK. It creates part of user's secret key SK_1 and the outsourced secret key for the OKGSP OSKOKGSP.

OKGSP.KenGen $(S, PK, SK_{OKGSP}, OSK_{OKGSP}) \rightarrow (SK_2)$**.** OKGSP runs the outsourced operation. It inputs user's attribute set S, OKGSP's secret key SK_{OKGSP}, and the outsourced secret key OSK_{OKGSP}. It generates another part of user's secret key SK_2 associated with S. Then, the user's secret key SK is generated, that is SK = $\{SK_1, SK_2\}$.

Phase 3. Data encryption: this phase includes two functions DO.Encrypt and OESP.Encrypt.

DO.Encrypt $(PK, T, M) \rightarrow$ **(CT_1).** Data owner (DO) runs the local algorithm. It inputs PK, an access policy T, and a file to be encrypted M. It outputs part of ciphertext CT_1 and the outsourced secret key for the OESP OSK_{OESP}.

OESP.Encrypt $(PK, OSK_{OESP}) \rightarrow (CT_2)$**.** OESP executes the outsourced operation. It inputs PK and OESP's secret key OSK_{OESP}. It generates another part of ciphertext CT_2. Then, the complete ciphertext CT is constructed, that is CT = $\{CT_1, CT_2\}$.

Phase 4. Data decryption: this phase includes three functions User.Delegate, ODSP.Decrypt, and User.Decrypt.

User.Delegate $(SK) \rightarrow \left(OSK_{ODSP}, SK_{delegate}\right)$**.** The decryption user runs the local algorithm. It inputs the user's secret key SK and outputs an outsourced secret key for the ODSP OSKODSP and the corresponding delegated key $SK_{delegate}$.

ODSP.Decrypt $(PK, CT, OSK_{ODSP}) \rightarrow IT$**.** ODSP operates the outsourced algorithm. It inputs PK, ciphertext CT which implicitly includes access policy T, and ODSP's secret key OSK_{ODSP}. It creates the intermediate result IT, only when the user's attribute set S matches the access policy T associated with CT.

User.Decrypt $\left(IT, CT, SK_{delegate}\right) \rightarrow M$**.** The decryption user runs the local operation. It inputs IT, CT, and $SK_{delegate}$. It outputs plaintext M.

Based on the proposed outsourcing model, most heavy CP-ABE-related costs are offloaded to the service providers responsible for each operation. Mobile users deal with lightweight decryption of random value in the element of CP-ABE ciphertext. However, high dependency on service providers is not flexible and impractical for real implementation. The maintenance and administrative cost for working with several service providers are also the burden. In the security point of view, the size of random value used for blinding the original message M in the intermediate ciphertext CT' must be large enough to make the CT' is resistant to Bruce force attack.

3.3 Outsourcing CP-ABE Decryption to a Proxy with Symmetric Encryption Combination

With this solution, there is another encryption layer such as symmetric encryption introduced to encrypt the data. Basically, a data file is encrypted with a symmetric encryption before it is outsourced to the cloud. CP-ABE is used to encrypt the symmetric key.

Here, CP-ABE is outsourced to the proxy server while mobile devices run partial CP-ABE decryption to get the symmetric key to decrypt the ciphertext. With the symmetric cryptography, mobile devices are able to decrypt the ciphertext efficiently.

Ruixuan Li et al. [3] proposed a framework of lightweight data sharing scheme (LDSS) for mobile cloud computing based on the integration of symmetric encryption and CP-ABE. The proposed scheme consists of four major algorithms as follows:

Setup (attribute set A, version attribute V) $\rightarrow$ (the master key MK, the public key PK)
KeyGen (attribute set A_u, master key MK) $\rightarrow$ (the attribute key SK_u that associated with attribute set A_u)
Encryption (symmetric key K, public key PK, access control tree T) $\rightarrow$ (ciphertext CT)
Decryption (ciphertext CT, access control tree T, attribute key of user SK_u) $\rightarrow$ (plaintext of symmetric key K)

In essence, the scheme applies symmetric cryptography algorithm such as AES, 3DES to encrypt a plaintext and then uses CP-ABE to encrypt the symmetric key. Then both ciphertext and encrypted symmetric key are stored on the cloud storage. For decryption, the mobile user gets CT from the cloud storage and decrypts encrypted symmetric key by using CP-ABE decryption key. Then, the symmetric key is obtained, and it will be used to decrypt the ciphertext. With this scheme, mobile devices are required to perform CP-ABE decryption. Also, if the file size is too large, the efficiency for the mobile devices in running decryption algorithm will be degraded. For the security point of view, LDSS is based on the traditional CP-ABE. Thus, the security strength is similar to CP-ABE approach. The difference is that it only changes the structure of access policy slightly for supporting in case the attribute is revoked.

4 Comparative Analysis

This section presents the comparative analysis of four works [1–4] that we have presented their concept in the previous section. The comparison of their cryptographic features is firstly presented in Table 1. Then, the detailed analysis is done through the comparison of their storage, computation, and communication cost shown in Table 2, 3, and 4 respectively. In order to simplify the representation of performance evaluation, we define the following symbol.

$|A_O|$: The number of attributes owned by data owner.
$|A_U|$: The number of attributes owned by data user.
$|T_A|$: The number of leaf nodes in access control policy.
E_{G0}, E_{G1}: The size of element in G_0 and G_1 groups.
G_0: Exponentiation and XOR operation in group G_0.
G_1: Exponentiation and XOR operation in group G_1.
G_e: Pairing operation in group G_0.
G_m: Multiplication operation in group G_0.
Z_p: The group {0, 1, …, p−1} multiplication modulo p.
IR: Intermediate result size of encryption and decryption process. The average size smaller than 10,000 bytes of CT.

4.1 Cryptographic Features

Table 1. Features comparison

Schemes	KP/CP-ABE/AES			Outsourced crypto operation
	CP-ABE	KP-ABE	AES	Enc/Dec/Key gen
[4]	✓	✓	x	Dec
[6]	✓	x	x	Enc/Dec
[8]	✓	x	✓	Enc/Dec
[9]	✓	x	x	Enc/Dec/Key gen

Table 1 presents the comparison of cryptographic settings of four schemes. In [1], decryption operation is solely outsourced to be done at the proxy server, while [2] and [3] were designed to support outsourcing encryption and decryption. In [2], the CP-ABE construct is divided to be executed by both a server and the mobile client. In [3], the authors introduced the symmetric encryption layer to be done at the client side. In addition to outsourcing encryption and decryption, Wang et al., also proposed a novel outsourcing scheme for key generation.

4.2 Storage Cost

Generally, the storage cost is one of the most significant overheads of the access control model in cloud computing. Big portions of storage overheads are the ciphertext, CP-ABE cryptographic components such as element size of random.

Table 2. Storage cost comparison

Schemes	Data Owner	AA	Data user	Cloud Storage
[4]	$3E_{G0}+E_{G1}$	E_{G0}	$(\|A_U\|+1)E_{G0}$	$(2\|T_A\|+1)E_{G0}+E_{G1}$
[6]	$3E_{G0}+E_{G1}$	E_{G0}	$(\|A_U\|+1)E_{G0}$	$(2[T_A]+1)E_{G0}+E_{G1}$
[8]	$3E_{G0}+E_{G1}$	E_{G0}	$(\|A_U\|+4)E_{G0}$	$(2[T_A]+3)E_{G0}+E_{G1}$
[9]	$4E_{G0}+E_{G1}$	E_{G0}	$(\|A_U\|+3)+4E_{G0}+E_{G1}$	$(2[T_A]+5)E_{G0}+E_{G1}$

As shown in Table 2, most works share the same overhead for storing CT on cloud storage. However, in all schemes, the data owner still needs to keep the PK for data encryption process while the data user needs to keep the SK for the decryption. In [4], the data owner needs to keep both *PK* and *SK* for data decryption process. Scheme [1] and [2] require least storage overhead because they do not require attribute versions and additional random as [3] and [4] did respectively. Attribute authority (*AA*) in all schemes retain element in G_0. Actually, ciphertexts and cryptographic elements stored in the cloud are the major portion of the storage overhead responsible by the cloud.

4.3 Communication Cost

Table 3. Communication cost comparison

Schemes	Encryption	Decryption
[4]	–	$(\|AU\|+1)E_{G0}+ \|IR\|$
[6]	$3E_{G0}+ E_{G1}+ \|IR\|$	$(\|AU\|+1)E_{G0}+ \|IR\|$
[8]	$3E_{G0}+ E_{G1}+ \|IR\|$	$(\|AU\|+4)E_{G0}+ \|IR\|$
[9]	$4E_{G0}+ E_{G1}+ \|IR\|$	$(\|AU\|+3) + 4E_{G0}+ EG1 + \|IR\|$

Table 3 presents the communication cost of each scheme. The communication overhead of each schemes happens when the data owner or the data user sends request with attached parameter to the proxy server. Generally, the proxy requires two parameters, PK and CT' for the encryption. Thus, communication cost of encryption process is propositional to the size of PK and CT'. For the decryption, most approaches require SK' and CT', but scheme [4] needs one more parameter PK. Hence, the communication cost of decryption of this scheme is higher than others.

4.4 Computation Cost

In essence, computation cost is the most important overhead all approaches have focused on. Table 4 presents the computation cost of each scheme.

Table 4. Computation cost comparison

Schemes	Encryption cost		Decryption cost	
	Mobile device	Provider or proxy	Mobile device	Provider or proxy
[4]	N/A	N/A	$2G_1$	$(2\|A_u\|+1)G_e+$ $(2\|T_A\|+1)G_1$
[6]	$(T_A\| + 2)G_0+ 2G1$	$(2\ \|TA\|\ +1)G_0$	$2G_1$	$(2\|A_u\|+1)G_e+$ $(2\|T_A\|+1)G_1$
[8]	$3G_0+ G_m$	$2\ \|T_A\|\ G_0$	$G_0+ G_m$	$(2\ \|\ A_u\|+1)\ G_0+$ $\|\ TA\ \|\ G_1$
[9]	$4G0 + 2G_1+ Zp$	$2\ \|T_A\|\ G_0$	$G_0+ 3G_1+ 2G_e$	$(2\|Au\|+3)Ge +$ $(2\|T_A\|+1)G_1$

As can be seen from Table 4, scheme [1] and [2] provide less decryption cost compared to [3] and [4]. This is because they used the same method where major bilinear paring operation is done by the proxy while mobile client did the exponential on the secret component on G_I. However, scheme [2] also outsourced the encryption operation to the proxy. In [3], symmetric encryption was used to improve the performance of ciphertext encryption and then used CP-ABE to encrypt the symmetric key. This scheme renders more computation for G_0 and G_m on mobile client. However, such complexity is a tradeoff for improved performance of cyphertext encryption.

In [4], there were more complexity in outsourcing provider as key generation, encryption, and decryption were offloaded to the provider. Mobile clients deal with the use of $SK_{delegate}$ to compute the random component of the intermediate ciphertext obtained from the proxy. Therefore, total computation cost is $G_0 + 3G_I + 2G_e$. Clearly, decryption cost consists of CP-ABE decryption, blind key generation $(G1 + G_0)$, and verification cost on G_e. The computation cost is also propositional to the number of key components that will be computed in each process as well. However, the computational overhead at mobile client is subject to the ciphertext size.

5 Discussion on Possible Research Direction

According to the review findings, we identify possible research directions as follows.

1. Security construct of the divided element of CP-ABE

Most existing solutions have focused on outsourcing the computation of encryption and decryption overheads by dividing the element of CP-ABE to be partially executed between the proxy and mobile device. Here, the fragmented ciphertext or secret key are vulnerable for brute force attack and collusion attack. The techniques such as randomization and zero-knowledge proof are worth to be examined and applied to support security and authenticity over the fragment cryptographic elements sent between the outsourcing server and mobile device.

2. Trust model of proxy

As the proxy is the untrusted or semi-trusted server, it is the primary target for the attack. Also, the secure communication between mobile clients and the proxy should be guaranteed by the trust model. Here, PKI is a palpable solution that can provide strong authentication and non-repudiation among the system entities.

3. Lightweight accountability and flexible verification

Accountability or auditing is crucial for any access control system. In mobile cloud environment, a lightweight public auditing scheme is needed to allow both data owner and third auditing parties to verify the integrity of the transactions occurred in the system. The capability in checking the audit log and ad-hoc verification via mobile devices are prominent. Strong data integrity techniques such as signature and HMAC are promising to be applied to the public auditing scheme.

4. Scalability and Fault tolerance of the outsourcing server

To serve a large number of mobile cloud users, the proxy or outsourcing server must be scalable and fault tolerance. Even though cloud system provides scale-out feature supporting a high volume of transaction processing, the failure can occur anytime. The system elements possessed by the proxy is critical. Hence, fault tolerance based on multiple proxy and secure data or system configuration replications should be available and they need to be designed in a secure manner. On real cloud system with several types of mobile devices.

6 Conclusion

We have presented the detailed review of a CP-ABE based access control model in mobile cloud computing. Existing schemes focus on minimizing the computation cost done at the mobile device by offloading this overhead to be executed by the proxy server or third party. Also, we provided comparative analysis of four schemes by analyzing their cryptographic constructs and their efficiency based on storage, communication, and computation overhead. Finally, we pointed out possible research directions that are currently required to enhance the practicality and security of the access control solution in mobile cloud environment.

References

1. Green, M., Hohenberger, S., Waters, B.: Outsourcing the decryption of ABE ciphertexts. In: Proceedings of the 20th USENIX Conference on Security (SEC 2011), p. 34. USENIX (2011)
2. Asim, M., Petkovic, M., Ignatenko, T.: Attribute-based encryption with encryption and decryption outsourcing. In: Proceedings of 12th Australian Information Security Management Conference, pp. 21–28. ECU Security Research Institute (2014)
3. Li, R., Shen, C., He, H., Gu, X., Xu, Z., Xu, C.: A lightweight secure data sharing scheme for mobile cloud computing. IEEE Trans. Cloud Comput. **6**(2), 344–357 (2018)
4. Wang, S., et al.: A fast CP-ABE system for cyber-physical security and privacy in mobile healthcare network. IEEE Trans. Ind. Appl. **56**(4), 4467–4477 (2020)
5. Fugkeaw, S., Sato, H.: Improved lightweight proxy re-encryption for flexible and scalable mobile revocation management in cloud computing. In: Proceedings of IEEE International Conference on Cloud Computing, 27 June–2 July, pp. 894–899. IEEE (2016)
6. Zhou, Z., Huang, D.: Efficient and secure data storage operations for mobile cloud computing. IACR Cryptol. ePrint Arch. IEEE (2012)
7. Li, Z., Li, W., Jin, Z., Zhang, H., Wen, Q.: An efficient ABE scheme with verifiable outsourced encryption and decryption. IEEE Access **7**, 29023–29037 (2019)
8. Goyal, V., Pandey, O., Sahai, A., Waters, B.: Attribute-based encryption for fine grained access control of encrypted data. In: Proceedings of ACM Conference on Computer and Communications Security (CCS 2006), pp. 89–98. ACM (2006)
9. Bethencourt, J., Sahai, A., Waters, B.: Ciphertext-policy attribute-based encryption. In: Proceedings of IEEE Symposium on Security and Privacy, pp. 321–334. IEEE (2007)
10. Fugkeaw, S.: A fine-grained and lightweight data access control model for mobile cloud computing. IEEE Access **9**, 836–848 (2021)
11. El Gamal, T.: A public key cryptosystem and a signature scheme based on discrete logarithms. IEEE Trans. Inf. Theor. **31**(4), 469–472 (1985)

Integration of MANET with In-Band SDN Controller

Thananop Thongthavorn(✉) and Sumet Prabhavat

Faculty of Information Technology, King Mongkut's Institute of Technology Ladkrabang, Bangkok, Thailand
{thananop,sumet}@it.kmitl.ac.th

Abstract. Mobile Ad Hoc Networks (MANETs) are Infrastructureless mobile networks that can be formed with a group of wireless devices. To establish communication, MANETs require a routing protocol to discover an end-to-end route under uncertain network topology due to node mobility. The traditional reactive routing protocol discovers the shortest route using only the node's local information and keeps utilizing the route until it collapses. This kind of method might not conform to the recent network condition. In this paper, we propose an SDN-based routing protocol for MANETs. By assigning a node in the network to act as an SDN controller, the route discovery action will be centralized. Since the SDN controller is not designed to support multi-hop wireless communication, we additionally design a mechanism to maintain connectivity between the controller and mobile nodes. The controller utilizes link-state information of every node to decide a route upon request. The simulation result shows that the proposed protocol overcomes the aforementioned routing problem. The generated route requests are decreased which indicating an increase in route stability, resulting in a better packet delivery ratio.

Keywords: MANETs · SDN · Routing protocol

1 Introduction

Mobile Ad Hoc Networks (MANETs) [1] are a self-organized mobile network that allows nodes to communicate or exchange data while infrastructure support is not necessary. MANETs can adapt to topology change according to the availability of mobile nodes. This made MANETs scalable and robust. Recently, MANETs is extended to utilize in vehicular networks (VANETs) [2], UAVs networks [3], and people-centric network (Opportunistic Networks: OppNets [4]).

The performance of MANETs depends on its routing protocol. The proactive routing protocol (e.g., OLSR, DSDV) discovers and maintains the route to every node in the network. On the other hand, in the reactive routing protocol (e.g., AODV, DSR), a source node only discovers the route to a destination when it has data to send. These conventional routing protocols operate decentrally and make a decision based on limited network information. When network topology changes, nodes in the network have to

P. Meesad et al. (Eds.): IC2IT 2021, LNNS 251, pp. 322–331, 2021.
https://doi.org/10.1007/978-3-030-79757-7_31

update their routing table accordingly. The network wide-flooding mechanism is erupted to allow nodes to recognize the new network conditions. Therefore, mobile nodes? bandwidth and energy will be wasted if the route break frequently.

Software-Defined Networking (SDN) [5, 6] is a network architecture designed to separate control plane from data plane. An SDN controller (in the control plane) has a role to orchestrate all devices in the data plane. The controller can be controlled by an administrator through an application. From this concept, SDN architecture has the potential to enhance MANETs performance. If an SDN controller can be deployed into MANETs, routing decisions are precisely made by depending on a global view of the network, which is provided by the controller.

In this paper, we propose an SDN-based routing protocol for MANETs. Mobile nodes report their link-state information to the SDN controller periodically. The controller will form the global network topology and discover the route when requested. Then, the controller modifies a routing table of every corresponding node to create an end-to-end route from source to destination. This method increases network adaptability while keeping wide-flooding activity under control.

The rest of the paper is organized as follows. Section 2 introduces the recent SDN-MANETs literature review. The detail of our proposed protocol is described in Sect. 3. The simulation and discussion are presented in Sect. 4. Section 5 summarizes the outcome of this paper.

2 Literature Reviews

Even though SDN architecture emerges to reduce configuration complexity in wired networks, many researchers looking for the possibility to extend SDN technology for supporting wireless networks [7]. Since wireless devices might intermittently disconnect from an SDN controller, network information collected at the controller might not match with actual network condition. To make matters even more difficult, in MANETs environment, node mobility adds more complication to the controller to associate with mobile nodes.

An integration between SDN controller and MANETs are proposed in some related works. Hans et al. [8] introduce SDN-MANETs architecture and practical implementation. A small-scale experiment highlights the topology change detection of SDN-MANETs routing protocol when compared with OLSR. Paolo et al. [9] present the middleware framework prototype for QoS management in SDN-MANETs. The conducted experiment reveals the SDN controller's ability to prioritize the data flow in the network. Konstantinos et al. [10] prove that the SDN-enabled mobile network can be deployed by installing SDN-related software into the commercial smartphone. Moreover, the hybrid SDN-MANETs conceptual architecture is also proposed in the paper to address open issues. These works confirm that an integration between the SDN controller and MANETs can be done with current technology.

Ian et al. [11] present SDN-MANETs architecture with various component designs including, the SDN wireless node components, the association method between SDN wireless node and controller, the flow table management according to flow-based prioritization. To prove these designs, the paper also presents the simulation results to compare the performance of SDN-MANETs with traditional MANETs routing protocols. Under the assumption that the controller always knows the network topology, the traditional routing protocols are outperformed. Klement et al. [12] propose an adaptation of the SDN controller to highly utilized MANETs which responsible for traffic scheduling tasks. A source node attaches bandwidth requirements to a route request message before sending it to the controller. The controller will use demanded bandwidth information to determine the best route at the current network condition. The wireless node in the data plane of [11, 12] is equipped with multiple wireless interfaces. One of the wireless interfaces is dedicated to connecting with an LTE access SDN controller. The SDN-based routing protocol, which is designed to support single wireless interface devices, is done by [13]. Unfortunately, the paper does not consider the impact of node mobility and the number of generated overheads in the simulation results.

3 Proposed Work

In the literature, almost all existing works implement SDN controllers in out-of-band mode. They improve network adaptability at an expense of network complexity which decreases applicability to real networks. To cope with the drawbacks, we propose MANETs routing protocol with an in-band SDN controller. Important functional components proposed in our work are (1) association mechanism between mobile nodes in data plane and an SDN controller in control plane (2) integration between SDN and MANETs routing protocols.

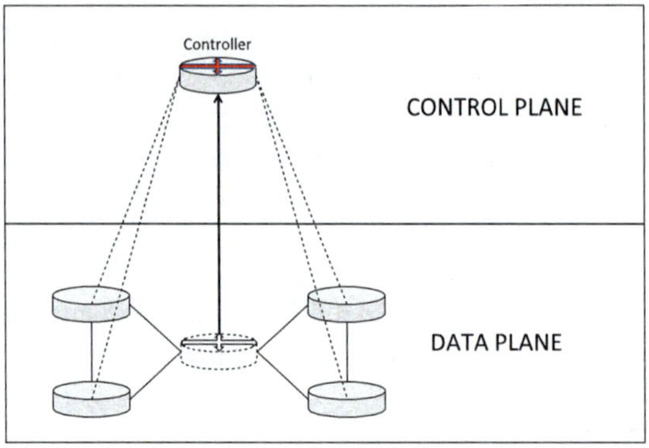

Fig. 1. System overview

3.1 System Components

System Overview. Figure 1 illustrates the system overview. One of the mobile nodes is assigned to be a controller. The others need to associate with the controller to exchange their network information. Afterward, the controller can manipulate the routing table of any mobile node to establish an end-to-end route when requested.

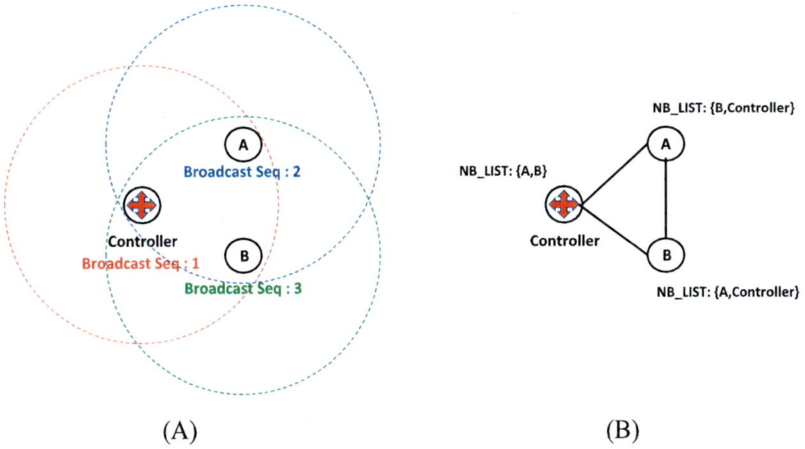

Fig. 2. Controller advertisement message dissemination

Mobile Nodes and Controller Association. A controller advertisement message (C_{adv}) is designed to inform the existence of an SDN controller to mobile nodes in the data plane. The controller periodically disseminates C_{adv} with an increasing sequence number over the network. When a node receives C_{adv}, it creates a reverse route to the controller before rebroadcasting C_{adv}. If the node receives C_{adv} with the same sequence number, it indicates that the node receives duplicate C_{adv}. The node discards duplicate C_{adv} but it will recognize the sender as its neighboring node. Figure 2 is used to explain the C_{adv} dissemination process. Figure 2A shows the transmission range of mobile nodes and an example of C_{adv} broadcast sequence, i.e., node B broadcast C_{adv} as the last node. Figure 2B presents a neighboring node list (NB_list) of every node when the C_{adv} disseminating process has been done. Since node B receives C_{adv} from the controller but has a queue to broadcast C_{adv} after node A, C_{adv} from node B is dropped by A and the controller. However, the NB_list of a and the controller is updated. Specifically, we utilize the C_{adv} messages to provide nodes a route to the controller and update the NB_list of mobile nodes at the same time.

Once a node recognizes a route to associates with the controller, the node starts to periodically report its NB_list. The NB_list packet includes its neighboring node list and a lifetime of each link. The link lifetime is set to 2 times of the controller advertisement interval. The lifetime will be updated after the node receives any packet from its neighboring node. The controller acknowledges the network topology by fabricating NB_list information. Since an activity to report the NB_list is triggered by C_{adv} dissemination, a global view of network topology will be updated synchronously with the C_{adv} advertisement interval. We will define a moment when the global view of network topology (stored at the controller) is updated as an "update period".

Route Discovery. When a node has data to send, it searches its routing table for a valid route to a destination. If the route is not found, the destination IP address will be attached to <route request field> of an NB_list packet before sending it to an SDN controller in the next reporting period. After the controller receives and analyzes the packet, the route discovery process is triggered. The controller applies Dijkstra's algorithm to discover the shortest route. The cost function of each link is presented in Eq. 1, where T is the link lifetime. The older the link lifetime the higher the cost of the link. Since the network collision may affect NB_list information accuracy on each node, the most recently updated link should be selected to avoid route breakage.

$$C = 1/T \tag{1}$$

In order to establish an end-to-end route, the controller generates an instruction packet (INS). INS is used to manipulate the routing table of each node that will be a part of the selected route. The controller unicasts the INS packets to each node in the selected route starting from the destination back to the source. When a node receives INS, a routing entry is created or updated according to information in the INS packet header. The routing entry informs the next hop to the destination. The source begins to send data after it receives routing information from INS. If the controller cannot find a route, it will retry the route discovery process after it receives a new NB_list from the source in the next update period. It is worth mentioning that every end-to-end route is tracked by the controller. The tracked routes are validated at the update period. If a better route is available, the controller generates INS to inform nodes according to the route rediscover result.

Route Maintenance. When a node detects the link failure, i.e., from any link-layer notification, the node invalidates the corresponding routing entry and then unicast NB_list to inform the controller. If the link failure occurs at any node in the active route, the node additionally generates a route error packet (RERR) and unicast it back to the source. RERR header has a field to specify the IP address of an unreachable destination. A node response to the RRER by invalidating its routing entry that stores the route to the unreachable destination.

In the case of a source node, it will request a new route from the controller by sending an NB_list with a special flag. the flag indicates that this NB_list is generated to respond with route breakage. the number of this special NB_list is limited by the ?request token? to prevent the source node from flooding NB_list to the controller. The controller then tries to maintain the connectivity by discovering an alternative route. If the route maintenance fails, the source node will wait until the next update period which its NB_list is automatically generated.

3.2 Proof of Work

In this section, we will prove our concept with a controlled simulation scenario. The simulation tool and parameters will be presented later in Sect. 4. From Fig. 3, A source node and a destination node are placed between a group of neighboring nodes. These neighboring nodes are split into two groups to form disjoint routes. At first, the source node is convinced to use <ROUTE A> to transmit data messages. In the middle of the simulation, a node in <ROUTE A> is moved away from its original position to break the route. We try to observe how fast the protocol will react to this situation to examine the network adaptability. We decide to compare our proposed protocol with AODV [14] because it has been proved to outperform the other traditional routing protocol [15]. An SDN controller node will be removed from the scenario when AODV is simulated.

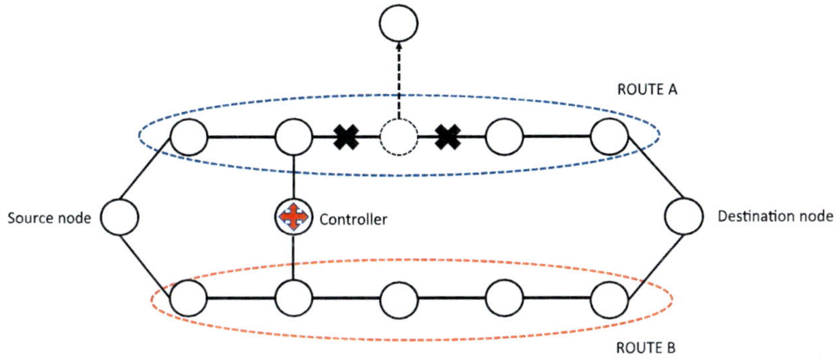

Fig. 3. Proof of work scenario setup

After the route break, we reckon the time from the point where the RERR packet is generated until the source node is informed to switch the route. The simulation result shows that our proposed protocol reacts to this situation 20% faster than AODV. While our protocol depends on the controller to rediscover the route, the source node operated with AODV needs to rediscover the route by itself. From this scenario, we can confirm that our proposed protocol can improve network adaptability in MANETs. Next, we will present the simulation results where the MANETs instability factors are included in the scenario.

4 Simulation and Analysis

We evaluated our proposed protocol by using Network Simulator 3 (NS 3.301). Table 1 shows simulation parameters which are also applied by the proof of work in Sect. 3.2. In this section, mobile nodes are randomly distributed over the simulated area. The random waypoint mobility model is installed on every node except for an SDN controller. Node maximum speed is set to 5m/s. Node maximum pause time is set to 10 s. These two parameters are varied under the uniform distribution. The controller is fixed at the center of the simulated area. To keep the randomness property of this scenario, the controller is prohibited from relay any data messages. The number of source nodes is varied from 1–5 nodes. We simulate each scenario for 300 s. The warm-up time is set to 15 s. Data packets are immediately generated after the warm-up period. Simulation results are summarized from 10 simulation trials. The following metrics are used to evaluate network performance:

- Route request attempt: the total number of route requests generated at the source node during the simulation. For our proposed protocol, we count the number of NB_list that has a route request field attached. If the protocol can avoid the route breakage

Table 1. Simulation parameters.

Parameters	Value
Simulate area	1 km^2
Wireless adaptor standard	802.11a (6 Mbps data rate)
Loss model	Range propagation loss model
Node's transmission range	250 m
Generated data traffic	CBR of 5 packets per second
Data packet size	512 bytes
Number of the controller node	1
Controller advertisement interval	10 s
Number of the request token	1

or successfully maintenance the route when it breaks, the number of route requests reduce accordingly. In other words, this metric implies an ability to adapt to the network condition of the routing protocol.

- Packet delivery ratio: the fraction of the total number of data messages received at the destination to the total number of generated data messages.
- Total routing overhead: the total number of every routing packet generated in the simulation.

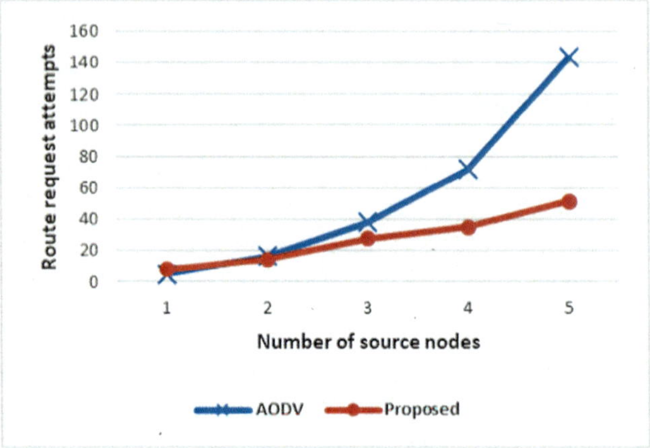

Fig. 4. Route request attempts

Figure 4 shows the number of route request attempts compared with AODV. Our proposed protocol achieves a better result. Since our protocol validates the route every update period, the route breakage problem is mitigated resulting in a lower number of route request attempts. On the other hand, AODV keeps utilizing a discovered route until its break. This is an important reason that increases route discovery activities. Even AODV has a route recovery mechanism, the route restoration cannot be guaranteed. When a node fails to recover a route, the route request process is triggered. Moreover, network collisions might occur during the route request process. If the source node cannot rediscover the route before its queue is overflowed, the buffered data messages are dropped.

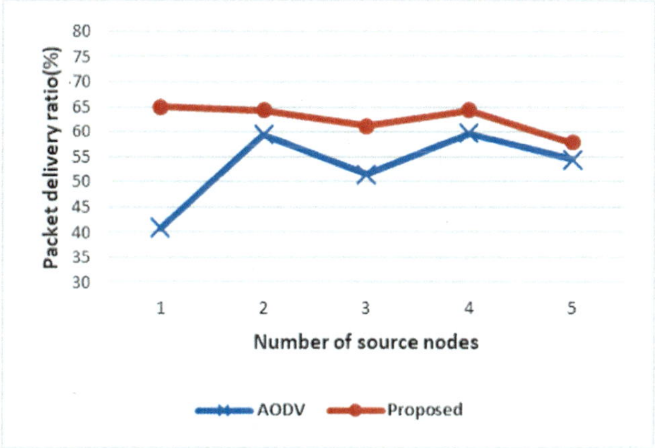

Fig. 5. Packet delivery ratio

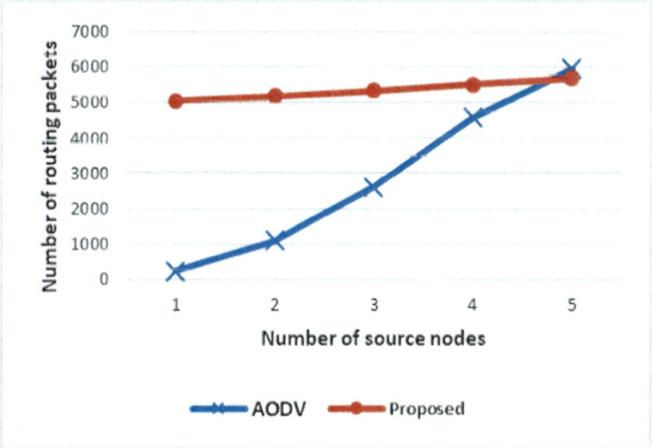

Fig. 6. Total routing overhead

Figure 5 presents the packet delivery ratio. Our proposed protocol achieves a better packet delivery ratio of up to 24% compared to AODV. Although we try to periodically update the route for every source-destination pairs, route breakage is inevitable. Therefore, we include the mechanism to let a source node ask for a new route from the controller between the update period. A source node will respond with network conditions faster this way. The reason we cannot further alleviate the packet delivery ratio is that the controller might lose a connection with some node in the network. In the worst case, the controller does not have enough information to form a route for some period. It can be observed that the packet delivery ratio drops when the number of source nodes increases. Since every node needs to associate with the controller, the bottleneck at the controller might be a cause of performance degradation. When there are more than 5 source nodes, the increasing collision rate will disconnect the controller from mobile nodes. The limitation presented by the simulation results may vary according to simulation scenarios.

Figure 6 shows the amount of routing packet requires to create an end-to-end route. Our proposed protocol has to broadcast the C_{adv}. Thus, the number of routing packets relates to the number of nodes in the network. In contrast, the route request activity at source nodes has a major impact on routing overhead in AODV. It can be observed together with Fig. 4 that these metrics increase with the same trend. An uncontrollable flooding activity of source nodes might affect the performance of the network by increasing the chance of network collision. In a scenario with 5 source nodes, the number of generated overheads in AODV has exceeded our proposed protocol.

5 Conclusion

In this paper, we propose a MANETs routing protocol which exploits the SDN architecture concept to centralize routing decision. An in-band SDN controller is deployed into the network to manage a communication route for every node. Since traditional SDN standard does not support mobile ad hoc communication, we design a mechanism that allows nodes to associate with the controller through a wireless medium. Mobile nodes report link-state information to the controller periodically or when a link failure is detected. Thus, the controller can adapt to a topology change. The simulation results show a positive outcome compared to AODV. By frequently updating the route to match with the current network condition, the route breakage is reduced. The packet delivery ratio consequently increases because source nodes achieve a more reliable connection. However, the routing overhead increases according to node density. It is a generic tradeoff for the flooding-related action initiate by the controller.

It can be summarized from simulation results that the bottleneck problem at the controller degrades the network performance. For future work, we will try to solve this problem. Deploying multiple controllers might be one of the possible solutions. Moreover, there are some open issues which can be further improved. The cost function can be tuned for another purpose or to achieve better results. The traditional protocol might be used as a fallback mechanism when a source node lost connection with the controller as mentioned in [11]. Nevertheless, this early-stage work evinces the possibility to centralize MANETs' routing decisions with an in-band SDN controller. This kind of SDN-based routing protocol requires further investigation.

Acknowledgments. This research project is financially supported by King Mongkut?s Institute of Technology Ladkrabang (KMITL).

References

1. Ververidis, C.N., Polyzos, G.C.: Service discovery for mobile Ad Hoc networks: a survey of issues and techniques. IEEE Commun. Surv. Tutorials **10**(3), 30–45 (2008). Third Quarter
2. Awang, A., Husain, K., Kamel, N., Aïssa, S.: Routing in vehicular ad-hoc networks: a survey on single- and cross-layer design techniques, and perspectives. IEEE Access **5**, 9497–9517 (2017)
3. Gupta, L., Jain, R., Vaszkun, G.: Survey of important issues in UAV communication networks. IEEE Commun. Surv. Tutorials **18**(2), 1123–1152 (2016). Second quarter
4. Trifunovic, S., Kouyoumdjieva, S.T., Distl, B., Pajevic, L., Karlsson, G., Plattner, B.: A decade of research in opportunistic networks: challenges, relevance, and future directions. IEEE Commun. Mag. **55**(1), 168–173 (2017)
5. Nunes, B.A.A., Mendonca, M., Nguyen, X., Obraczka, K. Turletti, T.: A survey of software-defined networking: past, present, and future of programmable networks. In: IEEE Communications Surveys & Tutorials, vol. 16, no. 3, pp. 1617–1634 (2014). Third quarter
6. Xia, W., Wen, Y., Foh, C.H., Niyato, D., Xie, H.: A survey on software-defined networking. IEEE Commun. Surv. Tutorials **17**(1), 27–51 (2015). First quarter
7. Kobo, H.I., Abu-Mahfouz, A.M., Hancke, G.P.: A survey on software-defined wireless sensor networks: challenges and design requirements. IEEE Access **5**, 1872–1899 (2017)
8. Yu, H.C., Quer, G., Rao, R.R.: Wireless SDN mobile ad hoc network: from theory to practice. In: 2017 IEEE International Conference on Communications (ICC), Paris (2017)
9. Bellavista, P., Dolci, A., Giannelli, C.: MANET-oriented SDN: Motivations, Challenges, and a Solution Prototype. In: 2018 IEEE 19th International Symposium on "A World of Wireless, Mobile and Multimedia Networks" (WoWMoM), Chania (2018).
10. Poularakis, K., Qin, Q., Marcus, K.M., Chan, K.S., Leung, K.K., Tassiulas L.: Hybrid SDN control in mobile Ad Hoc networks. In: 2019 IEEE International Conference on Smart Computing (SMARTCOMP), Washington, DC, USA (2019)
11. Ku, I., Lu, Y., Gerla, M.: Software-defined mobile cloud: architecture, services and use cases. In: 2014 International Wireless Communications and Mobile Computing Conference (IWCMC), Nicosia (2014)
12. Streit, K., Rodday, N., Steuber, F., Schmitt, C., Rodosek, G.D.: Wireless SDN for highly utilized MANETs. In: 2019 International Conference on Wireless and Mobile Computing, Networking and Communications (WiMob), Barcelona, Spain (2019)
13. Wei, D., Liu, Z., Cao, H.: SRPA: SDN-based routing protocol for Ad Hoc networks. In: 2018 9th International Conference on Information Technology in Medicine and Education (ITME), Hangzhou (2018)
14. Perkins, C., Belding-Royer, E., Das, S.: Ad Hoc on-demand distance vector (AODV) routing. RFC 3561 (2003)
15. Maan, F., Mazhar, N.: MANET routing protocols vs mobility models: a performance evaluation. In: 2011 Third International Conference on Ubiquitous and Future Networks (ICUFN), Dalian (2011)

Author Index

P. Meesad et al. (Eds.): IC2IT 2021, LNNS 251, pp. 333–334, 2021.
https://doi.org/10.1007/978-3-030-79757-7